Resiliency Models and Addressing Future Risks for Family Firms in the Tourism Industry

Anita Zehrer
MCI Management Center Innsbruck, Austria

Gundula Glowka
MCI Management Center Innsbruck, Austria

Katrin Magdalena Schwaiger
MCI The Entrepreneurial School, Austria

Victoria Ranacher-Lackner
MCI The Entrepreneurial School, Austria

A volume in the Advances in Hospitality, Tourism, and the Services Industry (AHTSI) Book Series

IGI Global
PUBLISHER of TIMELY KNOWLEDGE

Published in the United States of America by
IGI Global
Business Science Reference (an imprint of IGI Global)
701 E. Chocolate Avenue
Hershey PA, USA 17033
Tel: 717-533-8845
Fax: 717-533-8661
E-mail: cust@igi-global.com
Web site: http://www.igi-global.com

Library of Congress Cataloging-in-Publication Data

Names: Zehrer, Anita, editor. | Glowka, Gundula, 1991- editor. | Schwaiger, Katrin, 1989- editor. | Ranacher-Lackner, Victoria, 1987- editor.
Title: Resiliency models and addressing future risks for family firms in the tourism industry / Anita Zehrer, Gundula Glowka, Katrin Schwaiger, and Victoria Ranacher-Lackner, editor.
Description: Hershey PA : Business Science Reference, [2021] | Includes bibliographical references and index. | Summary: "This book provides an in-depth examine of tourism family firms since these firms tend to provide solutions for challenges such as dealing with uncertainty, becoming or staying resilient, and creating sustainable tourism destinations byconnecting knowledge from family business research to tourism research"-- Provided by publisher.
Identifiers: LCCN 2020057643 (print) | LCCN 2020057644 (ebook) | ISBN 9781799873525 (hardcover) | ISBN 9781799873532 (paperback) | ISBN 9781799873549 (ebook)
Subjects: LCSH: Tourism--Risk management. | Family-owned business enterprises--Risk management. | Small business--Management. | Organizational resilience. | COVID-19 (Disease)--Economic aspects.
Classification: LCC G155.A1 R4755 2021 (print) | LCC G155.A1 (ebook) | DDC 910.68--dc23
LC record available at https://lccn.loc.gov/2020057643
LC ebook record available at https://lccn.loc.gov/2020057644

This book is published in the IGI Global book series Advances in Hospitality, Tourism, and the Services Industry (AHTSI) (ISSN: 2475-6547; eISSN: 2475-6555)

British Cataloguing in Publication Data
A Cataloguing in Publication record for this book is available from the British Library.

Advances in Hospitality, Tourism, and the Services Industry (AHTSI) Book Series

ISSN:2475-6547
EISSN:2475-6555

Editor-in-Chief: Maximiliano Korstanje, University of Palermo, Argentina

MISSION

Globally, the hospitality, travel, tourism, and services industries generate a significant percentage of revenue and represent a large portion of the business world. Even in tough economic times, these industries thrive as individuals continue to spend on leisure and recreation activities as well as services.

The Advances in Hospitality, Tourism, and the Services Industry (AHTSI) book series offers diverse publications relating to the management, promotion, and profitability of the leisure, recreation, and services industries. Highlighting current research pertaining to various topics within the realm of hospitality, travel, tourism, and services management, the titles found within the AHTSI book series are pertinent to the research and professional needs of managers, business practitioners, researchers, and upper-level students studying in the field.

COVERAGE

- Service Management
- Casino Management
- Tourism and the Environment
- Customer Service Issues
- Sustainable Tourism
- Health and Wellness Tourism
- Leisure & Business Travel
- Cruise Marketing and Sales
- Service Training
- Food and Beverage Management

IGI Global is currently accepting manuscripts for publication within this series. To submit a proposal for a volume in this series, please contact our Acquisition Editors at Acquisitions@igi-global.com or visit: http://www.igi-global.com/publish/.

The Advances in Hospitality, Tourism, and the Services Industry (AHTSI) Book Series (ISSN 2475-6547) is published by IGI Global, 701 E. Chocolate Avenue, Hershey, PA 17033-1240, USA, www.igi-global.com. This series is composed of titles available for purchase individually; each title is edited to be contextually exclusive from any other title within the series. For pricing and ordering information please visit http://www.igi-global.com/book-series/advances-hospitality-tourism-services-industry/121014. Postmaster: Send all address changes to above address.

Titles in this Series

For a list of additional titles in this series, please visit:
http://www.igi-global.com/book-series/advances-hospitality-tourism-services-industry/121014

ICT Tools and Applications for Accessible Tourism
Celeste Eusébio (University of Aveiro, Portugal) Leonor Teixeira (University of Aveiro, Portugal) and Maria João Carneiro (University of Aveiro, Portugal)
Business Science Reference • © 2021 • 356pp • H/C (ISBN: 9781799864288) • US $215.00

Impact of New Media in Tourism
Maria Gorete Dinis (GOVCOPP, Polytechnic Institute of Portalegre, Portugal) Luís Bonixe (ICNOVA, Polytechnic Institute of Portalegre, Portugal) Sónia Lamy (ICNOVA, Polytechnic Institute of Portalegre, Portugal) and Zélia Breda (GOVCOPP, University of Aveiro, Portugal)
Business Science Reference • © 2021 • 389pp • H/C (ISBN: 9781799870951) • US $195.00

Growth of the Medical Tourism Industry and Its Impact on Society Emerging Research and Opportunities
Manjeet Singh (Central University of Jammu, India) and Subbaraman Kumaran (Alliance University, Bangalore, India)
Business Science Reference • © 2021 • 189pp • H/C (ISBN: 9781799834274) • US $195.00

Global Development of Religious Tourism
Emilia Alaverdov (Georgian Technical University, Georgia) and Muhammad Waseem Bari (Government College University, Faisalabad, Pakistan)
Business Science Reference • © 2021 • 366pp • H/C (ISBN: 9781799857921) • US $195.00

Innovation and Entrepreneurial Opportunities in Community Tourism
Jakson Renner Rodrigues Soares (Universidade da Coruña, Spain & Universidade Estadual do Ceará, Brazil)
Business Science Reference • © 2021 • 358pp • H/C (ISBN: 9781799848554) • US $195.00

For an entire list of titles in this series, please visit:
http://www.igi-global.com/book-series/advances-hospitality-tourism-services-industry/121014

IGI Global
PUBLISHER of TIMELY KNOWLEDGE

701 East Chocolate Avenue, Hershey, PA 17033, USA
Tel: 717-533-8845 x100 • Fax: 717-533-8661
E-Mail: cust@igi-global.com • www.igi-global.com

Table of Contents

Detailed Table of Contents

Section 1
Being Resilient Throughout Crises

The first section, 'Being Resilient Throughout Crises', addresses organizational resilience and consists of three chapters. The first and the second chapter showcase the COVID-19 pandemic in Europe, and the third chapter presents a longitudinal study of family firms in Croatia along the countries historic periods of communism, war, and access to the European Union. While the first chapter chooses a descriptive analysis of statistical data available in the European Union, Chapters 2 and 3 concentrate on family-firm owner-managers and their behavior. Therefore, both chapters conducted semi-structured qualitative interviews with family firm owner-managers.

The coronavirus is crippling the global economy. However, the economic impacts of the pandemic vary significantly across sectors of the economy. This outbreak and related lockdowns are putting the tourism industry under unprecedented pressure. Within months, world tourism went from over-tourism to non-tourism. Therefore, the overall objective of this chapter is to introduce the latest trends in tourism, with emphasis on family businesses. More precisely, the chapter encompasses (1) the theoretical background, which defines family business, highlights their characteristics, and summarizes the importance of leadership succession; (2) the role of family business resilience and their behaviour in times of crisis; (3) a review of relevant European Commission policy publications regarding EU tourism and statistical analyses of selected key tourism indicators; and (4) an analysis of family businesses environment in time of crisis. Hence, this chapter has documented the crucial role that family businesses tourism has within the European Union and on the front line of the current crises.

On March 11, 2020, the World Health Organization (WHO) announced a pandemic of COVID-19. The family-owned hospitality industry in the Tirol has been affected by governmental measures and travel bans in particular. The current study focuses on hospitality family business owners as the backbone of the regional tourism industry. Their dealing with the crisis and their perception of their own organizational resilience is explored. It uses a qualitative exploratory approach with semi-structured interviews to answer the research question. The most surprising result of the interviews was the optimistic state of mind visible among entrepreneurs. In this study, a better understanding of the tourism industry's dealing with a crisis such as the COVID-19 pandemic is created.

Through three dramatic periods—the period of communism, the period of the war and transition during the 1990s, and the period of EU admission—this longitudinal study investigates how small hotel owners in Croatia understand and practice resilience. Findings demonstrate that entrepreneurs have proactive understanding of resilience and practice different actions before the shock of a disruptive event/s occur/s. Interestingly, entrepreneurs perceive this disruption as opportunities to exploit for their business development. One of the most prevalent practices is diversification, where entrepreneurs start another firm in different, often unrelated, sectors.

Section 2
Creating Opportunities Through Crisis

The second section is about 'Creating Opportunities Through Crisis' and consists of two chapters. In this section, both chapters apply a community-oriented model of the destination. First, Chapter 4 takes the destination and its stakeholders and focuses on opportunities for developing resilience strategies at a destination-wide level. Exploring such opportunities is vital, as there might be a trade-off between firm resilience and regional resilience. Secondly, the fifth chapter returns to the perspective of organizational resilience of family firms, while also following a community-oriented destination model.

Against the background of its conceptual vagueness, practicing resilience is a complex task and a particular challenge for management. Resilience initiatives on the local and regional levels pursue participation by using different methods of stakeholder integration. In this contribution, the private sector is presented as a particularly difficult stakeholder group to attract for cooperative resilience building. Due to limited resources, corporate development and responsibility often remain in the spheres of an operational context. Thus, involvement in regional resilience building is often seen as desirable, but not an integral part of strategic corporate management. This contribution gives recommendations on how coordinated resilience initiatives should act in order to achieve the systematic participation of the private sector. If resilience practice defines measurable outcomes for entrepreneurship and positive experiences are made in the regional network, a sustainable participation of firms in the resilience development should be manifested.

Family firms in tourism and hospitality (FFTH) are often characterized by their long-term orientation and strong "familiness" and embeddedness in tourism destinations. Families however are often confronted with and have to adapt to global change and can sometimes struggle with drastic changes in the tourism industry. And while the current COVID-19 pandemic will mean even more challenges for FFTH, it may also lead to opportunities for responsible tourism development. This chapter attempts to shed light on FFTH's ability to manage change, discussing their specific characteristics through the lens of sustainability and responsibility. The following chapters will analyze FFTH characteristics to develop recommendations for FFTH in supporting responsible tourism, and present ideas for a tourism policy that promotes FFTH.

Section 3
Reacting in Times of Change

The third section, 'Reacting in Times of Change', consists of three chapters. All chapters focus on family firm performance during turbulent times. Chapters 6 and 7 therefore conduct interviews with family firm members of different generations working in the family business. Thus, the studies represent the perception of the owner-family towards the own family firm. Chapter 7 in contrast interviews non-family member employee to obtain a perception of the family firms' ambidexterity.

Successfully dealing with risk has become a major challenge for firms to stay competitive in the long run. The tourism industry is vulnerable to any crisis and therefore has to deal with a high degree of uncertainty including many risks. The Alpine tourism industry is shaped by small and medium-structured family businesses. Such firms have more informal decision-making processes and more generations involved, while risk taking depends on the decisions of an owner-manager. Understanding the risk perception owner-managers is thus a complex task. This study conducts n=12 qualitative interviews with senior and junior owner-managers of family SME in tourism. Understanding the risk perception for tourism family SME might therefore contribute to build resilience and long-term competitiveness.

Traditional Heurigen and Buschenschenken have been an important part of Austrian culture for centuries and are mostly run as small family firms. In those businesses, succession is inevitable. Several factors are key for the success of transition processes within a family. Traditional concepts are confronted with digitalization, new technologies, new demand, and social changes. For the chapter at hand, incumbents and heirs of four wine making and Heurigen businesses in the Northern Burgenland were interviewed. Its aim is to investigate the perception of the succession process itself as well as the risk of implementing innovations into traditional concepts. This study shows that innovations are directed by natural circumstances rather than customer demands. The owners rely on traditional concepts with incremental changes to keep customers attracted. Thanks to trust, open communication, and succession processes that spread over long periods of time, neither generation thinks of the handover as problematic.

This chapter examines the interplay of the fundamental constructs of perceived organizational resilience and organizational ambidexterity by providing insights from a case study of an integrated service company in the tourism and leisure industry located in the Swiss Alps. The authors theorize that organizational ambidexterity (i.e., the balance between adaptability and alignment) leads to higher levels of resilience by boosting the capability of the firm to mitigate and adapt to major disruptions in its environment. Data from a survey with non-family employees show that establishing structures of alignment and adaptability and imprinting an "and/or"-mindset into the company culture has enabled all employees to constantly navigate new challenges while staying aligned on priorities throughout the hardships of the COVID-19 pandemic. However, the positive effects of alignment and adaptability can only unfold if they are combined with clear communication and transparency by the company's leadership.

Preface

RESILIENCY MODELS AND ADDRESSING FUTURE RISKS FOR FAMILY FIRMS IN THE TOURISM INDUSTRY

Supporting infrastructure developments in remote regions and creating employment, tourism has developed as a major economic driver, generating more than USD 1.4 trillion globally in 2016 (UNWTO, 2017). In 2019, international arrivals increased by +7% compared to 2017 and another +6% to 2018 (UNWTO, 2021a). Against worldwide economic slowdown, the Middle East (+8%), Asia and the Pacific (+4%) and Europe (+4%) recorded an increase in arrivals in 2019 (UNWTO, 2021a). In 2020, the COVID-19 pandemic caused a drastic decline of 74% in international arrivals, i.e. over 1 billion less international arrivals than in the year before (UNWTO, 2021b). The World Tourism Organization estimates an USD 1.3 trillion loss in export revenues due to travel restrictions (UNWTO, 2021b). Thus, recent developments in the context of the COVID-19 pandemic appear to outpace all other tourism challenges.

However, tourism as an industry is in its nature predestinated for uncertainties that affect its market and its firms. Crises of various kinds, such as natural disasters, economic or political crises, have been affecting the hospitality sectors of many countries globally throughout time (Ritchie, Crotts, Zehrer, & Volsky, 2014; Hall, 2010; Glaesser, 2006; Pizam & Smith, 2000). The ISO 31000:2009 defines a risk as an effect of uncertainty on objectives. Tourism firms thus have to cope with challenges such as coping with terrorist attacks, political instability, economic recession, biosecurity threats or natural disasters (Ritchie, 2004). For hazards risks, like a terrorist attack, the probability of occurrence is relatively low for a single destination. In the case of occurrence, it is, however, followed by severe consequences, since political instability heavily influences the success of a destination (Liu & Pratt, 2017). Thus, destinations and its firms have to manage the risk perception of being the target of a terrorist attack and implement crisis management in the case of such an event (Veréb, Nobre, & Farhangmehr, 2018; Zehrer, 2013). In a digitalized world, tourism firms might therefore use social media as a communication tool to reach travelers. Many other risks such as declining snowfall, rising sea levels, more extreme weather events, over-tourism or natural resource pollution, provoked by tourism, urge the need for sustainable tourism development. Essential to sustainable development is identifying such risks and developing innovative opportunities to guarantee long-term firm survival.

Especially, small and medium-sized enterprises (SME) are hit by the pandemic and up to 120 million jobs are in danger (UNWTO, 2021b). Most SME are family-owned and run firms that differ from non-family firms in the willingness to pass the business to the next generation (Chua, Chrisman, & Sharma, 1999). Typically, more generations are involved in the daily operations of a family SME (Cabrera-Suarez, Saa-Perez, & Garcia-Almeida, 2001) and owner-managers shape the vision and mission of the business (Feltham, Feltham, & Barnett, 2005). The desire to maintain the business in the family over generations fosters a sustainable, long-term orientation of such firms (Gersick, Davis, Hampton, & Lansberg, 1997). To preserve the socio-emotional wealth of the family, e.g. reputation of the family (Berrone, Cruz, & Gomez-Mejia, 2012), family firms take corporate social responsibility (Dyer & Whetten, 2006) and adapt practices of sustainability (Le Breton-Miller & Miller, 2016). Gómez-Mejía, Haynes, Núñez-Nickel, Jacobson, and Moyano-Fuentes (2007), for example, highlight specific risk preferences for family businesses and conclude that while socio-emotional wealth constitutes one of the key goals for family businesses, strategies leading to a loss of this socio-emotional wealth are rather avoided even though they might bring some risk protection to the firm. Regarding the briefly outlined characteristics of family businesses, it is worthwhile analyzing this business type in particular when it comes to organizational resilience. Since family businesses play such an important role for the entire tourism industry, this book investigates resilience models of family firms, which could create useful learning effects for the future.

Considering that organizational resilience is able to weaken the negative impact of uncertain events, the Covid-19 pandemic might be seen as a wake-up call for the importance to study and develop resiliency models for the tourism industry and its family firms as addressed in this book. Organizational resilience "is the organization's capability to face disruptions and unexpected events in advance thanks to the strategic awareness and a linked operational management of internal and external shocks" (Annarelli & Nonino, 2016, p. 7). It does not only depend on internal or personal characteristics, but also structural or external factors (Zehrer & Leiß, 2019), while it is characterized by reflection on past critical events, coping with and adapting to them (Hedner, Abouzeedan, & Klofsten, 2011). Thus, if family firms are resilient, they will be able to handle challenging situations, without collapsing and to stabilize themselves by maintaining structures and functions (Innerhofer, Fontanari, & Pechlaner, 2018). The Covid-19 pandemic can therefore be seen as the opportunity to observe the patterns of long-term oriented family firms and their resilience, that otherwise might not have been revealed. In this manner, the authors stress the importance to develop resilience models that exceed the crisis-context and deal with different kinds of uncertainties.

ORGANIZATION OF THE BOOK

The book is organized into four main sections and 11 chapters. In the following, each of the four sections and its chapters are discussed. The first section, 'Being Resilient Throughout Crises', addresses organizational resilience and consists of three chapters. The first and the second chapters showcase the Covid-19 pandemic in Europe, the third chapter presents a longitudinal study of family firms in Croatia along the countries historic periods of communism, war and access to the European Union. While the first chapter chooses a descriptive analysis of statistical data available in the European Union, Chapter 2 and 3 concentrate on family-firm owner-managers and their behaviour. Therefore, both chapters conducted semi-structured qualitative interviews with family firm owner-managers.

In Chapter 1, Harc and Basarac Sertić discuss family business' characteristics and the role of resilience in such businesses. The chapter, 'Family Business Tourism Risk in the Context of Coronavirus Pandemic', presents statistical analyses of selected key tourism indicators in the European Union relevant to develop an understanding of the Covid-19 pandemic in this areas. Thus, the first chapter of this book introduces the characteristics of family firms and frames the current family business environment for the European Union.

The second chapter, 'The COVID-19 Pandemic and Organizational Resilience in Hospitality Family Firms: A Qualitative Approach', by Schwaiger and Zehrer explores the different aspects of organizational resilience of Alpine hospitality family firms. Data collection took place in May 2020, just after the first lockdown period was over in Austria. Interestingly, the study finds that family firms are rather optimistic and can manage a several month closing period without generating profits. Furthermore, the study reveals that owner-managers highly value family firm's employees, who play a vital role for the success and resilience of the firms during crisis.

In Chapter 3, 'How Small Hotel Owners Practice Resilience? Longitudinal Study Among Small Family Hotels in Croatia', Skokic conducts semi-structured interviews with family firms in Croatia. The first data selection took place during winter 2006/07 and the second, ten years later, during winter 2016/2017. This chapter therefore does not analyze resilience using the Covid-19 pandemic, but focuses on three dramatic periods that affected Croatian hospitality family firms: 1) the period of communism; 2) the period of the war and the following transition during the 1990s; 3) and the period of EU admission. The study presents traits, attributes and experiences of family entrepreneurs and finds a proactive approach towards resilience practices. Interestingly, this study finds that the entrepreneurs use diversification as a resilience practice.

The second section is about 'Creating Opportunities Through Crisis' and consists of two chapters. In this section, both chapters apply a community-oriented model of the destination. First, chapter four takes the destination and its stakeholders and focuses on opportunities for developing resilience strategies at a destination-wide level. Exploring such opportunities is vital, as there might be a trade-off between firm resilience and regional resilience. Secondly, the fifth chapter returns to the perspective of organizational resilience of family firms, while also following a community-oriented destination model.

In the fourth chapter, 'Resilience as an Opportunity Approach: Challenges and Perspectives for Private Sector Participation on a Community Level', Zacher and Pechlaner analyse possible resilience initiatives to integrate the private sector into a coordinated resilience-building process at a community level. Therefore, resilience is viewed as an active design and planning tool in tourism related contexts. As possible strategies to implement resilience building, three opportunities are discussed: 1) Community Resilience Toolkits and Resilience-Orientated Consultancy; 2) Establishing Networks and Organisational Structures for Resilience Building; 3) Objectives of National Resilience Strategies. Thus, Zacher, D. and Pechlaner, H. investigate the challenges and perspectives of resilience initiatives in integrating the private sector to ultimately provide recommendations on coordinated resilience initiatives, to enhance participation of the private sector.

In Chapter 5, 'Family Firm Management in Turbulent Times: Opportunities for Responsible Tourism', Eichelberger and Peters propose opportunities for responsible tourism development. Examining the special characteristics of family firms such as the construct of familiness and the long-term orientation through the lens of sustainability, this chapter identifies the opportunity for responsible tourism development as a possible positive aftermath of a crisis.

The third section, 'Reacting in Times of Change', consists of three chapters. All chapters focus on family firm's performance during turbulent times. Chapter 6 and 7 therefore conduct interviews with family firm members of different generations working in the family business. Thus, the studies represent the perception of the owner-family towards the own family firm. Chapter 7 in contrast interviews non-family member employee to obtain a perception of the family firms' ambidexterity.

In Chapter 6, 'The Risk Perception of Family Business Owner-Manager in the Tourism Industry: A Qualitative Comparison of the Intra-Firm Senior and Junior Generation', Glowka, Tusch, and Zehrer observe how the senior and the junior generation perceive risks differently by conducing twelve semi-structured interviews with both, family members of the senior and the junior generation. The results on differences in risk perception are then discussed along financing and investment management, marketing management, human resources management, cost management, eco-management, and succession.

In Chapter 7, 'Heurigen 2.0: Succession Processes of Family-Owned Wine Taverns in Northern Burgenland', Soleder attends to the influencing factors on the experience of introducing modern business strategies during succession processes in her investigation on the example of family-owned wine taverns and their modernized concepts after succession. She approaches the succession process with its concomitant modernization as either an anticipation of external threats and risks, or as a risk in itself. Considering the predominance of small and medium sized family enterprises in the Austrian tourism structure, successful integration of the younger generation is vital to ensure the survival of the firm. This process represents one of the potential risks and failures of small and medium sized enterprises, due to emotional attachment to the business, the uncertainty of modern business strategies and other factors, which can affect family life.

In Chapter 8, 'Crisis Favors the Prepared Firm: How Organizational Ambidexterity Relates to Perceptions of Organizational Resilience', Bechthold, Lude, and Prügl propose organizational ambidexterity as a necessary precondition for family firm resilience. The ability of the firm to run explorative and exploitative activities simultaneously, serves as a strategic framework to adapt to ever-changing situations continuously. Presenting a case study, the family firm, which is operating in the tourism industry, needs to cope with challenges imposed by climate change infectious crises and diverging strategic goals. The case study serves as study object to frame how family firms cope with uncertainty on different levels of the value chain. As a result, the authors propose that organizational ambidexterity is fruitful to firm resilience.

The fourth and final section of this book, 'Digitalizing the Tourism Product: Risk or Opportunity?' consists of three chapters exploring the communication of family firms to potential guests in a digitalized word. Therefore, Chapter 9 analyzes how family-run hotels can use social media as a communication tool in times of crisis. Chapter 10 addresses the resilience of the traveler and strategies for family firms to communicate with them. Finally, Chapter 11 examines the risk of information overload in a digitalized world and how holiday farms could address this risk in communication with their guests.

In the ninth chapter, 'Physically Apart, Emotionally Close: How Family-Run Hotels Use Instagram During the COVID-19 Pandemic – Social Media Crisis Communication During the COVID-19 Pandemic', Fecker, Nadegger, and Haselwanter present a mixed-method study analyzing the communication of family firms via the social media platform Instagram during the Covid-19 pandemic. Interestingly, results show that family-run hotels mainly posted positive attitudes, emotions and classic alpine marketing topics on their Instagram profiles. The authors suggests that family firms present themselves as resilient and anchored in their family values in times of crisis.

The tenth chapter, 'Travellers' Resilience to Major Travel Risks of Today, Global Terrorism, and Pandemic Disease: Opportunities and Challenges for Small Tourism Businesses', analyses the resilience of travellers and addresses small family firm's opportunities regarding traveller resilience. Veréb and Azevedo investigate the potential change of traveller's perception of travel risk, in the context of the current global challenges like infectious diseases (for example the COVID-19 pandemic), or global terrorism risk. As this angle of resilience research is still untouched, this contribution taps into an unexplored field by addressing crisis communication as well as guidelines to strengthen the resilience of travellers.

In Chapter 11, 'Dealing with the Risks of Information Overload: Promoting a Mindful Surprise Culture on Farm Holidays', Mangini and Quendler address the phenomenon of information overload. The authors argue that information overload can impact the touristic experience negatively, since visitors do not create sufficiently memorable and authentic experiences. The omnipresence of information, the amount and its complexity can result in overextension and feelings of overpowering. However, this chapter makes proposals on how family firms can address this the risk of information overload and create opportunities from the fact that information can be consumed and be received at all time.

Thus, this book provides an international perspective on contemporary issues and future challenges and risks of family firms in the tourism industry. While several books have been devoted to either tourism or family firms, this book represents a comprehensive attempt to address the discussions about the future of family firms in tourism. To deal with future challenges of tourism family firms, the authors of this book examine different perspectives at individual, firm and destination level. This book thus creates value for both, the tourism and family firm research community, as well as for practitioners in both spaces. Addressing the challenge of sustainability in an uncertain world, family firms might be a major contributor to manage and conserve tourism destinations successfully in the future. The objective is to connect knowledge from family business research to tourism research, in particular focusing on hospitality. Doing so, tourism practitioners and researchers might be able to use valuable linkages that can lead to sustainable solutions at individual, firm, and destination level. The same applies for tourism family firms that can benefit from the identification of future challenges for their businesses and points of departure for organizational resilience.

REFERENCES

Annarelli, A., & Nonino, F. (2016). Strategic and operational management of organizational resilience: Current state of research and future directions. *Omega*, *62*, 1–18. doi:10.1016/j.omega.2015.08.004

Berrone, P., Cruz, C., & Gomez-Mejia, L. R. (2012). Socioemotional Wealth in Family Firms: Theoretical Dimensions, Assessment Approaches, and Agenda for Future Research. *Family Business Review*, *25*(3), 258–279. doi:10.1177/0894486511435355

Cabrera-Suarez, K., Saa-Perez, P., & Garcia-Almeida, D. (2001). The Succession Process from a Resource- and Knowledge-Based View of the Family Firm. *Family Business Review*, *14*(1), 37–48. doi:10.1111/j.1741-6248.2001.00037.x

Chua, J. H., Chrisman, J., & Sharma, P. (1999). Defining the Family Business by Behavior. *Entrepreneurship Theory and Practice*, *23*(4), 19–39. doi:10.1177/104225879902300402

Dyer, W. G., & Whetten, D. A. (2006). Family firms and social responsibility: Preliminary evidence from the S&P 500. *Entrepreneurship Theory and Practice*, *30*(6), 785–802. doi:10.1111/j.1540-6520.2006.00151.x

Feltham, T. S., Feltham, G., & Barnett, J. J. (2005). The dependence of family businesses on a single decision-maker. *Journal of Small Business Management*, *43*(1), 1–15. doi:10.1111/j.1540-627X.2004.00122.x

Gersick, K. E., Davis, J. A., Hampton, M. M., & Lansberg, I. (1997). *Generation to generation: Life cycles of the family business*. Harvard Business Press.

Glaesser, D. (2006). *Crisis management in the tourism industry*. Routledge. doi:10.4324/9780080464596

Gómez-Mejía, L. R., Haynes, K. T., Núñez-Nickel, M., Jacobson, K. J. L., & Moyano-Fuentes, J. (2007). Socioemotional wealth and business risks in family-controlled firms: Evidence from Spanish olive oil mills. *Administrative Science Quarterly*, *52*(1), 106–137. doi:10.2189/asqu.52.1.106

Hall, C. M. (2010). Crisis events in tourism: Subjects of crisis in tourism. *Current Issues in Tourism*, *13*(5), 401–417. doi:10.1080/13683500.2010.491900

Hedner, T., Abouzeedan, A., & Klofsten, M. (2011). *Entrepreneurial resilience*. Taylor & Francis. doi:10.3402/aie.v2i1.6002

Innerhofer, E., Fontanari, M., & Pechlaner, H. (2018). *Destination resilience: Challenges and opportunities for destination management and governance*. Routledge. doi:10.4324/9780203701904

Le Breton-Miller, I., & Miller, D. (2016). Family firms and practices of sustainability: A contingency view. *Journal of Family Business Strategy*, *7*(1), 26–33. doi:10.1016/j.jfbs.2015.09.001

Liu, A., & Pratt, S. (2017). Tourism's vulnerability and resilience to terrorism. *Tourism Management*, *60*, 404–417. doi:10.1016/j.tourman.2017.01.001

Pizam, A., & Smith, G. (2000). Tourism and terrorism: A quantitative analysis of major terrorist acts and their impact on tourism destinations. *Tourism Economics*, *6*(2), 123–138. doi:10.5367/000000000101297523

Ritchie, B. W. (2004). Chaos, crises and disasters: A strategic approach to crisis management in the tourism industry. *Tourism Management*, *25*(6), 669–683. doi:10.1016/j.tourman.2003.09.004

Ritchie, B. W., Crotts, J. C., Zehrer, A., & Volsky, G. T. (2014). Understanding the effects of a tourism crisis: The impact of the BP oil spill on regional lodging demand. *Journal of Travel Research*, *53*(1), 12–25. doi:10.1177/0047287513482775

UNWTO. (2017). UNWTO Tourism Highlights: Vol. 2017. Madrid, Spain: World Tourism Organization.

UNWTO. (2021a). International Tourism Highlights, 2020 Edition. Madrid, Spain: World Tourism Organization (UNWTO).

UNWTO. (2021b). *2020: Worst year in tourism history with 1 billion fewer international arrivals*. Retrieved from https://www.unwto.org/taxonomy/term/347

Veréb, V. N., Nobre, H., & Farhangmehr, M. (2018). The fear of terrorism and shift in cosmopolitan values. *International Journal of Tourism Cities*.

Zehrer, A. (2013). The culture of facing crisis in tourism – an empirical study on crisis competencies of future destination managers. In H. Pechlaner, D. Abfalter, & G. Dal Bò (Eds.), *Cultural entrepreneurship – Challenges and strategies. Culture Meets Economy* (pp. 43–62). EURAC Book.

Zehrer, A., & Leiß, G. (2019). Family entrepreneurial resilience–an intergenerational learning approach. *Journal of Family Business Management*.

Section 1

Being Resilient Throughout Crises

The first section, 'Being Resilient Throughout Crises', addresses organizational resilience and consists of three chapters. The first and the second chapter showcase the COVID-19 pandemic in Europe, and the third chapter presents a longitudinal study of family firms in Croatia along the countries historic periods of communism, war, and access to the European Union. While the first chapter chooses a descriptive analysis of statistical data available in the European Union, Chapters 2 and 3 concentrate on family-firm owner-managers and their behavior. Therefore, both chapters conducted semi-structured qualitative interviews with family firm owner-managers.

Chapter 1
Family Business Tourism Risk in the Context of Coronavirus Pandemic

Martina Harc
https://orcid.org/0000-0002-7207-9032
Croatian Academy of Sciences and Arts, Croatia

Martina Basarac Sertić
Croatian Academy of Sciences and Arts, Croatia

ABSTRACT

The coronavirus is crippling the global economy. However, the economic impacts of the pandemic vary significantly across sectors of the economy. This outbreak and related lockdowns are putting the tourism industry under unprecedented pressure. Within months, world tourism went from over-tourism to non-tourism. Therefore, the overall objective of this chapter is to introduce the latest trends in tourism, with emphasis on family businesses. More precisely, the chapter encompasses (1) the theoretical background, which defines family business, highlights their characteristics, and summarizes the importance of leadership succession; (2) the role of family business resilience and their behaviour in times of crisis; (3) a review of relevant European Commission policy publications regarding EU tourism and statistical analyses of selected key tourism indicators; and (4) an analysis of family businesses environment in time of crisis. Hence, this chapter has documented the crucial role that family businesses tourism has within the European Union and on the front line of the current crises.

DOI: 10.4018/978-1-7998-7352-5.ch001

INTRODUCTION

The coronavirus outbreak has paralysed the world economy. However, the economic impact of the pandemic differs greatly between various sectors of the economy. The impact and recovery potential depend, among other things, on the economic structure of each country, for example, those with a large number of small and medium-sized enterprises (SMEs) are more affected by the crisis (EC, 2020a). In that context, the exfoliation of the recession facing Europe is huge. For millions of Europeans, this will be a life-changing year. Never before, except in wartime, have whole industries lock down and raised up new problems so fast (McKinsey Global Institute, 2020a). It appears that the most affected economies are the ones highly dependent on tourism. Namely, this novel coronavirus (SARS-CoV-2), that causes COVID-19, is putting the tourism industry under unprecedented pressure. Within months, global tourism had shifted from over-tourism to non-tourism.

More precisely, five years prior to 2019, the tourism sector was responsible for one in four of all net new jobs created globally (WTTC, 2020a). Last year one in ten people worked in travel and tourism-related jobs (World Economic Forum, 2020), contributing with $8.9 trillion and generating 10.3%, to the global economy (WTTC, 2020a). That was before current pandemic crisis. While many popular destinations are cautiously welcoming visitors again, the industry is continuing to lose jobs and revenues. A huge drop in tourism will have a major influence on countries that depend on the tourism industry and could have large-scale effects on their national gross domestic product. The United Nations World Tourism Organisation (UNWTO, 2020) estimated in May that the number of international tourists could fall by 60 to 80 per cent in 2020 and that some 197.5 million jobs could be lost in this sector.

When it comes to European Union, tourism represents the third largest socioeconomic activity in the EU after the trade and distribution and construction sectors (EC, 2010). Moreover, taking into account the sectors linked to it, tourism's contribution to GDP is even greater (EC, 2010). Besides, tourism is the fourth largest EU export category and brings spill-over benefits to the European economy as a whole: 1 EUR of value added generated by tourism results in additional 56 cent of value added in indirect effect on other industries (EC, 2020b). Because of that, Europe was confronted by a public health challenge that quickly became the most drastic economic crisis in its history (EC, 2020c). With its long-term vision, governance and values-driven strategy, the family businesses can be role model for the global solution.

In that context, the importance of family businesses is unquestionable. Thus, what is clear beside that family businesses represent a vital part of the economy, are the implications that involve their commitment to local communities, the long-term stability they bring, the responsibility they bear as owners and the values they stand for (EC, 2009:22). The covid outbreak has led to major revenue drops for most of small and medium sized enterprises and threaten their survival. While at the same time the majority of global employment is depending on the pace of recovery of small and medium-sized enterprises, the future of the global economy will very much depend on how SMEs manage to get through and emerge from the crisis (ITC, 2020). Hence, it is important to analyse the environment in which family businesses operates. For countries such as Spain and Italy, the contribution of tourism to their GDP is even higher than the global average, 14.3% and 13.0% of their GDP in 2019, respectively (WEF, 2020). In Croatia for example, it contributes with around 20% to GDP, confirming the economy's high reliance on the tourism sector, which makes it particularly vulnerable to external shocks (Orsini and Ostojić, 2018). Therefore, this paper analyses the family business's tourism risk in the context of coronavirus in the European Union and introduce the latest trends in tourism, with emphasis on family businesses. According to the results of the conducted analysis, unlike other industry sectors, the loss of tourism revenue resulting in major economic and job losses followed by an uncertain future in the sector, particularly for family businesses and tourism related stakeholders. It is important to be analysed because family businesses think about next generations, succession and legacy in accordance with longevity improves prosperity. When well governed and operating with long-term perspectives, possesses the right competences for developing a resilience business models. Therefore, we ask: How family businesses (un)successfully deal with the crisis, and what future generations can learn from the crisis?

The rest of chapter encompasses: (1) the theoretical background which defines family business, highlights their characteristics and summarize the importance of leadership succession, (2) the role of family business resilience, and their behaviour in times of crisis, (3) a review of relevant European Commission policy publications regarding EU tourism and statistical analyses of selected key tourism indicators, and (4) an analysis of family businesses environment in time of crisis. The chapter also outlines the main recommendations and suggests future and emerging trends. It can be concluded that this chapter on tourism risks deals with the coronavirus epidemic which is testing Europe in a way that would have been unthinkable a few months ago and analyses the fact that the pandemic has changed economic prospects for the coming years in the Union.

THEORETICAL BACKGROUND

This section defines family business, highlights their characteristics and summarize the importance of family businesses. The section also analyses the role of family business resilience, and their behaviour in times of crisis.

Defining Family Businesses

The family businesses play an important role in dynamism and strength of the European economy, long-term stability and sustainability (Siakas et al., 2014). Family businesses are the cornerstone of most national economies because they create jobs, foster innovation, grow faster and generate superior returns (Credit Suisse Research Institute, 2018). Particularly, family businesses that operates within tourism and hospitality industry are very important in regional development (Zapalska and Brozik, 2007) and responsible for destination development (Glowka et al., 2020). Although many of the challenges facing family businesses are also concerning and SMEs (EC, 2009), they differ from other types of businesses (Siakas et al., 2014). The fact that family businesses involve three overlapping elements, namely the family, the business and the ownership, makes them different from other types of businesses (Taquiry and Davis, 1982).

Most of SMEs (especially micro and small enterprises) are family businesses and a large majority of family companies are SMEs (EC, 2009). More than 60% of all companies in Europe are family owned. They range from companies with one owner to large international enterprises. Large or small family-run companies, whether listed or unlisted, play an important role in the EU economy.

The family businesses are one of the most prevalent, persistent and common forms of business organization in the world (Hiebl et al., 2019, Nordqvist and Jack, 2020). Family businesses perform better and enjoy a stronger financial structure than non-family businesses suggesting that family businesses recover better or more easily from different kind of crisis and persist in stronger performance (Salvato et al., 2020; Hansen et al., 2018; Amann and Jaussaud, 2012; Siakas et al., 2014; Chrisman et al., 2011; Gueye and Simon, 2010). They are particularly resilient, referring to the ability of organizations to avoid, absorb, respond to, and recover from situations that could threaten their existence (Lengnick-Hall and Beck, 2005).

Miller and Le Breton-Miller (2003, p. 127) define a family business as 'one in which a family has enough ownership to determine the composition of the board, where the CEO and at least one other executive is a family member, and where the intent is to pass the firm on to the next generation'. Villalonga and Amit (2006) identified three key dimensions that characterise family businesses: (1) family members hold top management positions, (2) family members retain significant control over the company through the distribution of capital among non-family shareholders and, (3) a significant part of the capital is held by one or several families (from Amann and Jaussaud, 2012). Family businesses have long-term focus on their businesses rather than on short-term profits and prefer succession through generational changes in ownership and management, relative to non-family owned businesses. Thus, family businesses are on average older than non-family businesses (KMU Forschung Austria, 2008).

Family firms are ambitious, aiming to grow and ensure long-term success and stability to their businesses. They are vital to all economies, offering stability as well as commitment and responsibility to their communities and employees for the long term. Family businesses differ from non-family businesses in a way they stand for stronger culture and values, measure success in broader terms than just profit and growth, make decisions faster than their peers, endure entrepreneurial spirit and take a longer-term approach to decision making (PWC Family Business Survey, 2016).

Succession is an important and often challenging subject for any family business (KPMG, 2019). Although 70% of current family business leaders do not have a succession plan in place (Calabro and Valentino, 2019), family business can be considered as natural lasting and long-lived development (Gueye and Simon, 2010). In line with that, Zehrer and Leiβ (2019) explored the leadership succession in family business by using an intergenerational learning approach. Authors discussed how intergenerational learning develops resilient families in business during leadership succession. In the final phase, authors identified five practical implications (namely: reduce stress level, define competences and responsibilities, create a knowledge-oriented culture, build distance and establish adaption structures) explaining how families can improve their understanding of intergenerational learning and develop resilience at the same time during the leadership succession. Pittino et al. (2019) argue that increased generational involvement and the participation in ownership and management of the family firm enhance family firm's growth. Precisely, authors confirmed that with involvement of new generations of family members in family business, wealth of firm increases while strategic renewal becomes more important. Furthermore, with each succession, new family members introduce fresh knowledge and insight to the firm, positively affecting business to innovate and grow.

Family Business Resilience

Resilience, defined as ability to respond effectively and adopt positively to adverse events, have an important role in the sustainable tourism management and can help organisations to better cope with turbulences (Jiang et al., 2019). Despite the challenges posed by current crisis, family-owned businesses are proven as more resilient than non-family-owned businesses. Further, family owned companies offer safety in periods of tough times. In first six months of this year they outperformed non-family-owned companies, proving the family alpha factor is still present (Credit Suisse Research Institute, 2020).

The strong performance of family businesses aligns with several theoretical perspectives. In particular, the theory of organizational resilience suggests that a resilient company can take situation-specific, robust and transformative actions when confronted with unexpected events (Lengnick-Hall and Beck 2009, from Amann and Jaussaud, 2012). According to Nordqvist and Jack (2020) 'the ability of a company or an organization to persist the substantial changes in the external business and economic environment or the ability to withstand disruptions and catastrophic events – resilience – has become one of the most important factors to be able to manage the crisis caused by the COVID-19 pandemic'. The term 'resilience' has been used at the organizational level to describe how some companies are more successful in dealing with and responding to the unfamiliar. It refers to those organizations that are able to respond quicker, recover faster, or develop more successful ways of doing business under various types of adversity than others (Linnenluecke, 2017). In other words, resilience is a progressive state of organisational evolution representing organisation's constant change and adaptability (Jiang et al. 2019). Coutu (2002) highlighted three key dimensions that characterise resilient organisations: (1) facing reality, meaning don't lose sense of reality, (2) the search for meaning, either don't lose core direction in times of crisis and (3) ritualized ingenuity, referring to ability of sufficient using of available resources. Further, referring on resilience, Nordqvist and Jack (2020) claimed that we can learn about resilience in time of crisis and through leveraging specific advantages common in long-standing family businesses, namely: strong commitment to the family business, long-term orientation of family businesses and strong independence in the business. Also, authors discuss how to successfully deal with disadvantages of family businesses that threaten the organisational resilience such as dealing with dominant leader, coping with conflicts and limiting opportunities. Salvato et al. (2020) revealed that family firms display greater resilience when facing adversity and greater ability to turn adversities into entrepreneurial opportunities than their non-family counterparts. Authors concluded that resilience is main reason for family firm's comparatively greater longevity in regard to non-family firms.

Family Business in Times of Crisis

For any kind of tourism organisations, crises and disasters can be observed as severe occurrences that endanger the business activities of tourism organisations and damage the overall destination reputation by affecting tourists' perceptions, disabling organisations to recover using its own resources (Jiang et al., 2019). According to previous experiences, disasters and crises appear unexpected and occur regularly, present a shock for the tourism industry and are almost always poorly handled (Prideaux, 2003). In line with that, tourism industries are not well prepared to manage crises and disasters (Wang and Ritchie, 2012).

In recent publication Deloitte (2020) propose three phases that resilient leader must face in order to overcome the Covid outbreak: respond - in which s company deals with the present situation and manages continuity, recover – during which a company learns and emerges stronger and thrive -where the company prepares for and shapes the "new normal". Recovery phase includes workforce-related strategies: reflect, recommit, re-engage, rethink, and reboot. These five actions can help organizations to lead, prepare and support their workforce through the recovery phase while laying the foundation for the next phase – thriving in the new normal. Although some moments during the crisis will call for immediate action, in order to learn to emerge stronger, leaders need to find moments to stop, reflect, and think ahead before making further moves (McKinsey Global Institute, 2020b). In line with overcome the crisis, Wenzel et al. (2020) identify four types of strategic responses to crises: retrenchment, persevering, innovating and exit. Retrenchment is observed as response to crisis which refers to reduction in costs, assets, and product related activities. With potential to narrow the scope of a business activities, it is recognized as short term response to crisis. Second strategy, persevering refers to measures aimed at sustaining a firm's activities, preserving the status quo and mitigating the adverse impacts of crisis. It refers to sustaining firm survival in the medium run. In order to recognize crisis as an opportunity, innovating is observed as opportunity for strategic renewal in response to crisis. As valuable strategy, innovating refers to sustaining firm survival in the longer run if the crisis continues for a longer period. And finally, exit, as response to crisis, is unavoidable when the other responses fail. It refers to discontinuation of a firm's activities in response to crisis. Using the Wenzel's framework, Kraus et al. (2020) provided important and timely implications for family firm owners and managers by proposing an advanced model for changes during a crisis for short-term adaption and long-term firm positioning. Suggesting that first, companies must ensure liquidity, reducing their fixed costs through implementation of shortened work hours or any other appropriate government support. These short-term effects should be extended by identifying inefficiencies in the company, which could also help achieving long-term effects. Further, family firms that have the opportunity to continue operating in the wake of the crisis should take advantage of this opportunity and create the appropriate framework conditions for future actions. The aim is not only to survive the crisis but also to emerge from it stronger. Further, Omar et al. (2020) identified business challenges that SMEs are coping with during the immediate quarantine as operational problems (i.e. operation disruption; supply chain disruption; problems in future business direction) and financial related problems (i.e. cash flow problems; access to stimulus packages; risk of bankruptcy). Authors concluded that company's financial reserved and emergency funds are very efficient tools for business sustainability during the crisis period as well as other assets and skills allowing firms to react to the new business opportunities. Schwaiger and Zehrer (2020) argue that emotional attachment of the family affects the performance of family firms during a crisis. The importance of employees also highlighted as asset. Moreover, the long-established businesses, according to authors should be able to cope with several months of closure or zero revenue for a short period. Finally, authors concluded that intrinsic features of family businesses indicate a higher resilience than other organizational forms.

TOURISM IN THE EUROPEAN UNION AND IN THE CONTEXT OF THE CORONAVIRUS PANDEMIC

This section aims to improve the understanding of the recent development of tourism in the European Union, and to identify and signalize the issues associated with it. The section involves a review of relevant European Commission policy publications regarding EU tourism and statistical analyses of selected key tourism indicators. Hence, this section has documented the crucial role that tourism has in general and within the European Union.

Tourism in the European Union

Tourism made up 10 percent of global GDP in 2019 and was worth almost $9 trillion, making the sector nearly three times larger than agriculture (McKinsey Global Institute, 2020c). In that sense, "tourism is one of the largest and fastest growing sectors of the global economy" (Weston et al., 2019). According to the World Travel & Tourism Council (WTTC) it was worth $8.8 trillion in 2018 (around €7.9 trillion), outperforming the global economy for the eighth consecutive year (WTTC, 2019). When it comes to Europe, it is the leading tourist destination in the world whereby tourism is at the centre of a huge ecosystem of businesses, contributing substantially to prosperity and jobs in all Member States (EC, 2020d). As an activity that impinges on cultural and natural heritage and on traditions and contemporary cultures in the European Union, tourism provides a textbook example of the need to reconcile economic growth and sustainable development, including an ethical dimension (EC, 2010). Besides, tourism is an economic activity capable of generating growth and employment in the EU, while contributing to development and economic and social integration, particularly of rural and mountain areas, coastal regions and islands, outlying and outermost regions or those undergoing convergence (EC, 2010). Tourism is also a phenomenon with an important territorial dimension, uneven spatial distribution between and within countries and localised impacts, with many EU regions heavily dependent on this sector (Batista et al., 2019). Moreover, for the developed economies analysed by Zurub et al. (2015), tourism is a fundamental industry to EU's economic growth because it contributes significantly to the gross domestic product.

Considering the diversity of the activities and effects described above, it can be assumed that tourism is not "just" a sector per se, but rather a booster for other relevant sectors and activities (EC, 2020e). Tourism related activities tackle challenges which, if properly addressed, can become opportunities to boost the competitiveness of EU regions (EC, 2020e). European Commission (2020e) highlights that tourism initiatives can be instrumental in fostering additional results across complex value-chains: from its interlinkages with creative industries and upstream sectors such as agriculture or construction, to its essential role in fostering sustainable infrastructure and circular models for local economies, up to the increased embeddedness of tourism products/services into ICT and social media (Tourism 4.0).

Tourist services provide an important source of export earnings. According to Dupeyras and MacCallum (2013), the tourism services are an important driver in the shift towards a service economy. The positioning of tourism as an export and the ability to bring in payments to a country can change quickly (Dupeyras and MacCallum, 2013). This rate of change has been is further accelerated by globalisation, economic crisis, strong competition and fourth industrial revolution. Policy makers recognise that the long-term growth of tourism services depends not only on being able to satisfy customers' needs and desires, but to develop a global competitiveness position that is supported by evidence in the balance of payments (Dupeyras and MacCallum, 2013). Strategic vision, support for quality resources and promotion of services become vital for the growth in tourism services (Dupeyras and MacCallum, 2013).

Considering that tourism-related activity contributes to several sectors of the economy, the WTTC has coined the term 'visitor economy' to capture economic effects at all levels: direct, indirect, and induced effects (EC, 2020e). Namely, when the number of visitors increases, this directly results in increased consumption which is reflected throughout the entire supply chain (for instance maintenance services, equipment and food producers, transport services, etc.) (EC, 2020e). This also contributes to increased economic activity of workers in those sectors (who then spend the earned money in the economy). Thus, the effects of the tourism sector on the EU economy and employment extend further and, indirectly, its contribution to EU GDP is estimated at around 10% (EC, 2020e).

From the given review, it is evident that tourism is an important driver of economic policy of the European Union. Figure 1 provides evidence on total contribution of travel & tourism to GDP and total employment in European Union members, in 2019, thus points to a conclusion that the impacts of tourism on the EU economies are heterogeneous, with differences in the importance of tourism to domestic output, as well as employment effects. Hence, as illustrated in Figure 1, the importance of tourism also stems from the fact that as many as twelve-member states record a double-digit dependence of their GDP on tourism, with Croatia leading the way. In addition, if we consider the contribution of tourism to GDP/employment in the various countries, Croatia, Cyprus, Greece and Portugal lead the charts in terms of contribution to GDP and employment with values greater than 20%. According to EC (2020e) they are closely followed by a lot of 8 countries with values ranging between 10% and 20% for both indicators (such as Italy, Spain, Austria, United Kingdom, etc.) and an additional group of countries (e.g., Finland, the Czech Republic, Hungary, etc.) for which the contribution to GDP and employment is situated between 5% and 10%. On the other hand, Poland and Lithuania have the lowest percentages in the EU, but they are still around 5%, reflecting the important overall contribution of the sector to the economy and employment in all Member States (EC, 2020e). Generally, larger countries tend to have slightly higher percentages of tourism contribution to GDP (with Portugal and Austria being the exceptions) rather than smaller ones due to larger multiplicative effects in their economy (EC, 2020e).

Figure 1. Total contribution of travel & tourism to GDP and total employment in European Union members, in 2019
Source: WTTC

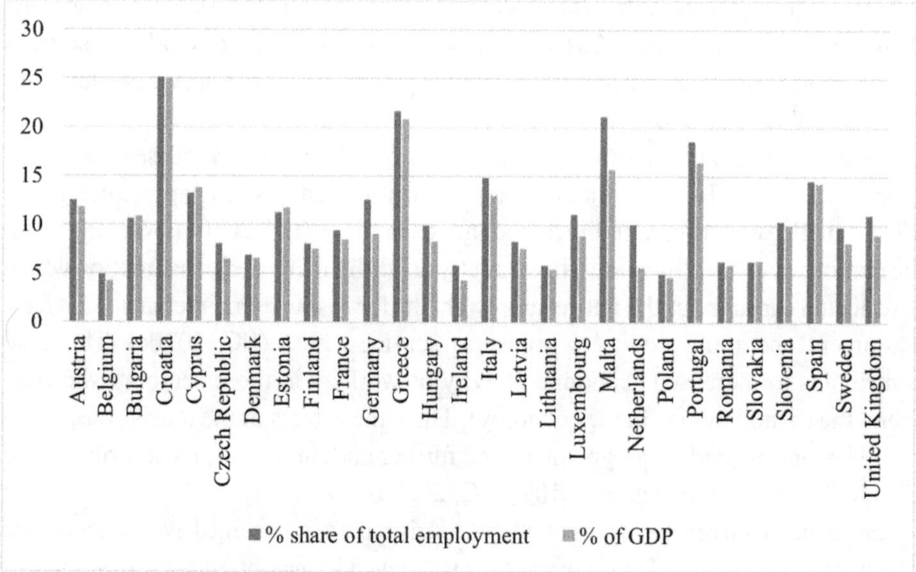

Figure 2 shows the number of international tourist arrivals in Europe from 2010 to 2019. The number of international tourist arrivals rose from 490.6 in 2010 to 745.2 million arrivals in 2019.

Figure 2. Number of international tourist arrivals in Europe from 2010 to 2019 (in millions)
Source: Statista (UNWTO World Tourism Barometer July 2020, page 6)

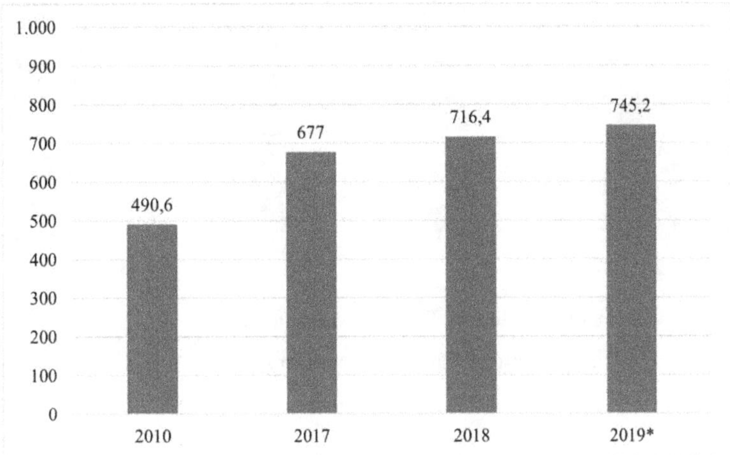

Figure 3 provide additional evidence on international tourist arrivals in Europe from 2010 to 2019, by region. According to Statista (2020), in Europe, the region with the highest number of tourist arrivals in 2019 was the Southern/Mediterranean region, with 304.1 million. Comparatively, the region with the least international tourists was Northern Europe, with 80.2 million visitors in 2019 (Statista, 2020).

Moreover, tourism has the potential to play a significant role in the economic aspirations of many European Union regions and can be of particular importance in remote/peripheral regions, such as the EU's coastal, mountainous or outermost regions (Eurostat, 2020). Infrastructure that is created for tourism purposes contributes to local and regional development, while jobs that are created or maintained can help counteract industrial or rural decline (Eurostat, 2020).

Tourism in the European Union in the Context of Coronavirus Pandemic

More than 1.5 million lives have been lost to COVID-19 since the beginning of the year and the number of victims continues to grow. Consequently, the literature

Figure 3. International tourist arrivals in Europe from 2010 to 2019, by region (in millions)
Source: Statista (UNWTO World Tourism Barometer July 2020, page 6)

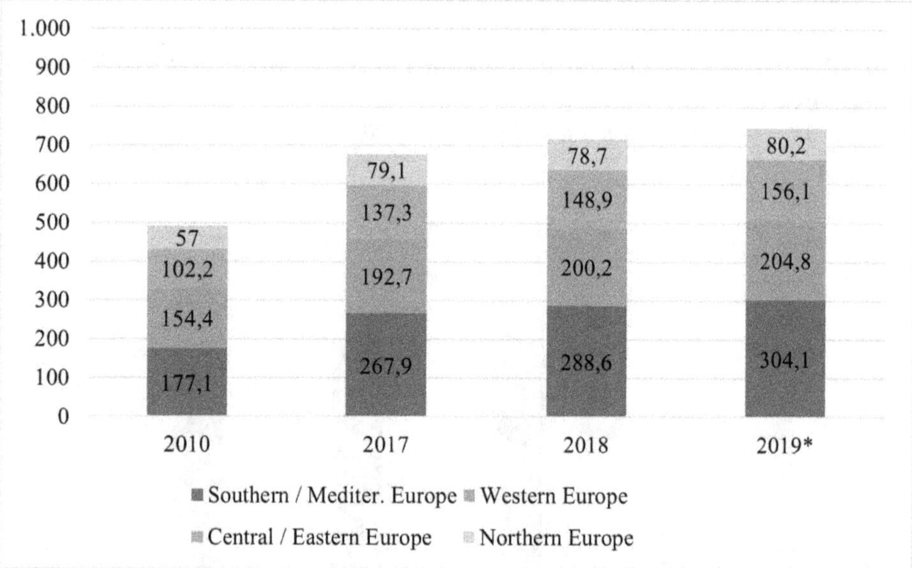

on the economic crisis triggered by the coronavirus pandemic has been expanding at a very rapid pace (IMF, 2020). When it comes to tourism, it is a sector in the European Union that makes a significant direct and indirect positive contribution to EU economic growth, job creation and safeguarding, and SME activities. Although it is resilient, the sector faces shared EU-wide challenges, including the need to ensure sustainability and digital transformation (EC, 2020e). However, in the context of the current crisis, EU tourism is one of the most affected. Hence, besides the difficult background in the tourism industry highlighted in a number of documents in previous years, now the European tourism faces new major challenges. Namely, before COVID-19, most tourism ministries and authorities focused on destination marketing, industry promotions, and research (McKinsey Global Institute, 2020c). Many are now dealing with a raft of new regulations, stimulus programs, and protocols (McKinsey Global Institute, 2020c).

This is supported by the last year's WEF analysis: *Travel and Tourism at a Tipping Point* (WEF, 2019) which showed the healthy growth of the industry, with increased competitiveness worldwide set against the slower improvement and adoption rates of necessary infrastructure and sustainable tourism management practices respectively. However, the outbreak of the COVID-19 pandemic has altered economic prospects for the years ahead in the Union.

Chronologically, on 30 December 2019 the first case of coronavirus infection was confirmed in Wuhan, People's Republic of China. This novel coronavirus, known as covid-19, has spread from the People's Republic of China to other countries. On 30 January 2020 following the recommendations of the Emergency Committee, the World Health Organization (WHO) declared that the outbreak constitutes a Public Health Emergency of International Concern (PHEIC). Further, on 11 March 2020 coronavirus confirmed as pandemic by WHO. The singularity of the current situation lies in the spread of the virus, its geographical coverage and the measures implemented by governments to stop its dissemination (mobility and travel restrictions, lockdown, confinement, closure of shops and hotels, etc.) (Marques Santos et al., 2020). Currently, the world is struggling for vaccine and proper treatment for this pandemic. Unprecedented lockdown, travel restrictions, economic crisis, lock on import-export activities and others have raised a serious challenge before the world to explore the solutions for revival from it (Kumar, 2020).

The contraction of the economy would begin with the diminishing of the travel and tourism sectors. According to Kenneth Rogoff (Professor of Economics, Harvard University) "The short-term collapse in global output now underway already seems likely to rival or exceed that of any recession in the last 150 years" (World Bank, 2020a). Moreover, according to Jerome Powell (Chair, The U.S. Federal Reserve System) "The scope and speed of this downturn are without modern precedent, significantly worse than any recession since World War II" (World Bank, 2020a).

Namely, as can be seen from the Figures 4 and 5, since 1870, the global economy has experienced 14 global recessions[1] and in each of these episodes, there was a contraction in global real per capita GDP (World Bank, 2020b). Current projections imply that the covid-19 global recession will be the fourth deepest in this period and the most severe since the end of World War II (World Bank, 2020b). According to World Bank (2020b) analysis, economic activity collapsed in the first quarter as a result of the coronavirus outbreak and related lockdowns and closures, although there is evidence of a bottoming out.

In particular, as shown in Figure 6, the fall in activity caused by the coronavirus has been concentrated in services sectors that are typically stable. Travel restrictions and concerns about coronavirus have led to a precipitous fall in tourism, a sector that in recent years has accounted for about 7 percent of global exports of goods and services, with sharp declines in economies with the most severe outbreaks (World Bank, 2020b). Moreover, according to Figure 7 which shows an employment loss in the travel and tourism industry due to the coronavirus pandemic worldwide in 2020, by region (in millions), it is evident that more than 100 million people have lost their jobs (with the highest number of redundancies recorded in Asia and Europe).

Aside from the negative impact on the labour market, the coronavirus has consequently had a negative impact on GDP. As can be seen from the data in Table

Figure 4. Global GDP (Index, 1870 = 100)
Source: World Bank (2020b)

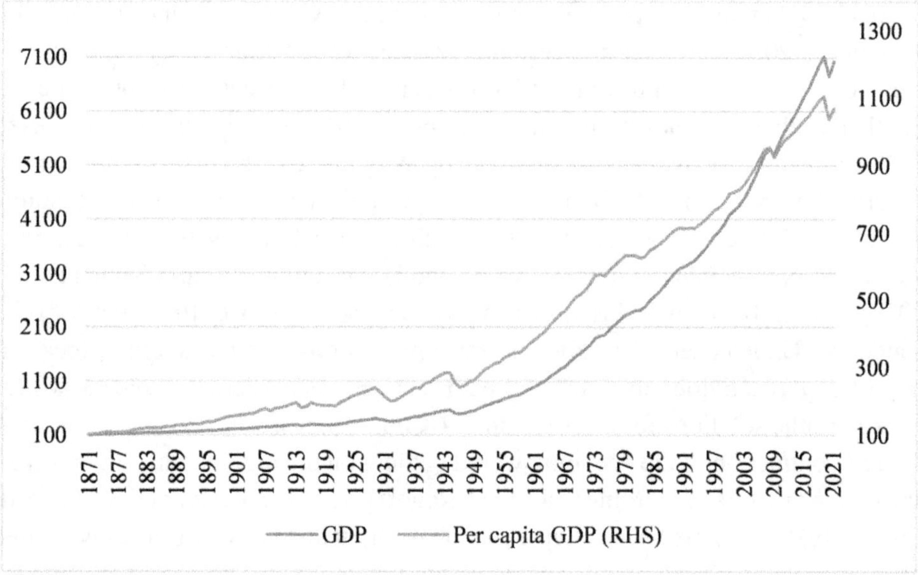

Figure 5. Global GDP growth
Source: World Bank (2020b)

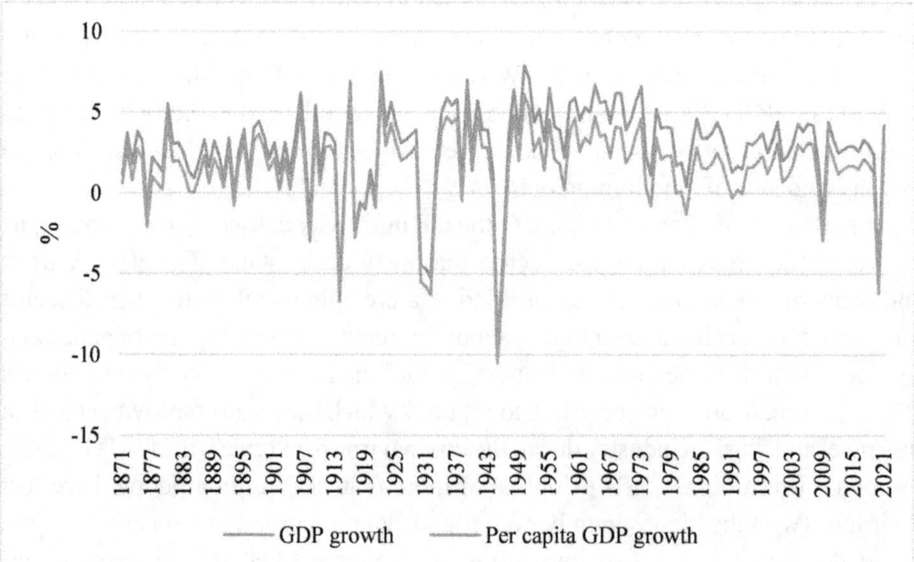

Figure 6. Global manufacturing and services PMI
Note: Purchasing Managers' Index (PMI). PMI readings above (below) 50 indicate expansion (contraction) in economic activity.
Source: World Bank (2020b)

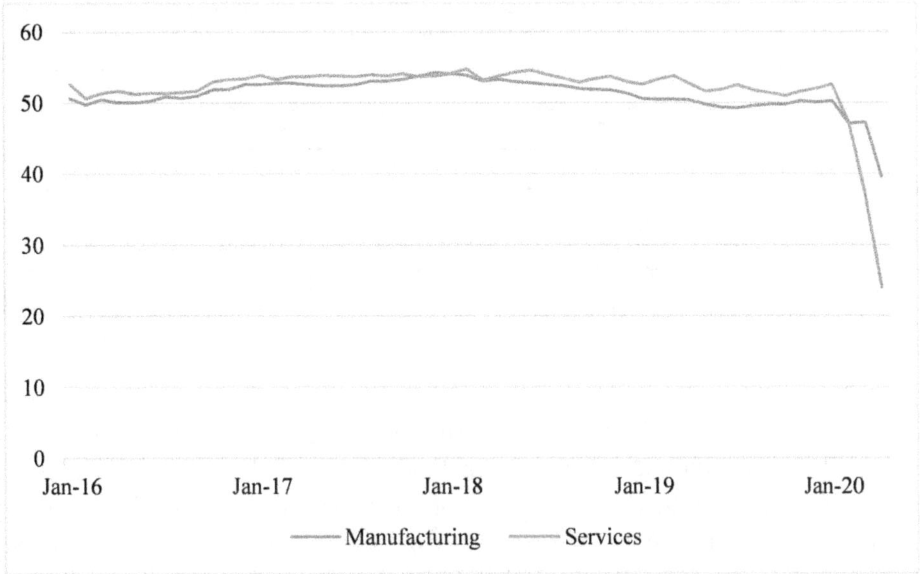

Figure 7. Employment loss in the travel and tourism industry due to the coronavirus (COVID-19) pandemic worldwide in 2020, by region (in millions)
Source: WTTC

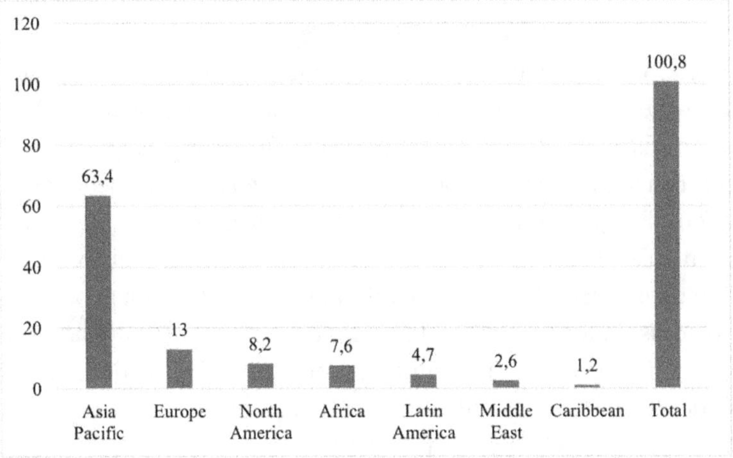

Table 1. Loss in GDP from jobs at risk in the travel and tourism industry due to the coronavirus (COVID-19) pandemic worldwide in 2020, by region (in billion U.S. dollars)

	Jobs at Risk	GDP Loss in Billion U.S. Dollars
Asia Pacific	63,4	1.041
Europe	13	708,5
North America	8,2	680,7
Middle East	2,6	96,2
Latin America	4,7	83,8
Africa	7,6	52,8
Caribbean	1,2	26,4

Source: WTTC

1, the loss of GDP in billions of American dollars is recorded on all continents, especially in Asia, Europe and North America.

It is clear therefore that the tourism sector is facing an unprecedented crisis, with highly uncertainty in its recovery. According to availability of the second quarter data, the impact of pandemic-related travel restrictions has become apparent, including the vulnerability of destinations that are heavily dependent on the tourism sector and on international source markets (ETC Quarterly report (Q2/2020). International, regional and local travel restrictions immediately affected not only tourism sector but national economies (Gossling et al., 2021). Due to the latest World Economic Outlook (WEO) forecast, global growth is predicted at –4.9% in 2020. The World Travel & Tourism Council (WTTC) estimated that the global economic impact on the tourism industry would be more than 5 times larger than the impact of the 2008 global financial crisis (WTTC, 2020b). To date, if new coronavirus cases continue to rise and Europe experiences a severe second wave of infections, the potential scenarios for the next coming months (from June to December 2020), foresee a decrease of tourist arrivals in EU up to 68% in 2020 (Marques Santos et al., 2020). Less severe scenarios foresee a drop between 38% and 52%. Subsequently, between 6.6 and 11.7 million jobs could be at risk of reduction in working hours or permanent jobs losses in 2020. These jobs at risk represent between 3.2% and 5.6% of the total active population of EU (Marques Santos et al., 2020). Croatia as well as Austria, Slovenia, Cyprus, Malta, Greece and Spain are the countries with the highest share of jobs at risk, based on the contribution of tourism to their national employment, their dependence on international tourists and the share of temporary employees in related sectors (Marques Santos et al., 2020).

The outbreak of current crisis has impacted countries at different times, in different ways and in varying degrees (UNWTO, 2020). Therefore, certain countries and regions will be more affected than others. Regions with an ecosystem strongly dependent on the tourism sector will be probably the most affected (Marques Santos et al., 2020). In the second quarter of 2020, total services turnover was reduced by 16.5% in the EU-27 and by 18.1% in the euro area. The highest decline during the second quarter was recorded in Croatia (-35.0%). Accommodation and food services recorded a decline of 55.4% in the EU-27 and 57.1% in the euro area. The strongest decline was registered in Malta (-94.8%), Spain (-78.2%), Portugal (-66.4%), Croatia (-66.3%), and Italy (-62.6%) (Eurostat, 2020). Inability in maintaining its core activity in the period of lockdown - accommodation services, resulted in strongest decline within total services sector. Alternative such as teleworking was not possible in accommodation sector.

According to McKinsey Global Institute (2020c) forecasts, it will take four to seven years for tourism demand to return to 2019 levels, which means that overcapacity will be the new normal in the medium term. This prolonged period of low demand means that the way tourism is financed needs to change. The aforementioned types of policies are expensive and will be difficult for governments to sustain over multiple years. They also might not go far enough. A recent OECD survey[2] of SMEs in the tourism sector suggested more than half would not survive the next few months, and the failure of businesses on anything like this scale would put the recovery far behind even the most conservative forecasts (McKinsey Global Institute, 2020c).

FAMILY BUSINESSES RESILIENCE IN TOURISM IN THE CONTEXT OF THE CORONAVIRUS PANDEMIC

The severity of the pandemic for the family businesses is obvious. Around 80% of the family-owned companies reported that their business had been negatively impacted by the pandemic. Moreover, the family businesses ranked pandemic as the second-biggest challenge in the next five years after the need to innovate and retain staff (Credit Suisse Research Institute, 2020). According to IMF 2020, small and medium enterprises are more vulnerable than their larger counterparts due to, among other factors, their limited access to credit. The effects of the current crisis on SMEs are likely to be even more severe than in previous crises because of the prevalent of SMEs in the hardest-hit sectors such as restaurants and hotels. (IMF, 2020: 57). Furthermore, small companies that provide 70% of jobs in countries around the world and about half of economic activity with workers and customers under lockdown, and supply chains tested by shutdowns, have been put under stress (ITC, 2020:12). A recent study (McKinsey Global Institute, 2020d) on the employment impact of the pandemic in Europe articulated that "at least two of three jobs at risk are in an SME, and more than 30 percent of all jobs at risk are found within microenterprises consisting of nine employees or fewer". Furthermore, it includes 70 percent of SME jobs at risk in the accommodation and food sector.

According to Welter and Kranzusch (2020) in the past, the German "Mittelstand" businesses (SMEs) has proven their resilience in times of crisis, due to their stronger long-term orientation in relation to larger companies, as well as due to their overall lower dependence on foreign markets. Though the current crisis may be different from the previous ones, regarding the crisis resilience of Mittelstand businesses, the IfM Bonn developed two scenarios to assess the impact of the crisis on SMEs. In the first scenario (lockdown limited to 2-2.5 months) for most Mittelstand businesses, a short-term crisis would not lead to liquidation. But the existence of companies particularly at risk are in consumption-oriented and manufacturing sectors. In the case of second scenario (lasting more than six months) a higher rate of liquidation and consequently spreads to other economic sectors can be expected. A first estimates assumed that a significant loss of jobs between 850 000 and 1.6 million could be expected among SMEs, especially among microenterprises.

Nevertheless, the ITC (ITC, 2020) analysed 4,467 companies in 132 countries in order to gather the evidence of the pandemic impact on SMEs. Analysis shows that the pandemic has strongly affected the majority (55%) of business respondents. Further, according to collected data, nearly two-thirds of micro and small firms reported that the crisis strongly affected their business operations, while 40% of large companies reported as affected. One-fifth of SMEs responded that they are under the risk of shutting down businesses permanently within three months. As presented in Figure 8, the analysis shows as well that the service companies have been hit the hardest. According to sectoral impact, in accommodation and food services, 76% of surveyed companies responded that lockdowns strongly affected their business operations. More than three quarters of companies in this sector experienced a reduction in sales. (ITC, 2020:14)

According to ITC analysis, the businesses diverged in their business approaches to pandemic crisis. Some companies adopted retreating strategies as responses to the crises, undermining the long-term competitiveness of the company. Many companies adopting this strategy laid off employees, sold off assets or get on new debt, at the same time disrupting their long-term viability. About 20% of the businesses that participated in the ITC analysis accepted this type of approach. Analysis shows as well that small exported oriented companies were significantly less likely to take the retreating approach than domestically oriented companies.

Many other companies (64% of large and 59% of SMEs) adopted resilience as business strategy to come over the crisis, as can be seen from the Figure 9 below. Following resilience strategy, companies use win-win strategy scaling down or

Figure 8. The scale of business operations affected by the coronavirus pandemic
Source: ITC, 2020.

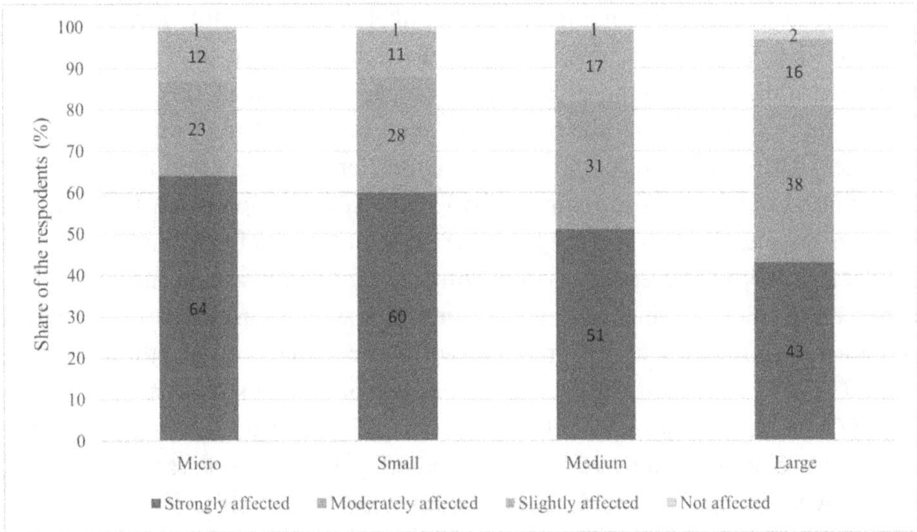

adjusting the business temporarily in order to emerge from the crisis as good as before, if not even stronger. Many companies being resilient use resilient approaches such as, moving their sales towards online channels, finding new suppliers or teleworking. As a result, resilient companies survive the crisis.

An even more constructive response to the crisis is agility. The most agile firms transformed themselves to fit the new situation, creating new products according to

Figure 9. The scale of business approaches to crisis
Source: ITC, 2020.

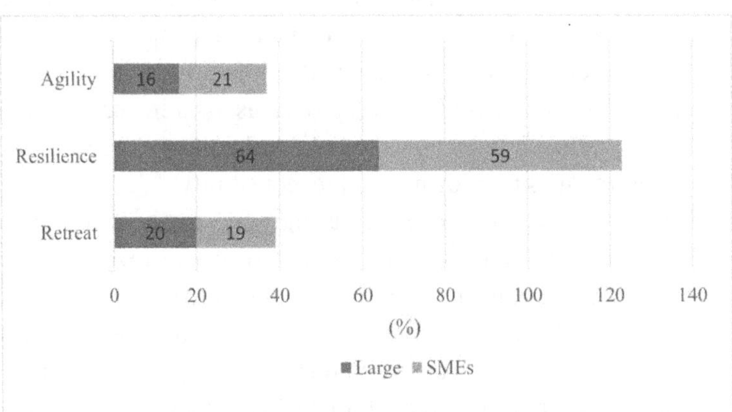

the newest trends. It includes new business models in response to current situation such as designer masks or rapid testing technologies. The analysis reveals that smaller firms (21%) were significantly more likely to adopt agility as responses to the crisis than larger enterprises. At the same time, however, they also tended to adopt retreating strategies more than bigger companies (Figure 9). (ITC, 2020:15,16)

The study conducted by the Suisse Research Institute (Credit Suisse Research Institute, 2020) explored through the lens of the current pandemic how family-owned companies differ from those companies where the family is no longer at the heart of decision making. Prior, the severity of the crisis is obvious for family and non-family businesses alike, both agreeing that there will be no return to business as usual. Family-owned companies are characterized as companies with a longer-term focus when running their businesses, which provides them more robust returns on a through-cycle basis. The study clarified that during the periods of market stress (such as crises) family-owned companies tend to have above-average defensive characteristics that allow them to perform well and generate superior growth, regardless their size. The family-owned companies, as declared not only generate stronger revenue growth, but also tend to be more profitable than their non-family-owned peers. Using cash flow return on investment as an indicator in order to review profitability, the study showed a clear and consistent degree of outperformance by family-owned companies on a global level as well as for each of the regions covered by database in the study. Furthermore, in the study there have been calculated net-debt-to-EBITDA levels for family-owned companies and compared to their non-family-owned companies. The data clearly suggested that, on average, family-owned companies do run their operations with lower gearing levels. Moreover, the study revealed that during the previous Financial Crisis, family-owned companies reduced debt levels more rapidly than their non-family-owned peers, both smaller and larger family-owned companies. The results confirmed their previous work (Credit Suisse Research Institute, 2018) on family-owned companies which highlighted that the stronger through-cycle financial performance might be the result of a longer-term focus of family companies having on their businesses compared to non-family-owned companies.

Finally, the more conservative financial model of family-owned companies built on lower leverage and strong cash flow has proved as asset in current and previous crisis. Notably, they relied less on government employment support to retain their workforce, conceiving it as their own social responsibility.

Although the latest forecasts show an increasingly negative impact of the pandemic on global GDP growth, the most recent business surveys show some confidence improvements, possibly due to the significant policy efforts to address the SME liquidity gap and the lifting of lockdown measures. Consequently, the previous support measures for SME survival are gradually shifting from liquidity support measures to recovery support measures (OECD, 2020a:1,2).

RECOMMENDATIONS AND MANAGERIAL IMPLICATIONS

Therefore, the solution is likely to be on national levels, as each economy is differently affected by COVID-19. In addition, European members rely differently on tourism revenues. Reshaped strategies, new publications adopted by the European Commission, SMEs guidelines and help packages and NextGeneration EU are some of the strategic frameworks that could play a key role in the post Covid-19 recovery process. The measures and their implementation should start immediately, and they should relate to the short, medium and long-term. These actions are required so that tourism sector can continue in its fight and returned to pre-crisis levels as soon as possible. It is clear that a one-size-fits-all solution does not exist, as the challenges of a tourism comeback will be different depending on the unique characteristics of each destination within and between countries (OECD, 2020b).

Hence, in order to help EU countries to start gradually lift travel restrictions and allow tourism businesses to reopen, the European Commission in May 2020 released a package of guidelines, recommendations and actions (EC, 2020b). One of the first directions of this package is a Re-open EU – an interactive tool providing the essential information to travellers to plan European travel and holidays while respecting necessary health precautions. Further, to address the crisis the Commission's efforts continue with recovery plan including support to EU countries for investment and reforms (EC, 2020a).

Next, in May 2020, "*Coronavirus response in relation to tourism: 2020 Tourism and transport in 2020 and beyond*", adopted by the Commission, provides "guidance to the Member States so that they can gradually lift restrictions to free movement and internal border controls in a phased and coordinated manner", in line with epidemiological and public health criteria.

A new initiative, REACT-EU known as initiative that continues and extends the crisis response and crisis repair measures, will provide support to EU countries, with a budget of €47.5 billion (EC, 2020c). The budget is available from 2020 and distributed according to the impact of the crisis. The objective of the budget is to ensure the ongoing financing of key measures to address crises, while supporting countries most in need. Beside main purpose, the initiative aims in support workers and SMEs, health systems and the green and digital transitions across sectors – from tourism to culture (EC, 2020f). The overall recovery plan also includes a Recovery and Resilience Facility of €672.5 billion of loans and grants in frontloaded financial support for the crucial first years of the recovery (EC, 2020g). The Facility will help repair the damage from the crisis and prepare a better future for the next generations (EC, 2020g).

Furthermore, since June 2020, all 27 EU countries have been progressively opening their borders in order to allow people to travel in and around the Europe.

As an additional incentive, on 1 July 2020, the European Commission adopted the European Skills Agenda for sustainable competitiveness, social fairness and resilience. This Agenda aims to improve the relevance of skills for all working age people in the EU to strengthen sustainable competitiveness, ensure social fairness and build resilience through 12 “actions”. A first building block of the Skills Agenda is to foster cooperation through a Pact for Skills (EC, 2020f). The Pact will initially focus on the industrial ecosystems heavily affected by the current crisis and the priority areas identified in the European Green Deal. Tourism has been identified as one of the four priority sectors (EC, 2020f).

In the family business context, companies require dynamic and flexible risk management to navigate an unpredictable future in which change comes quickly (McKinsey Global Institute, 2020d). Moreover, according to McKinsey Global Institute (2020d) nearly all organizations need to refresh and strengthen their approach to risk management to be better prepared for the next normal. Dynamic risk management has three core component activities: detecting potential new risks and weaknesses in controls, determining the appetite for risk taking, and deciding on the appropriate risk management approach (McKinsey Global Institute, 2020d). However, today many firms see enterprise risk management as a dreary necessity but hardly a source of dynamism or competitive advantage. It can suffer from being static and separate from the business. But dynamic and integrated risk management, which includes the ability to detect risks, determine appetite, and decide on action in real time, is growing ever more critical (McKinsey Global Institute, 2020d).

Additionally, the Davos Agenda 21 highlighted four ways family businesses can lead the pandemic recovery (Ye, 2021): i. longevity improves prosperity (because public companies operate on a quarterly basis; while families businesses operate under the principle of generational legacy), ii. exercise responsible ownership (because families with a controlling interest in their business have the ability to implement radical changes throughout the organization, quickly and efficiently), iii. business for good (because with a holistic ownership approach, families can ensure a positive impact in business, family office and family giving) and iv. embrace stakeholder capitalism. The family business model has its virtues and weaknesses, but these examples reveal that a better model of capitalism is possible (Ye, 2021). When shareholders, the board and management are well governed and aligned under strong values, businesses can drive positive change with profound impact (Ye, 2021).

It is therefore clear that policymakers in many countries are being pressured to respond to the coronavirus epidemic (Ozili and Arun, 2020). Due to the European Union’s and national measures such as special lay-off schemes, special bank loans or subsidies to support liquidity, short-time work schemes, working from home or teleworking, most companies have survived the pandemic (Marques Santos et al. 2020).

FUTURE RESEARCH DIRECTIONS

When it comes to future research, after the end of the pandemic, it will be necessary to look at the full economic and other consequences caused by the coronavirus in the tourism sector, how many companies survived, and how many went out of business. Then it would be desirable to explore new directions that would ensure the sector's fastest recovery. For instance, we find it important to include in future research questions similar to: how improving risk management in the companies can help to keep pace with the changing environment characterized by an unexpected and catastrophic events which are expected to occur more frequently in the future. In near future, companies need to rethink, refresh and strengthen their approach to risk management to be better prepared for the next normal.

CONCLUSION

This chapter highlights the importance of tourism. Moreover, tourism is becoming one of the fastest growing sectors in the European economy, thereby demonstrating its remarkable resilience and flexibility (WTO, 2018). For many Member States, European regions and cities, tourism is a key contributor to the economic and social prosperity. It also provides much needed employment and revenue. Tourism contributes 10% of European Union GDP, it is a global leader with 40% of international arrivals, 2.4 million businesses, over 90% of which SMEs, and € 0.56 of added value for every €1 generated in tourism (EC, 2020d).

However, the coronavirus pandemic has directly resulted in huge economic contraction, GDP and job losses. This disease has caused the collapse of air traffic and travel, and tourism in general. The coronavirus outbreak is testing Europe in a way that we couldn't imagine just a few months ago. The outbreak of the COVID-19 pandemic has altered economic prospects for the coming years in the EU. Based on descriptive analysis, the implications of pandemic travel restrictions and national lockdown have become evident, including the vulnerability of destinations which rely heavily on the tourism sector and foreign markets. As family-owned enterprises are an important part of the tourism industry, they simultaneously play an important role in the EU economy. Unfortunately, a recent OECD study of SMEs in the tourism sector suggested more than half would not outlive the following few months, and the failure of businesses on anything like this scale would put the recovery far behind even the most conservative forecasts. Hence, this chapter provides theoretical background by defining family business, highlighting their characteristics and summarizing the importance of leadership succession. In that context, it is emphasized the fact that despite the challenges posed by current crisis, family-owned businesses are proven as more resilient than non-family-owned businesses. Further, family owned companies offer safety in periods of tough times. The Credit Suisse Research Institute (2020) research shows that in first six months of this year family-owned business outperformed non-family-owned companies, proving the family alpha factor is still present. Therefore, their characteristics of resilience in times of crisis have proven to be particularly important.

REFERENCES

Amann, B., & Jaussaud, J. (2012). Family and non-family business resilience in an economic downturn. *Asia Pacific Business Review*, *12*(1), 202–223. doi:10.1080/13602381.2010.537057

Batista e Silva, F., Kavalov, B., & Lavalle, C. (2019). Socio-economic regional microscope series - Territorial patterns of tourism intensity and seasonality in the EU. *Publications Office of the European Union, JRC117669*. Advance online publication. doi:10.2760/061885

Calabro, A., & Valentino, A. (2019). *STEP2019 Global Family Business Survey - REPORT The impact of changing demographics on family business succession planning and governance*. Deloitte Development LLC.

Chrisman, Chua, J. H., & Steier, L. P. (2011). Resilience of Family Firms: An Introduction Entrepreneurship. *Theory into Practice*, *35*(6), 1107–1119. doi:10.1111/j.1540-6520.2011.00493.x

Coutu, D. L. (2002). How resilience works. *Harvard Business Review*, *80*, 46–56. PMID:12024758

Credit Suisse Research Institute. (2018). *The CS Family 1000 in 2018. Credit Suisse AG*. Research Institute.

Credit Suisse Research Institute. (2020). *The Family 1000: Post the pandemic. Credit Suisse AG*. Research Institute.

Deloitte. (2020). *Workforce strategies for post COVID-19 recovery*. https://www2.deloitte.com/content/dam/Deloitte/us/Documents/about-deloitte/workforce-strategies-for-post-covid-19-recovery.pdf

Dupeyras, A., & MacCallum, N. (2013). *Indicators for Measuring Competitiveness in Tourism: A Guidance Document*. OECD Tourism Papers, 2013/02, OECD Publishing. https://www.oecd.org/cfe/tourism/indicators%20for%20measuring%20competitiveness%20in%20tourism.pdf

European Commission. (2001). *Working together for the future of European tourism.* COM(2001) 665 final. https://eur-lex.europa.eu/LexUriServ/LexUriServ.do?uri=COM:2001:0665:FIN:en:PDF

European Commission. (2007). *Agenda for a sustainable and competitive European tourism.* COM(2007) 621 final. https://eur-lex.europa.eu/legal-content/EN/TXT/PDF/?uri=CELEX:52007DC0621&from=EN

European Commission. (2009). *Overview of family–business–relevant issues: research, networks, policy measures and existing studies.* Final report of the expert group.

European Commission. (2010). *Europe, the world's No 1 tourist destination – a new political framework for tourism in Europe.* COM(2010) 352 final. https://eur-lex.europa.eu/legal-content/EN/TXT/PDF/?uri=CELEX:52010DC0352&from=EN

European Commission. (2012). *Implementation and development of the common visa policy to spur growth in the EU.* COM(2012) 649 final. https://eur-lex.europa.eu/legal-content/EN/TXT/PDF/?uri=CELEX:52012DC0649&from=EN

European Commission. (2014). *A European Strategy for more Growth and Jobs in Coastal and Maritime Tourism.* COM(2014) 86 final. https://eur-lex.europa.eu/LexUriServ/LexUriServ.do?uri=COM:2014:0086:FIN:EN:PDF

European Commission. (2020a). *Europe's moment: Repair and Prepare for the Next Generation.* COM(2020) 456 final. https://eur-lex.europa.eu/legal-content/EN/TXT/PDF/?uri=CELEX:52020DC0456&from=EN

European Commission. (2020b). *Tourism and transport in 2020 and beyond.* COM(2020) 550 final. https://eur-lex.europa.eu/legal-content/EN/TXT/PDF/?uri=CELEX:52020DC0550&from=EN

European Commission. (2020c). *Commission welcomes the political agreement on REACT-EU.* European Commission - Press release. https://ec.europa.eu/commission/presscorner/detail/en/IP_20_2140

European Commission. (2020d). *The EU helps reboot Europe's tourism.* Factsheet, FS/20/851. https://ec.europa.eu/commission/presscorner/detail/en/FS_20_851

European Commission. (2020e). *Study on the Contribution of Tourism to Local and Regional Development - Evidence from the European Structural and Investment Funds 2012-2018 EASME/COSME/2018/034.* Final report. Luxembourg: Publications Office of the European Union.

European Commission. (2020f). *Tourism and employment: how the severity of future coronavirus waves could impact jobs*. https://ec.europa.eu/jrc/en/news/tourism-and-employment-how-severity-future-coronavirus-waves-could-impact-jobs

European Commission. (2020g). *Annual Sustainable Growth Strategy 2021*. COM(2020) 575 final. https://eur-lex.europa.eu/legal-content/EN/TXT/PDF/?uri=CELEX:52020DC0575&from=EN

European Travel Commission Quarterly Report. (2020). *European tourism: trends & prospects, Quarterly report (Q2/2020)*. https://etc-corporate.org/uploads/2020/07/ETC-Quarterly-Report-Q2-2020_Final-Public.pdf

Eurostat. (2020). *Tourism statistics at regional level, Statistics Explained*. https://ec.europa.eu/eurostat/statistics-explained/pdfscache/1945.pdf

Glowka, G., Kallmünzer, A., & Zehrer, A. (2020). Enterprise risk management in small and medium family enterprises: The role of family involvement and CEO tenure. *The International Entrepreneurship and Management Journal*. Advance online publication. doi:10.100711365-020-00682-x

Gössling, S., Scott, D., & Michael Hall, C. (2021). Pandemics, tourism and global change: A rapid assessment of COVID-19. *Journal of Sustainable Tourism*, *29*(1), 1–20. doi:10.1080/09669582.2020.1758708

Gueye, S., & Simon, E. (2010). Is the family business a safer type of governance in time of crisis? *Problems and Perspectives in Management*, *8*(4).

Hansen. (2018). *Family Firm Performance over the Business Cycle: A Meta-Analysis*. Research Papers in Economics No. 6/18. https://ssrn.com/abstract=3288094

Hiebl, Duller, C., & Neubauer, H. (2019). Enterprise Risk Management in Family Firms: Evidence from Austria and Germany. *The Journal of Risk Finance*, *20*(1), 39–58. doi:10.1108/JRF-01-2018-0003

International Monetary Fund. (2020). *World Economic Outlook: A Long and Difficult Ascent*. https://www.imf.org/en/Publications/WEO/Issues/2020/09/30/world-economic-outlook-october-2020#Full%20Report%20and%20Executive%20Summary

International Trade Centre. (2020). *SME Competitiveness Outlook 2020: COVID-19: The Great Lockdown and its Impact on Small Business*. ITC. https://www.intracen.org/uploadedFiles/intracenorg/Content/Publications/ITCSMECO2020.pdf

Jiang, Y., Ritchie, B. W., & Verreynne, M.-L. (2019). Building tourism organizational resilience to crises and disasters: A dynamic capabilities view. *International Journal of Tourism Research, 21*(6), 882–900. doi:10.1002/jtr.2312

KMU Forschung Austria. (2008). *Overview of Family Business Relevant Issues.* Final Report. SME Research Austria. https://www.kmuforschung.ac.at/wp- content/uploads/2018/02/EPS_Familienunternehmen.pdf

KPMG. (2019). *European family business barometer. Succession top of mind as business families eye their legacy.* https://assets.kpmg/content/dam/kpmg/xx/pdf/2019/11/european-family-business-barometer.pdf

Kraus, S., Clauss, T., Breier, M., Gast, J., Zardini, A., & Tiberius, V. (2020). The economics of COVID-19: Initial empirical evidence on how family firms in five European countries cope with the corona crisis. *International Journal of Entrepreneurial Behaviour & Research, 26*(5), 1067–1092. doi:10.1108/IJEBR-04-2020-0214

Kumar, P. (2020). Impact assessment of COVID-19: In tourism perspective. *Dogo Rangsang Research Journal, 10*(6).

Lengnick-Hall, C. A., & Beck, T. E. (2005). Adaptive fit versus robust transformation: How organizations respond to environmental change. *Journal of Management, 31*(5), 738–757. doi:10.1177/0149206305279367

Lengnick–Hall, C. A., & Beck, T. E. (2009). *Resilience capacity and strategic agility: prerequisites for thriving in a dynamic environment.* Working Paper, University of Texas.

Linnenluecke, M. K. (2017). Resilience in Business and Management Research: A Review of Influential Publications and a Research Agenda. *International Journal of Management Reviews, 19*(1), 4–30. doi:10.1111/ijmr.12076

Marques Santos, A., Madrid, C., Haegeman, K., & Rainoldi, A. (2020). *Behavioural changes in tourism in times of Covid-19.* Publications Office of the European Union. https://publications.jrc.ec.europa.eu/repository/bitstream/JRC121262/report_covid_tour_emp_final.pdf doi:10.2760/00411

McKinsey Global Institute. (2020a). *How European businesses can position themselves for recovery.* https://www.mckinsey.com/industries/public-and-social-sector/our-insights/how-european-businesses-can-position-themselves-for-recovery

McKinsey Global Institute. (2020b). *Leadership in a crisis: Responding to the coronavirus outbreak and future challenges.* https://www.mckinsey.com/business-functions/organization/our-insights/leadership-in-a-crisis-responding-to-the-coronavirus-outbreak-and-future-challenges

McKinsey Global Institute. (2020c). *Reimagining the $9 trillion tourism economy—what will it take?* https://www.mckinsey.com/industries/travel-logistics-and-transport-infrastructure/our-insights/reimagining-the-9-trillion-tourism-economy-what-will-it-take

McKinsey Global Institute. (2020d). *Meeting the future: Dynamic risk management for uncertain times.* https://www.mckinsey.com/~/media/McKinsey/Business%20 Functions/Risk/Our%20Insights/Meeting%20the%20future%20Dynamic%20 risk%20management%20for%20uncertain%20times/Meeting-the-future-Dynamic-risk-management-for-uncertain-times.pdf?shouldIndex=false

Miller, D., & Le Breton–Miller, I. (2003). Challenge versus advantage in family business. *Strategic Organization, 1*(1), 127–134. doi:10.1177/1476127003001001222

Nordqvist, M., & Jack, S. (2020). *Family businesses can tell us a lot about what makes a company resilient in times of crisis.* Stockholm School of Economics Institute for Research. https://www.hhs.se/contentassets/12721b1ec98e41f3b9ff3 47d02b18b3e/a40.pdf

OECD. (2020a). *Coronavirus (COVID-19): SME Policy Responses.* https://read.oecd-ilibrary.org/view/?ref=119_119680-di6h3qgi4x&title=Covid-19_SME_Policy_Responses

OECD. (2020b). *OECD Tourism Trends and Policies 2020.* OECD Publishing., doi:10.1787/6b47b985-

Omar, A. R. C. (2020). The impact of Covid-19 Movement Control Order on SMEs' businesses and survival strategies. *Geografia OnlineTM Malaysian Journal of Society and Space, 16*(2), 139–150. doi:10.17576/geo-2020-1602-11

Orsini, K., & Ostojić, V. (2018). *Croatia's Tourism Industry: Beyond the Sun and Sea.* Economic Brief 036. Luxembourg: Publications Office of the European Union. https://ec.europa.eu/info/sites/info/files/economy-finance/eb036_en.pdf

Ozili, P. K., & Arun, T. (2020). *Spillover of COVID-19: impact on the Global Economy.* https://mpra.ub.uni-muenchen.de/99850/

Pittino, Chirico, F., Henssen, B., & Broekaert, W. (2019). Does Increased Generational Involvement Foster Business Growth? The Moderating Roles of Family Involvement in Ownership and Management. *European Management Review, 17*(3), 785–801. Advance online publication. doi:10.1111/emre.12366

Prideaux, B. (2003). The need to use disaster planning frameworks to respond to major tourism disasters: Analysis of Australia's response to tourism disasters in 2001. *Journal of Travel & Tourism Marketing, 15*(4), 281–298. doi:10.1300/J073v15n04_04

PWC Family Business Survey. (2016). *The 'missing middle': Bridging the strategy gap in family firms.* www.pwc.com/fambizsurvey2016

SalvatoC.SargiacomoM.AmoreM. D.MinichilliA. (2020). Natural disasters as a source of entrepreneurial opportunity: family business resilience after an earthquake. https://ssrn.com/abstract=3693817

Schwaiger, K., & Zehrer, A. (2020). *Unternehmensresilienz von Familienunternehmen in Krisenzeiten.* Eine Analyse von Resilienz-Faktoren im Gastgewerbe am Beispiel der COVID-19-Krise.

Siakas, K., Naaranoja, M., Vlachakis, S., & Siakas, E. (2014). Family Businesses in the New Economy: How to Survive and Develop in Times of Financial Crisis. *Procedia Economics and Finance, 9*, 331–341. doi:10.1016/S2212-5671(14)00034-3

Statista. (2020). *International tourist arrivals in Europe from 2010 to 2019, by region.* https://www.statista.com/statistics/261702/international-tourist-arrivals-in-europe-by-region/

Statista. (n.d.). https://www.statista.com/

Tagiuri, R., & Davis, J. (1982). Bivalent attributes of the family firms. Working paper, Harvard Business School, Cambridge, MA. *Family Business Review, 9*(2), 199-208.

UNWTO. (2020). *International tourist numbers could fall 60-80% in 2020.* UNWTO Reports. https://www.unwto.org/news/covid-19-international-tourist-numbers-could-fall-60-80-in-2020

Villalonga, B., & Amit, R. (2006). How do family ownership, control, and management affect firm value? *Journal of Financial Economics, 80*(2), 385–417. doi:10.1016/j.jfineco.2004.12.005

Wang, J., & Ritchie, B. W. (2012). Understanding accommodation managers' crisis planning intention: An application of the theory of planned behaviour. *Tourism Management, 33*(5), 1057–1067. doi:10.1016/j.tourman.2011.12.006

Welter, F., Wolter, H., & Kranzusch, P. (2020). *Preliminary assessments of the IfM Bonn on the economic impact of the coronavirus pandemic on the "Mittelstand" businesses in Germany. IFM-Background Paper*. IFM.

Wenzel, M., Stanske, S., & Lieberman, M. B. (2020). Strategic responses to crisis. *Strategic Management Journal, 42*(2), O16–O27. doi:10.1002mj.3161

Weston. (2019). *Research for TRAN Committee – European tourism: recent developments and future challenges, European Parliament*. Policy Department for Structural and Cohesion Policies.

World Bank. (2020a). *Chapter 3: Lasting Scars of the COVID-19 Pandemic*. Global Economic Prospects. https://pubdocs.worldbank.org/en/112641588788257004/Global-Economic-Prospects-June-2020-Topical-Issue-1.pdf

World Bank. (2020b). *Global Outlook – Pandemic, Recession: The Global Economy in Crisis*. Global Economic Prospects. https://openknowledge.worldbank.org/bitstream/handle/10986/33748/211553-Ch01.pdf

World Economic Forum. (2019). *The Travel & Tourism Competitiveness Report 2019: Travel and Tourism at a Tipping Point*. Geneva: World Economic Forum. http://www3.weforum.org/docs/WEF_TTCR_2019.pdf

World Economic Forum. (2020). *COVID-19: These countries are most at risk from falling tourism*. https://www.weforum.org/agenda/2020/07/coronavirus-covid19-travel-tourism-gdp-economics/

WTTC. (2019). *Travel & Tourism continues strong growth above global GDP*. https://www.wttc.org/about/media-centre/press-releases/press-releases/2019/travel-tourism-continues-strong-growth-above-global-gdp/

WTTC. (2020a). *Travel & Tourism: Global Economic Impact & Trends 2020*. https://wttc.org/Portals/0/Documents/Reports/2020/Global%20Economic%20Impact%20Trends%202020.pdf?ver=2021-02-25-183118-360

WTTC. (2020b). *Travel & Tourism Economic Impact from COVID-19: Global Data*. World Travel & Tourism Council. https://wttcweb.on.uat.co/Research/Economic-Impact

WTTC Data Gateway. (n.d.). https://tool.wttc.org/

Ye, J. (2021). *4 ways family businesses can lead the pandemic recovery*. World Economic Forum. https://www.weforum.org/agenda/2021/01/4-ways-family-businesses-can-lead-the-pandemic-recovery/

Zapalska, A. M., & Brozik, D. (2007). Managing family businesses in the tourism and hospitality industry: The transitional economy of Poland. *Proceedings of Rijeka Faculty of Economics: Journal of Economics and Business*, *25*(1), 141–165.

Zehrer, A., & Leiß, G. (2019). Family entrepreneurial resilience – an intergenerational learning approach. *Journal of Family Business Management.* https://www.emerald.com/insight/content/doi/10.1108/JFBM-09-2018-0037/full/html doi:10.1108/JFBM-09-2018-0037

Zurub, Ionescu, A., & Constantin, V. D. (2015). Measuring the Economic Impact of Tourism in European Emerging Markets. *Procedia Economics and Finance*, *32*, 95–102. doi:10.1016/S2212-5671(15)01369-6

ENDNOTES

1 Since 1870, the world economy has experienced 14 global recessions: in 1876, 1885, 1893, 1908, 1914, 1917-21, 1930-32, 1938, 1945-46, 1975, 1982, 1991, 2009, and 2020, and in each of these episodes, there was a contraction in global real per capita GDP (World Bank, 2020).

2 See Tourism policy responses to the coronavirus (COVID-19), OECD, June 2020, oecd.org.

Chapter 2
The COVID-19 Pandemic and Organizational Resilience in Hospitality Family Firms:
A Qualitative Approach

Katrin Magdalena Schwaiger
MCI Management Center Innsbruck, Austria

Anita Zehrer
https://orcid.org/0000-0003-0325-7170
MCI Management Center Innsbruck, Austria

ABSTRACT

On March 11, 2020, the World Health Organization (WHO) announced a pandemic of COVID-19. The family-owned hospitality industry in the Tirol has been affected by governmental measures and travel bans in particular. The current study focuses on hospitality family business owners as the backbone of the regional tourism industry. Their dealing with the crisis and their perception of their own organizational resilience is explored. It uses a qualitative exploratory approach with semi-structured interviews to answer the research question. The most surprising result of the interviews was the optimistic state of mind visible among entrepreneurs. In this study, a better understanding of the tourism industry's dealing with a crisis such as the COVID-19 pandemic is created.

DOI: 10.4018/978-1-7998-7352-5.ch002

INTRODUCTION

Tourism crises may negatively affect the visitor's perception of a destination and therefore means a reduction in tourist arrivals (Sönmez et al. 1994). Especially the tourism industry is susceptible to negative events and the preparation to deal with natural and human-made catastrophes is challenging (Pforr & Hosie, 2008). The nature of crises is manifold: natural disasters, economic or political crises, have been impacting the hospitality industry on a global level (Hall, 2010; Glaesser, 2006; Pizam & Smith, 2000; Pizam et al., 1997). However, the whole world is simultaneously struck by one crisis and to an equally strong extent. On March 11, 2020, the World Health Organization (WHO) finally announced a pandemic of COVID-19 (WHO, 2020). Immediately, measures were taken to halt the rapid spread of the new corona virus by governments throughout the globe, while these measures have been influencing not only people's everyday lives, but entire economies. The resulting measures have particularly influenced the tourism sector (Karim et al., 2020) with challenges remaining even after reopening. The question arises how to combine health requirements and the long-term resilience plans necessary after reopening businesses. This requires academic research providing the hospitality businesses with strategies how to become or stay resilient and capable of acting in these demanding times of COVID-19.

With regard to the industry's structure, small and medium-sized businesses play a vital role in the accommodation sector of the hospitality industry, while family businesses are the central business form (Märk et al., 2010; Thomas et al., 2011; Buhalis and Peters, 2006). 85% of all businesses in the Tirol are family-owned and in the hospitality industry, the share of family businesses is 92% making this type of business the vital backbone of the hospitality industry (Mandl, 2008). Family businesses differentiate themselves from non-family businesses by a number of factors concerning human resources management. The employer image of family businesses includes respect and appreciation, an integrative communication behavior towards employees, and other non-monetary values in general (Leiß & Zehrer, 2018; Mooney et al., 2016; Shekhar & Kumar, 2012; Stewart, 2003). These non-monetary values include social working conditions, loyalty and particularly, long-term working relationships; this employee-oriented approach leads to an increase in organizational resilience (Hauswald et al., 2015; Krappe et al., 2011). This may indicate a higher resilience for family businesses than other organizational forms (Amann & Jaussaud, 2012). On the other hand, business failure research proposes a risk coming along with emotional attachment of the owner-manager to the business in small family-run hospitality venues: lack of a marketing background, insufficient management skills or a potentially low level of innovation might increase the likelihood of business collapse (Glowka & Zehrer, 2020; Zehrer, & Leiß, 2016; Pikkemaat &

Zehrer, 2016). Therefore, this type of business needs to be focused on in particular regarding the current crisis.

In the Alpine tourism region of the Tirol, where large part of the local value added is gained in hospitality, the COVID-19 pandemic poses a particularly harsh challenge. In 2019, the Tirol generated a gross value added of around € 32 billion, while 15% (€ 4.9 billion) were generated by the hospitality sector (Garbislander et al., 2020). The Tirolean Economic Chamber proposes three different scenarios concerning the development of the hospitality industry after the pandemic. Loss in gross value added will reach numbers from € 2.25 billion to € 4.85 billion depending on the scenario (Garbislander et al., 2020). However, losses remain in a scale of billions regardless of the scenario, which underlines the devastating effects of the crisis.

We contribute to preliminary research in three different ways. First, we contribute to a better understanding of family businesses dealing with a global crisis by including the rather under-researched stakeholder group of hospitality employees (Baum, 2018) in addressing their respective employers. Second, we add to the literature on the new COVID-19 pandemic while extending research on crisis management and resilience. Third, this study takes on a mixed methods approach addressing several levels of resilience in connection with organizational resilience: governmental, regional and individual (family business owner perceptions). Consequently, we answer the following leading research question using a qualitative research design:

RQ: How do business owners deal with the crisis and how do they view different aspects of organizational resilience?

In the following, we first present relevant research concerning crises in the hospitality industry, family businesses and resilience. We further outline the mixed methods research design and present the results of our qualitative study parts, highlight contributions and limitations. Finally, we outline implications for further research and practice, as well as derive recommendations for action in the affected businesses.

LITERATURE REVIEW

Crises and Disasters in the Hospitality Industry

A crisis is defined as an "undesired, extraordinary, often unexpected and timely limited process with ambivalent development possibilities. It demands immediate decisions and countermeasures in order to influence the further development again positively for the organization (destination) and to limit the negative consequences as much as possible" (Glaesser, 2006, p. 14). A crisis is therefore synonymous with an event, which disrupts the pre-existing state of affairs (Laws et al., 2007). Plenty of research has been carried out on political risks, terrorism and crime and its impacts on travel behavior in general or destination choice in particular (Tarlow & Santana, 2002; Goodrich, 2002; Pizam & Smith, 2000). In general, a crisis will occur because of life events disturbing the preexisting psychological equilibrium (Jacobson, 1980).

Ritchie (2004) has highlighted the need for disaster and crisis phenomena research with specific reference to the tourism industry. Literature so far has mainly focused on natural disasters and their impact on tourism communities, vulnerability and resilience. For example, in a developing country setting, Tsai et al. (2016) focus on communities with hospitality businesses. The authors assess community leaders and residents in natural disaster-prone areas in Taiwan. They find that respondents' perceptions of tourism impacts are positively correlated with community resilience. They concluded that for enhancing community resilience in high-risk tourism areas, it is important to address tourism as part of the overall community development. This is closely connected with community resilience. In this respect, the essential role of the existing social capital and a functioning network in a region to increase resilience is attested in the literature (Aßheuer et al., 2013; Braun & Aßheuer, 2011). While these studies assessed slum dwellers' vulnerability and resilience natural disasters in Bangladesh, the role of networks as a measure of resilience in the hospitality industry may not be underestimated regardless of the destination (Tibay et al., 2018; Tsao & Ni, 2016).

Notwithstanding the type of crisis, the worst organizational consequences of a crisis may be avoided by an active crisis management (Zehrer, 2009). The hospitality industry tends to lack adequate preparation against crises even though existing shortages in resilience are known (Becken & Hughey, 2013). This is also true for the current crisis, because nobody in the industry would have anticipated a global pandemic with complications like the COVID-19 crisis. It becomes obvious that effective crisis tools and systematic and complete documentation about crisis processes are essential to benchmark and learn from comparable examples.

Family Business Characteristics

As a large part of hospitality businesses are family-owned (Märk et al., 2010), it is worthwhile analyzing this type of business regarding the current COVID-19 crisis. Three key dimensions of family businesses have been brought forward by Villalonga and Amit (2004): family members hold top management positions, family members retain control over the firm through the distribution of capital among non-family shareholders, and an essential part of financial capital is held by one or several families. Family businesses can generally be defined as businesses "governed and/or managed with the intention to shape and pursue the vision of the business held by a dominant coalition controlled by members of the same family or a small number of families in a manner that is potentially sustainable across generations of the family or families" (Chua et al., 1999, p.25).

Any business is striving for a positive and distinguishable employer image (Agostini et al., 2015). The employer image of family businesses depends on different variables than non-family businesses (Danler & Zehrer, 2017). It is mainly the coexistence of family and business as well as a general overlapping of family and business issues, alongside with non-monetary values, that make family businesses stand out (Chrisman et al., 2008; Märk et al., 2010). Specific integrative communication behaviors, long-term employment, particular sustainability efforts and particularly social working conditions are typical constituents of a family business employer image likewise (Krappe et al., 2011; Leiß & Zehrer, 2018). These image aspects are even more important, as monetary factors alone do not "motivate any action other than the purely passive action of remaining in the organization" (Kovach, 1996, p. 5). The more non-monetary attributes, typical for family businesses, the more likely is career longevity (Chang et al., 2010; Sokro, 2012), and thus the more likely is resilience of an enterprise with long-term oriented employees (Zehrer & Leiß, 2019). Altogether, this contributes to a specific value canon typical for family businesses determined by respect and honesty (Stewart, 2003). These values are included in the employer image in order to achieve a competitive advantage for the family business (Presas et al., 2014) and thus, creating organizational resilience.

Organizational Resilience in the Context of Family Businesses

The concept of organizational resilience has been brought forward disaster management literature (McManus et al., 2008). While there are many definitions, resilience can be called a multifaceted concept involving the ability of subjects (e.g. cities, countries, regions, industrial sectors) to resist the impact of a shock, while, at the same time, the ability to recover from the adverse consequences of this shock (Cellini & Cuccia, 2015). According to the so called Positive Appraisal Style Theory Of Resilience (PASTOR) by Kalisch et al. (2014), resilience is defined as the ability to maintain and adapt the essential function and structure in the case of disturbance (Holling, 1973).

Seven principles of a resilient organization are pointed out by Mallak (1989): perceive experiences constructively, perform positive adaptive behaviors, ensure adequate external resources, expand decision-making boundaries, practice bricolage (creating an order of whatever is available), develop tolerance for uncertainty, and build virtual role systems. A resilient organization is one that designs and implements effective strategy to advance the organization while increasing the chances of its survival (Mallak, 1989). Annarelli's and Nonino's (2016) definition serves as the working definition for our study: "Organizational resilience is the organization's capability to face disruptions and unexpected events in advance thanks to the strategic awareness and a linked operational management of internal and external shocks (p. 7). In a business context, resilience is a key factor and is correlated to internal (e.g. development of a desirable personal identity, experiences of power and control, adherence to cultural traditions, experiences of social justice and experiences of a sense of cohesion with others) and external contextual component (e.g. access to material resources and access to supportive relationships) (Hedner et al., 2011). Thus, organizational resilience depends on personal characteristics as well as structural factors (Zehrer & Leiß, 2019; Hedner et al., 2011) and is characterized by reflections on past, critical events, coping with and adapting to them (King, 2016). The resilient factor is crucial to perform effectively during the crisis and successfully adapt to the environment after a crisis (Dartey-Baah, 2015).

Unexpected challenges typically occur unexpectedly and require fast and decisive decision-making (Heath, 1995; Ritchie, 2004). According to Dartey-Baah (2015), resilience is both performance-oriented and change-oriented in the "current volatile, uncertain, complex and ambiguous (VUCA) business environment" (p. 101). Regarding small family businesses, it has been shown that increased family ownership reduces the likelihood for formal crisis procedures in the management (Faghfouri et al., 2015) and that the emotional attachment of the family impacts the performance of family firms during a crisis (Arrondo-García et al., 2016). Furthermore, family businesses usually are financially more stable than non-family businesses and it is suggested that family businesses recover better from crises, for example from an economic downturn (Amann & Jassaud, 2012). Therefore, links between family businesses and the characteristics of resilient organizations become evident.

METHODOLOGY

The study was directed to generate findings without limiting a-priori hypotheses (Mehmetoglu & Altinay, 2006). Therefore, to answer the research question, we undertook a qualitative study to gain a deeper insight into the processes and perceptions of family business owners regarding their own organizational resilience during the COVID-19 crisis. It was intended to identify emergent themes in the data. This made grounded theory epistemologically and ontologically the most suitable approach (Glaser & Strauss, 1967). Data were collected through semi-structured in-depth interviews with family business owners. These key informants were contacted among those who disclosed their e-mail addresses in a quantitative survey taken prior to this study. The interviews were conducted in May 2020 and lasted between 30 and 50 minutes. The interviews were translated from German into English to present key findings and quotes. While guaranteeing anonymous treatment of respondent data, the interviews were transcribed and coded with the coding software MAXQDA-12. Inter-coder reliability was provided by the separate coding of the information by the two authors of this paper. We ensured that themes emerge from data analysis rather than the researchers' presumptions (Javadi & Zarea, 2016). Another criterion for increasing validity was the authors' continuous self-reflection during the coding process (Leiß & Zehrer, 2018).

RESULTS

Data were collected by means of six semi-structured in-depth interviews with business owners in the hospitality industry, who participated in an a-priori quantitative study. A detailed description of the respondent characteristics is depicted in Table 1.

Table 1. Interview respondent profile (n = 6)

Participant	Age	Gender	Organization	Employees	Size (rooms)	Seasonal
1	28	F	hotel	15	30	yes
2	24	F	4-star hotel & restaurant	18	25	yes
3	56	M	4-star superior hotel	80	93	yes
4	25	M	3-star hotel & restaurant	12	15	no
5	N/A	M	restaurant	7	does not apply	no
6	49	M	hotel	50	29	no
7	25	M	pension & restaurant	18	30	no

Table 2 provides an overview of the themes and subthemes within the interviews. Five key themes concerning business owners' dealing with the crisis were prevalent in the data: Employees as key resilience factor, personal characteristics and strategies, network resilience, dealing with finances and communication behavior.

Table 2. Theme and sub-theme overview of interview content analysis

Themes	Sub-Themes
Employees as key resilience factor	core resilience factor key to a successful enterprise part of the family cooperativeness of employees
Personal charateristics & strategies	shock-like state fear of future immersing into work intensive problem-solving discovering new fields for action
Network resilience	cooperation with other businesses upon reopening networking and consulting with other owners networks with the destination viewed less important cooperation for image upgrade
Dealing with finances	governmental subsidies enough own capital and liquidity investments in the right place
Communication behaviors	open and honest towards employees continuous communication and appreciation of employees honest and informing towards guests social media / staying alive virtually

Employees as Key Resilience Factor

The data shows consent in the fact that employees play a vital role regarding the success and resilience of an enterprise. A loyal and cooperative workforce is considered essential in the maintenance of a resilient business, for example "*The most important (role). Employees constitute the success of the business* (Participant 5,), or "*If we did not have such a great consensus with employees, we would not have been able to manage everything so well*" (Participant 1). The role of employees receives another focus that increases resilience: a specific and strong tie with the owner family: "*In such a small family business, you have a strong connection; the staff is part of the family! […] Partly, they have been in the business for more than 35 years*" (Participant 7).

Personal Characteristics and Strategies

In general, employer personality is essential in how business owners deal with crises. There was consent among participants that this type of crisis has caught them completely off-guard and they have never expected such a thing. In fact, the abrupt closure of businesses led a shock-like state of mind for some participants. After overcoming this first paralyzing phase of perceived intractability, different personalities led to different ways of dealing with the crisis. Fear of the future arose in particular due to the lack of predictability with the situation. The main reaction among entrepreneurs clearly was an immediate getting back to work and improve their businesses, such as renovating or finishing construction work. All participants portray a rather optimistic perspective in dealing with the crisis and keeping energy in the system for the phase of going back to normal: "*We immersed into work and are diving even further into work now [...]. Staying fit, eat healthy, that was a challenge, but I say it again, working and solving one problem after the other*" (Participant 6). "*There are always possibilities, it is always going on, and you don't have to be pessimistic, and somehow it is always going on*" (Participant 3).

Network Resilience

Network resilience is a valuable tool concerning the dealing with crises and disasters. Interestingly, regional networks are not strongly in the perception of employers. For example, "*You only meet in certain forums, but they are superficial. The tourism association is quite weak here*" (Participant 4). However, there are some cooperative structures with other businesses, particularly during re-opening. Moreover, it is perceived meaningful to stay in contact with fellow owners in the region to find creative solutions and build an alliance of knowledge: "*It is very important to be connected on a local level. [...] For example, we have experts in marketing, active people regarding research and law in our business; this is a great advantage*" (Participant 3). In general, respondents agree that networking and exchanging information as well as cooperative strategies with other employees and the destination are vital for a resilient future development.

Dealing with Finances

Questions of financial issues arise due to the enforced closure of businesses. There is consent that governmental mitigation measures are to some extent helpful when it comes to a crisis. However, this is not without some criticism regarding the rather bureaucratic proceedings and delays in notifications and distributing money. Furthermore, there is a strong focus on previously saved own financial capital in the family business. Most of the participants underline that a well-managed business must survive a closure of some months. Investments should only be taken where necessary and only with a solid budgeting plan, for example: "*You need liquidity. [...] and then you have to manage it, you have to. In our opinion, it is very advisable to have cash on the side*" (Participant 1). However, the optimistic view of some respondents leads to certain investments to add upgrades to their business or finishing the long-postponed constructions: "*In my opinion, you should invest and do something; I improved the garden in the corona break, embellished the house, and so on, that's what people like and something is happening*" (Participant 5).

Communication Behaviors

Participants portray how communication structures are based on honesty, openness and understanding towards employees. Transparency and compassion are described as important aspects in this communication: "*Employees were so thankful that we reacted this clearly. [...] to look how employees feel and that compassion helped me on the other hand, keeping my energy*" (Participant 6). Appreciation and gratitude for employees' cooperation in difficult times need to be provided and also shown. Regarding guests, being transparent is viewed as a valuable strategy to remain trustworthy after re-opening and conveying a feeling of safety to potential guests, for example "[...] *Simply being in direct communication with the guest without pretending something. Simply communication, honest communication with the guests*" (Participant 2). Also social media plays an important role in communicating with guests.

DISCUSSION

Contribution to Literature

The current study intends to contribute to previous research on organizational resilience in three ways. First of all, we are adding to the growing body of literature on the new COVID-19 pandemic with a specific respect to family businesses in the hospitality industry. Second, we create a better understanding of family businesses in their dealing with a global crisis, such as the pandemic. By including the rather under-researched stakeholder group of hospitality employees through the lens of their respective employers we respond to another gap in the literature. Third, this study takes on a holistic mixed methods approach addressing several levels of resilience in connection with organizational resilience based on family business owners' perceptions: governmental, regional and individual.

The results of the qualitative analysis are in accordance with a statement by Arrondo-García (2016) that emotional attachment of the family affects performance of family firms during crises. For example, human resources in a family business context are not sole numbers, but employees are viewed as part of the family and try to integrate their staff as much as possible in handling the crisis. During the entire crisis likewise, family business supported the classic employer image of family businesses, such as specific integrative communication behaviors, long-term focus, sustainability efforts and social working conditions (Krappe et al., 2011; Leiß & Zehrer, 2018). The qualitative part of the study shows that an honest conversation or communicating problematic issues in a meaningful way is essential in overcoming a crisis. This goes in line with the fact that apart from external structures, it is specifically the internal and personal characteristics of businesses that foster organizational resilience (Zehrer & Leiß, 2019). Moreover, the importance of employees in connection with organizational resilience has been highlighted by the qualitative results. This matches the above-mentioned employee and appreciation-focused management typical for a family business employer image (e.g. Mooney et al., 2016).

Entrepreneurs portrayed an open-minded and very positive or optimistic view of the future portrayed by participants in the qualitative part of the study, which was a very surprising result. In this respect, the results from the qualitative research match to a great extent. The crisis has unambiguously taken its toll. However, particularly concerning finances, business owners shared an optimistic perspective. It was highlighted that long-established businesses should always be able to cope with several months of closure or lack of revenue for a short period of time. This supports a study by Amann & Jassaud (2012) showing the specific features of family businesses lead to a higher resilience than other organizational forms and are more cautious with regards to debt.

Practical Implications

The results indicate the importance of contingency plans including definite financial resources planning. For practitioners, it may be helpful to think about various measures to be able to pay the bills during a crisis. A combination of own capital alongside with the usefulness of specific governmental subsidies are a first step to financially tackle the precarious situation. It may be worthwhile thinking about what investment is meaningful and may be helpful after reopening.

Furthermore, the results of the qualitative interviews reveal that employees, significantly influence organizational resilience. There is consent among the interview participants that employees are the key business success factor and inevitable for a resilient business. A continuous open communication and appreciation policy towards their employees may be interesting for owners. This would foster resilience development, but also overcoming the rather negative image of the industry. Finally, networking and social media as a resilience driver has only partially used by practitioners so far, but may be an interesting thought for owners regarding networks with colleagues to face crises more creatively and efficiently.

The study may also prove valuable for politics and the destination management. Valuable insights into how family business owners view the support granted by institutions and how they may be helped better regarding crises are presented. Building up networks within and with the destination management is viewed as a key resilience factor and should be considered. The industry image remains a central issue. The qualitative results suggest this must be tackled directly after overcoming the crisis. Governmental and destination institutions may want to pay attention to the rather difficult area of skilled labor shortage in connection with and largely caused by an unfortunate employer image.

LIMITATIONS AND CONCLUSION

Several limitations of this study are related to the methodology. One limitation of the paper relates to the fact that we were solely using subjective data for the analysis, asking family business owners on their perception of organizational resilience. Furthermore, the study might not be generalizable to other geographic regions with different business structures and political systems.

To sum up, we created a better understanding of the constituents of organizational resilience in the context of family firms by identifying factors exerting influence on organizational resilience of family business owners in the hospitality industry during the COVID-19 pandemic. Through our research question, by using a qualitative approach, we studied how business owners deal with the crisis and how they perceive organizational resilience. Here, among business owners, there is a common sense of careful optimism. The answers regarding future perspectives, financial safety and general perceived organizational resilience led to a surprising conclusion: the rather optimistic, yet carefully optimistic, mind-set.

Regarding future research, we suggest to draw on bigger sample sizes and also other geographical areas for cross-cultural comparisons of crisis recovery. Future research should thus include various sectors and industries to generate more profound results. Moreover, one could also undertake a longitudinal study to capture resilience behavior at the beginning, during and at the end of the COVID-19 pandemic to see how the attitude and behavioral traits might have changed over time. Hence, more research is necessary to analyze organizational resilience of family businesses in times of crises to help them better deal with crisis events such as the current COVID-19 pandemic.

REFERENCES

Agostini, L., Filippini, R., & Nosella, A. (2015). Brand-building efforts and their association with SME sales performance. *Journal of Small Business Management*, *53*, 161–173. doi:10.1111/jsbm.12185

Amann, B., & Jaussaud, J. (2012). Family and non-family business resilience in an economic downturn. *Asia Pacific Business Review*, *18*(2), 203–223. doi:10.1080/13602381.2010.537057

Annarelli, A., & Nonino, F. (2016). Strategic and operational management of organizational resilience: Current state of research and future directions. *Omega*, *62*, 1–18. doi:10.1016/j.omega.2015.08.004

Arrondo-García, R., Fernández-Méndez, C., & Menéndez-Requejo, S. (2016). The growth and performance of family businesses during the global financial crisis: The role of the generation in control. *Journal of Family Business Strategy*, *7*(4), 227–237. doi:10.1016/j.jfbs.2016.11.003

Aßheuer, T., Thiele-Eich, I., & Braun, B. (2013). Coping with the impacts of severe flood events in Dhaka's slums—The role of social capital. *Erdkunde*, 21–35.

Baum, T. (2018). Sustainable human resource management as a driver in tourism policy and planning: A serious sin of omission? *Journal of Sustainable Tourism*, *26*(6), 873–889. doi:10.1080/09669582.2017.1423318

Becken, S., & Hughey, K. F. (2013). Linking tourism into emergency management structures to enhance disaster risk reduction. *Tourism Management*, *36*, 77–85. doi:10.1016/j.tourman.2012.11.006

Braun, B., & Aßheuer, T. (2011). Floods in megacity environments: Vulnerability and coping strategies of slum dwellers in Dhaka/Bangladesh. *Natural Hazards*, *58*(2), 771–787. doi:10.100711069-011-9752-5

Bryman, A., & Bell, E. (2011). *Business Research Methods*. Oxford University Press.

Buhalis, D., & Peters, M. (2006). SMEs in tourism. In D. Buhalis & C. Costa (Eds.), *Tourism Dynamics, Challenges and Tools: Present and Future Issues* (pp. 116–129). Butterworth-Heinemann.

Cellini, R., & Cuccia, T. (2015). The economic resilience of tourism industry in Italy: What the 'great recession' data show. *Tourism Management Perspectives*, *16*, 346–356. doi:10.1016/j.tmp.2015.09.007

Chang, C. C., Chiu, C. M., & Chen, C. A. (2010). The effect of TQM practices on employee satisfaction and loyalty in government. *Total Quality Management*, *21*(12), 1299–1314. doi:10.1080/14783363.2010.530796

Chrisman, J. J., Steier, L. P., & Chua, J. H. (2008). Toward a Theoretical Basis for Understanding the Dynamics of Strategic Performance in Family Firms. *Entrepreneurship Theory and Practice*, *32*(6), 935–947. doi:10.1111/j.1540-6520.2008.00264.x

Chua, J. H., Chrisman, J. J., & Sharma, P. (1999). Defining the family business by behavior. *Entrepreneurship Theory and Practice*, *23*(4), 19–39. doi:10.1177/104225879902300402

Danler & Zehrer. (2017). *Die Wahrnehmung der Arbeitgebermarke von Famlienunternehmen am Beispiel von Studierenden*. Forschungsforum der Österreichischen Fachhochschulen 2017.

Dartey-Baah, K. (2015). Resilient leadership: A transformational-transactional leadership mix. *Ecology and Systematics*, *4*, 1–23.

Dillman, D. A., Smyth, J. D., & Christian, L. M. (2009). *Internet, mail, and mixed-mode surveys: The tailored design method*. John Wiley.

Dörflinger, C., Dörflinger, A., Gavac, K., & Vogl, B. (2013). Familienunternehmen in Österreich – Status quo 2013. Wien.

Faghfouri, P., Kraiczy, N. D., Hack, A., & Kellermanns, F. W. (2015). Ready for a crisis? How supervisory boards affect the formalized crisis procedures of small and medium-sized family firms in Germany. *Review of Managerial Science*, *9*(2), 317–338. doi:10.100711846-014-0158-0

Faulkner, B. (2001). Towards a framework for tourism disaster management. *Tourism Management*, *22*(2), 135–147.

Frank, H., Lueger, M., Nosé, L., & Suchy, D. (2010). The concept of 'Familiness': Literature review and systems theory-based reflections. *Journal of Family Business Strategy*, *1*(3), 119–130.

Garbislander, S., Haigner, S., & Jenewein, S. (2020). Wertschöpfungseffekte des *Tourismusbedingten Nachfrageausfalls für Tirol im Zuge der Corona-Pandemie*. Studie Wirtschaftskammer Tirol. Available at: https://www.wirtschaft.tirol/wp-content/uploads/2020/04/Details-zur-Studie_-Wertschöpfungseffekte-des-tourismusbedingten-Nachfrageausfalls-im-Zuge-der-Corona-Pandemie-in-Tirol.pdf

Glaesser, D. (2006). *Crisis management in the tourism industry*. Butterworth-Heinemann.

Glowka, G., & Zehrer, A. (2019). Tourism Family-Business Owners' Risk Perception: Its Impact on Destination Development. *Sustainability, 11*(6992), 1–16.

Gómez-Mejía, L. R., Haynes, K. T., Núñez-Nickel, M., Jacobson, K. J., & Moyano-Fuentes, J. (2007). Socioemotional wealth and business risks in family-controlled firms: Evidence from Spanish olive oil mills. *Administrative Science Quarterly, 52*(1), 106–137.

Goodrich, J. (2002). September 11, 2001 Attack on America: A Record of the Immediate Impacts and Reactions in the USA Travel and Tourism Industry. *Tourism Management, 23*, 573–580.

Habbershon, T. G., & Williams, M. L. (1999). A resource-based framework for assessing the strategic advantages of family firms. *Family Business Review, 12*(1), 1–25.

Hair, J. F., Anderson, R. E., Tatham, R. L., & Black, W. C. (1995). *Multivariate Data Analysis*. Prentice-Hall Inc.

Hall, C. M. (2010). Crisis events in tourism: Subjects of crisis in tourism. *Current Issues in Tourism, 13*(5), 401–417.

Hauswald, H., Hack, A., Kellermanns, F. W., & Patzelt, H. (2016). Attracting new talent to family firms: Who is attracted and under what conditions? *Entrepreneurship Theory and Practice, 40*(5), 963–989.

Holling, C. S. (1973). Resilience and Stability of Ecological Systems. *Annual Review of Ecology and Systematics, 4*, 1–23.

Jacobson, G. F. (1980). Crisis Theory. *New Directions for Mental Health Services, 6*, 1–10.

Javadi, M., & Zarea, K. (2016). Understanding thematic analysis and its pitfall. *Journal of Client Care, 1*(1), 33–39.

Jiang, Y., Ritchie, B. W., & Verreynne, M. L. (2019). Building tourism organizational resilience to crises and disasters: A dynamic capabilities view. *International Journal of Tourism Research, 21*(6), 882–900.

Kalisch, R., Müller, M., & Tuescher, O. (2014). A conceptual framework for the neu-robiological study of resilience. *Behavioral and Brain Sciences, 27*, 1–49.

Karim, W., Haque, A., Anis, Z., & Ulfy, M. A. (2020). The Movement Control Order (MCO) for COVID-19 Crisis and its Impact on Tourism and Hospitality Sector in Malaysia. *International Tourism and Hopitality Yournal*, *3*(2), 1–07.

Kovach, K. A. (1996). *Strategic Human Resource Management.* University Press of America. Inc.

Krappe, A., Goutas, L., & von Schlippe, A. (2011). The "family business brand": an enquiry into the construction of the image of family businesses. *Journal of Family Business Management.*

Leiß, G., & Zehrer, A. (2018). Intergenerational communication in family firm succession. *Journal of Family Business Management*, *8*(1), 75–90.

Mallak, L. (1998). Putting organizational resilience to work. *Industrial Management*, 8-13.

Mandl, I. (2008). *Overview of Family Business Relevant Issues*. Final Report. KMU Forschung Austria.

Märk, S., Kraus, S., & Peters, M. (2010). Der Einfluss der Familie auf den Unternehmer: Eine qualitativ-empirische Untersuchung von Familienunternehmen auf Basis der Stewardship-Theorie. *ZfKE–Zeitschrift für KMU und Entrepreneurship*, *58*(1), 31–59.

Matteucci, X., & Gnoth, J. (2017). Elaborating on grounded theory in tourism research. *Annals of Tourism Research*, *65*, 49–59.

McKercher, B., & Hui, E. L. L. (2004). Terrorism, economic uncertainty and outbound travel from Hong Kong. In C. M. Hall, D. J. Timothy, & D. T. Duval (Eds.), *Safety and Security in Tourism: Relationships, Management and Marketing* (pp. 99–116). Haworth Press.

McManus, S., Seville, E., Vargo, J., & Brunsdon, D. (2008). Facilitated process for improving organizational resilience. *Natural Hazards Review*, *9*(2), 81–90.

Mehmetoglu, M., & Altinay, L. (2006). Examination of grounded theory analysis with an application to hospitality research. *International Journal of Hospitality Management*, *25*(1), 12–33.

Mooney, S. K., Harris, C., & Ryan, I. (2016). Long hospitality careers–a contradiction in terms? *International Journal of Contemporary Hospitality Management.*

Ogundimu, E. O., Altman, D. G., & Collins, G. S. (2016). Adequate sample size for developing prediction models is not simply related to events per variable. *Journal of Clinical Epidemiology*, *76*, 175–182.

Pforr, C., & Hosie, P. J. (2008). Crisis Management in Tourism: Preparing for Recovery. *Journal of Travel & Tourism Marketing*, *23*(2–4), 249–264.

Pikkemaat, B., & Zehrer, A. (2016). Innovation and service experiences in small tourism family firms. *International Journal of Culture, Tourism and Hospitality Research*, *10*(4), 343–360.

Pizam, A., & Smith, G. (2000). Tourism and Terrorism: A quantitative Analysis of major Terrorist Acts and their Impact on Tourism Destinations. *Tourism Economics*, *6*(2), 123–138.

Pizam, A., Tarlow, P. E., & Bloom, J. (1997). Making Tourists feel safe: Whose Responsibility is it? *Journal of Travel Research*, *36*(1), 23–28.

Presas, P., Guia, J., & Munos, D. (2014). Customer's Perception of Familiness in Travel Experiences. *Journal of Travel & Tourism Marketing*, *31*(2), 147–161.

Ritchie, B. W. (2004). Chaos, crises and disasters: A strategic approach to crisis management in the tourism industry. *Tourism Management*, *25*(6), 669–683.

Shekhar, D., & Kumar, D. N. S. (2011). *Perspectives Envisaging Employee Loyalty-A Case Analysis*. Available at SSRN 1961430.

Siddiqui, K. (2013). Heuristics for sample size determination in multivariate statistical techniques. *World Applied Sciences Journal*, *27*(2), 285–287.

Sokro, E. (2012). Impact of employer branding on employee attraction and retention. *European Journal of Business and Management*, *4*(18), 164–173.

Sönmez, S. F., Backman, S. J., & Allen, L. (1994). *Managing tourism crises: A guidebook*. Department of Parks, Recreation and Tourism Management, Clemson University.

Stewart, A. (2003). Help one another, use one another: Toward an anthropology of family business. *Entrepreneurship Theory and Practice*, *27*(4), 383–396.

Sydnor-Bousso, S. B. (2009). *Assessing the impact of industry resilience as a function of community resilience: the case of natural disasters* (Doctoral dissertation). The Ohio State University.

Tarlow, P. E. (2003). The Tenth Annual Las Vegas Tourism Security and Safety Conference, Las Vegas, Nevada. *International Journal of Tourism Research*, *4*, 251–252.

Thomas, R., Shaw, G., & Page, S. J. (2011). Understanding small firms in tourism: A perspective on research trends and challenges. *Tourism Management*, *32*(5), 963–976.

Tibay, V., Miller, J., Chang-Richards, A. Y., Egbelakin, T., Seville, E., & Wilkinson, S. (2018). Business resilience: A study of Auckland hospitality sector. *Procedia Engineering*, *212*, 1217–1224.

Tsai, C. H., Wu, T. C., Wall, G., & Linliu, S. C. (2016). Perceptions of tourism impacts and community resilience to natural disasters. *Tourism Geographies*, *18*(2), 152–173.

Villalonga, B., & Amit, R. (2004). *How do family ownership, control and management affect firm value?* EFA 2004 Maastricht Meetings Paper no. 3620. Available at: https://papers.ssrn.com/sol3/papers.cfm?abstract_id=556032

WHO. (2020). *Coronavirus disease 2019 (COVID-19). Situation Report – 51.* Available at: https://www.who.int/docs/default-source/coronaviruse/situation-reports/20200311-sitrep-51-covid-19.pdf?sfvrsn=1ba62

Zehrer, A. (2009). The culture of facing crisis in tourism – an empirical study on crisis competencies of future destination managers. Culture Meets Economy Conference "Culture and Crisis – opportunity or threat?"

Zehrer, A., & Leiß, G. (2019). Family Entrepreneurial Resilience - An intergenerational learning approach. *Journal of Family Business Management*. Doi:10.1108/JFBM-09-2018-0037

Chapter 3
How Small Hotel Owners Practice Resilience:
Longitudinal Study Among Small Family Hotels in Croatia

Vlatka Skokic
Faculty of Economics, Business, and Tourism, University of Split, Croatia

ABSTRACT

Through three dramatic periods—the period of communism, the period of the war and transition during the 1990s, and the period of EU admission—this longitudinal study investigates how small hotel owners in Croatia understand and practice resilience. Findings demonstrate that entrepreneurs have proactive understanding of resilience and practice different actions before the shock of a disruptive event/s occur/s. Interestingly, entrepreneurs perceive this disruption as opportunities to exploit for their business development. One of the most prevalent practices is diversification, where entrepreneurs start another firm in different, often unrelated, sectors.

INTRODUCTION

The literature on tourism and hospitality (T&H) entrepreneurship has grown significantly over the last decade (Fu, Okumus, Wu and Köseoglu, 2019; Solvoll, Alsos and Bulanova, 2015; Thomas, Shaw and Page, 2011). Although widely recognised that family firms dominate the T&H industry and are characterised by longevity (Peters and Buhalis, 2013), their resilience is left somewhat under-researched (Engeset, 2020). One explanation can be related to the T&H literature

DOI: 10.4018/978-1-7998-7352-5.ch003

which largely focuses on entrepreneurial agency and motivation to start a business (Skokic, Lynch and Morrison, 2016; Thomas et al., 2011) and entrepreneurial outcomes (Fu et al., 2019). A further explanation is related to the perception of resilience, 'which has become a useful tool for understanding how the tourism industry responds to challenges' (Engeset, 2020, p. 69). Therefore, more attention has been given to the tourist destinations themselves rather than the individual firms or entrepreneurs. For the purposes of this study we understand resilience as the ability of organizations to avoid, absorb, respond to, and recover from, situations that could threaten their existence (Lengnick-Hall and Beck, 2005). The generic management literature investigating resilience converges on two streams. On the one hand, the individual-oriented stream explores the traits, attributes and experiences of entrepreneurs. This stream assumes that individual traits shape and create resilience, however, they exclude actual resilience practices. On the other hand, the firm-oriented stream investigates resilience at the firm level, overlooking the owners/managers. Such a dichotomy between the individual and the firm neglects the role of owners'/managers' understandings of resilience and how they practice resilience (Conz, Lamb and De Massis, 2020).

This study aims to address this gap and investigates how small hotel owners in Croatia understand resilience and how they practice it. Croatia is considered as an appropriate setting since the country underwent three dramatic periods: the period of communism; the period of the war and transition during the 1990s; and the period of EU admission. To investigate resilience as a situational and context-dependent phenomenon (Bamberger, 2008) it is necessary to employ a qualitative longitudinal methodology. Our findings reveal that entrepreneurs have a proactive understanding of resilience and practice different actions before the shock of a disruptive event/s occur/s. Interestingly, entrepreneurs perceive this disruption as opportunities to exploit for their business development. One of the most prevalent practices is diversification, where entrepreneurs start another firm in different, often unrelated, sectors.

LITERATURE REVIEW

Resilience of Family Firms

It is established in the literature that family firms (FF) are a resilient form of business organization because they demonstrate continuity and capability to survive over long time periods (Conz et al., 2020). For instance, James (2006) argues how the FF are 'uniquely resilient during downturns because of [their] ability to mobilise social as well as financial capital' and that such resilience is 'a reflection of profound continuity' (p.340). He further argues that family dynasties, in essence, represent a depth of tradition, which promotes loyalty, commitment and resilience, and create powerful corporate brands (p.75). However, few studies address the question of resilience among FF (Chrisman, Chua and Steier, 2011; Conz et al., 2020).

The concept of resilience has been explored among a range of disciplines, from ecology to social sciences, leaving the conceptualisation of the term and definitions fragmented (Conz and Magnani, 2020). Besides definitional issues, the mainstream literature is divided across two themes (Conz et al., 2020). On the one hand, the firm-oriented approach, which deals with the concept at the firm level and assumes that the practice of resilience is a firm attribute exclusively. Studies investigating FF resilience through a firm-oriented approach (eg Acquaah, Amoako-Gyampah and Jayaram, 2011; Christman et al., 2011), observe the firm, the family and the individual as separate units. For instance, Acquaah et al. (2011) compares the relationship between manufacturing strategy and firm performance while Mzid, Khachlouf and Soparnot (2019) investigate how family capital influences the resilience of family firms. On the other hand, an individual-oriented approach focuses on the attributes, characteristics, traits and the pre-acquired knowledge of owner managers. These studies assume that characteristics of those who operate and run the business will be the main predictors of their strategic decisions. For instance, studies show how demographic attributes and characteristics of top management teams influence opportunity exploitation (De Massis, Eddleston and Rovelli, 2020) or how CEOs risk-taking propensity impacts new product portfolio innovativeness (Kraiczy, Hack and Kellermanns, 2015).

However, FF are complex systems, comprising of both the family and the business systems and can be described as cross-institutional organizations (Paul, Winter, Miller and Fitzgerald, 2003). Brewton, Danes, Stafford and Haynes (2010) recognise family firm resilience as the 'reservoir of both individual and family resources that cushions the family firm against disruptions and is characterized by individual and collective creativity used to solve problems and get work done' (p. 156). Danes et al. (2009) argue through investigation of sustainability as a consequence of natural disasters, that the resilience of FF can be understood as a stored capacity which will be called upon when disruptions occur. Therefore, FF resilience is also about accumulating resources which will be accessed when crises occur. To bridge different perspectives and to conceptualise FF resilience, Conz et al. (2020) propose that future research should seek different experiences and practices of entrepreneurs in FF and how they perceive and practice resilience. The main premise in this view is that resilience is embedded into a specific socio-economic context. Following this premise in a phenomenological study among longstanding family owned/managed wineries in Italy and Australia, Conz et al. (2020) four types of resilience understandings determined by resilience approach are revealed, as well as the time frame in which resilience is practiced. Controllers and developers are those entrepreneurs who practice resilience before a disruption occurs. The difference between these two types is that the former aims to either avoid or absorb the change while the latter seeks to adapt to change. Perpetuators and consolidators are those entrepreneurs who practice resilience while the shock is occurring. The difference is again in resilience understanding, where the former group aims to absorb the change ensuring for the business to stay at the same level of equilibrium while the latter aims to adapt to change to reach a new level of equilibrium, understanding resilience as an adaptive consolidation. Their results provide evidence that entrepreneurs' understandings of resilience is one of the main factors which determines the way it is practiced, and their understandings are dependent on the wider context. This approach can be useful when examining resilience of FF in turbulent settings, such as former transition economies, since entrepreneurs operated in hostile environments, with almost non-existent institutional frameworks (Aidis, Estrin and Mickiewicz, 2008).

Resilience of Family Firms in Tourism

Chowdhury and Prayag (2019) argue that resilience among T&H firms is important within the mainstream discussion of organisational resilience, as the T&H industry differs from other industries. For instance, the T&H industry is highly sensitive to global trends: it is vulnerable to natural disasters and highly reliant on inbound visitors, increasing the exposure of tourism businesses to sudden shocks. Although the T&H sector is dominated by small- and medium-sized (SMS) family firms worldwide (Getz, Carlsen and Morrison, 2004), far too little attention has been paid to the resilience of T&H firms.

The T&H literature does not ignore the concept of resilience, but a vast majority of the studies apply the concept to the destination level and investigate how the tourism destinations respond to various challenges, for example, revealing the adaptive resilience of a living cultural heritage destination (Bui, Jones, Weaver and Le, 2020), the 'overtourism' problem (Butler, 2017; Cheung and Li, 2019) environmental crisis-prone tourism destinations (Tsao and Ni, 2016) or humanitarian crises (Pappas and Papatheodorou, 2017). On the other hand, a growing section of the T&H entrepreneurship literature focuses largely on the characteristics of the entrepreneurs at the start up stage alongside the operating characteristics of T&H firms (Fu et al., 2019; Skokic et al., 2016). A significant number of studies argue that many T&H firms are lifestyle-oriented, implying that entrepreneurs are driven by specific lifestyle factors (such as to live in a certain area or to be one's own boss) when establishing a business (Ahmad, 2015; Ateljevic and Doorne, 2000; Banki and Ismail, 2015; Williams, Shaw and Greenwood, 1989). Lifestyle orientation affects the attitudes of entrepreneurs towards the operational characteristics of a business, such as business growth; which is not considered as a priority and sometimes it is also purposively rejected (Ioannides and Petersen, 2003). In addition, many FF in the T&H sector are driven by family-centred goals, such as preserving the socioemotional wealth of family members, family harmony or family legacy (Getz and Carlsen, 2000, Morrison and Teixeira, 2004). Therefore, the preservation of family values and interests rather than profit maximisation goals is the fundamental motivation for FF (Morrison and Teixeira, 2004).

Family business research puts significant emphasis on the importance of non-financial goals, especially socioemotional wealth (SEW). Berrone, Cruz & Gomez-Mejia (2012) define SEW through 5 dimensions: family control and influence, identification of family members with the firm, binding social ties, emotional attachment of family members, and renewal of family bonds to the firm through dynasties (p. 259). Memili et al. (2020) conducted one of the first studies investigating the impact of SEW on firm performance among Turkish family hotels. Their findings show that SEW preservation can limit the performance of small and medium sized FF in H&T. However, their results also suggest that FF psychological capital, which involves hope, optimism, efficacy and resilience, can buffer those negative effects. Memili, Welsh and Luthans (2013) argue that psychological capital may be more prevalent in FF, because of unique characteristics of FF, such as long–term orientation and the family's lasting involvement. Engeset (2020) attempts to examine this issue and investigates how family ownership contributes to business resilience among rural hotel owners in Norway. His findings demonstrate that the resilience of rural hotels depends significantly on family capital and identifies the most important components of human and social capital that build resilience: an entrepreneurial mind and the willingness to work hard; complementary skills and respect; and commitment. Although the study employed a small sample (10 hotels), it shows interesting results in terms of hotel owners' responses to the crisis, where most of them bounce forward after the crisis by innovating or adapting to the changes.

In contrast to the mainstream business and entrepreneurship literature, a majority of the existing studies which examine resilience practices among T&H firms focus on organizations in a post-disaster context (Biggs, Hall and Stoeckl, 2012; Chowdhury et al., 2019; Dahles and Susilowati, 2015; Orchiston, Prayag and Brown, 2016; Prayag et al., 2020). For instance, Chowdhury et al. (2019) examine the relationship between social capital and organisational resilience as predictors of business performance among tourism firms in the post disaster context (earthquake) of Christ Church, New Zealand. The study demonstrates the importance of adaptive resilience, which has a major effect on business performance. Applying ethnographic methods in a longitudinal study among non-rated accommodation providers in Yogyakarta, Indonesia, Dahles and Susilowati (2015) findings demonstrate significant resilience during the decade of crisis (economic, political and natural). Their research reveals that entrepreneurs use survival, adaptation and innovation responses to the crisis, where adaptation is the most common. For instance, entrepreneurs oriented towards the domestic market and directed marketing efforts accordingly, or they started new businesses in sectors which serve both the domestic and international market, such as restaurants and bike rentals. Williams, You and Joshua (2020) examine the resilience of small tourism firms in a remote island through the prism of instrumental stakeholder theory. Their findings reveal that close relationship capabilities with stakeholders have a strong effect on business resilience. Also their study identifies a range of disruptive events, mostly specific to a remote island, all of which disturb the operational strategies of businesses.

Although limited in number, those studies investigating resilience practices among T&H firms clearly demonstrate that small businesses are particularly responsive to exogenous shocks as they are innovative and adaptable (Williams and Vorley, 2014). Russell and Faulkner (2004 p.561) note that 'it is often through times of upheaval, where there exists an edge-of chaos state, that entrepreneurs spot opportunities in the environment and use their creativity to bring about innovation'. They add that unpredictable triggers are very often the main catalyst for entrepreneurial innovation, in line with Orchiston, Prayag and Brown's (2016) results. However, the apparent dichotomy between the individual- and the firm-level of analysis neglects the role of entrepreneurial understanding of resilience and how entrepreneurs practice resilience.

STUDY CONTEXT

The study context has to be understood by reflecting on three periods which have had substantive effects on the overall socio-economic development and entrepreneurial activity in Croatia. Croatia was a part of the former Yugoslavia (1945 - 1990). Private ownership and entrepreneurship were, to a large extent, forbidden during this time and entrepreneurs were portrayed rather negatively (Kornai, 2006).

In this context, tourism and agriculture were among the few industries where private initiatives were allowed. However, by constitution the government restricted the size of businesses in terms of employees (maximum five personnel permitted) and was reluctant to legalize private property (Dana, Palaić and Ramadani, 2018). Within the coastal Croatia, a number of people started to offer summer houses for rent. Other forms of entrepreneurship were highly limited as the government restricted the key enabling conditions for entrepreneurial activity. During the Yugoslav period, Croatia was the most popular tourist destination in Yugoslavia, where two thirds of all accommodation capacities were located (Kobasic, 1987). In 1991, Croatia declared independence, after many political and economic controversies.

The period of transition to the market economy was extremely challenging for Croatia (Alpeza, Oberman Peterka and Has, 2018). This was further aggravated by the Homeland War with Serbia (1991-1995), which caused significant human and financial losses. Additionally, many sectors of the manufacturing industry were either closed or destroyed, while tourism activity was almost non-existent (Schönfelder, 2005). In terms of tourism accommodation capacity, properties were either bombed or were used as temporary homes for many refugees. After the end of the war in 1995, tourism revenues and arrivals showed only minor improvements. Moreover, up to 2002 the tourism sector did not experience any significant changes, since large hotel enterprises did not adjust to new market trends and their position in the tourism market deteriorated even further due to the inadequate privatization model.

New policy initiatives implemented after 2002 encouraged entrepreneurs to invest in the tourism sector which significantly improved the quality and the capacity of accommodation, properties which were more tailored to changing tourist needs. The number of tourists' arrivals and tourism revenues continued to increase and the tourism industry grew exponentially, especially after 2013, when Croatia joined the European Union (Marić, Samardžić and Protrka, 2017). Over the last decade, the tourism industry has grown rapidly, reaching its peak in 2019 when tourism contributed more than 25% of GDP and 25% employment (WTTC, 2020). In 2019 a record 19.6 million tourists arrived in Croatia (DZS, 2020).

Therefore, Croatia offers an appropriate context to investigate the resilience of FF in tourism over a longitudinal period since it has experienced several shocks which in turn have provoked entrepreneurial responses. In addition, the tourism sector is dominated by small and family owned firms, providing a homogeneous sample pool of participants through which to understand the research problem (Miles and Huberman, 1994).

METHOD

Sampling

The first study was conducted during winter 2006/07[1] through semi-structured interviews. At that time, Croatia did not have appropriate statistical databases of small firms and no official record of tourism entrepreneurs. The National Hotel Classification (NHC), which lists all hotels in Croatia by size and the ownership, was used to identify small hotel owners. Due to the strong geographical diversity and the level of tourism development, which differs considerably within the country, it was decided to focus on the Splitsko - Dalmatiska county. The identified area has a long established tourist tradition and strong entrepreneurial activity in the tourism sector which enabled the researcher to find a representative sample population (Miles and Huberman, 1994). Following the NHC guidelines which define a small hotel as one having up to 40 rooms, 64 small hotels are identified. The entrepreneurs were contacted personally to establish initial contact and to schedule an interview. The research followed the theoretical saturation principle (Eisenhardt, 1989) and stopped after the 37th interview was conducted. To distinguish the family firms in this sample, Chua et al.'s definition (1999, p. 25) is used. A family firm is 'a business governed and/or managed with the intention to shape and pursue the vision of the business held by a dominant coalition controlled by members of the same family or a small number of families in a manner that is potentially sustainable across generations of the family or families'. Out of 37, 22 were identified as family firms following Chua et.al.'s criteria.

Since the resilience is understood as situational and context dependent (Bamberger, 2008), interview questions were structured around four key themes to capture individual circumstances and contextual features. The themes were structured around entrepreneurial biographies (life histories) and members' motivation to start a business (including the role in the FF); socio-economic environment for entrepreneurship; formal and informal networks; business lifecycle and operations. These themes enabled the researcher to understand a variety of entrepreneurial circumstances and how entrepreneurs understand and interpret different aspects of their lives. The themes offered a broad understanding into the specific Croatian context, including the developments of the tourism industry, turbulent transitional environment and promising EU setting. More importantly, these insights provided a context for understanding entrepreneurial resilience, as it was possible to situate participants' perceptions in place and time and in specific contextual circumstances. The order and the style of questions to investigate entrepreneurial resilience were flexible, due to significant variations in the participants' life and business experience. For instance, the primary question to capture subjects' understanding of resilience was 'How do you perceive (for example, the fall of communism) and the way it affected your business? Entrepreneurs talked openly about their experiences and in few cases probing questions were raised to further clarify the issue, for instance 'are there any examples of …' or 'can you tell me more about …'. The primary question to capture their responses to the idea of resilience was 'Which activities enabled your business to endure over time?' or 'How those challenges were experienced and met?'

The second study was conducted ten years later, during winter 2016/2017. Longitudinal design offers an opportunity to investigate entrepreneurial resilience in time and whether participant understandings and practices change over time. All entrepreneurs were in business and accepted an invitation to participate in the second study. Due to travelling clashes, 17 interviews with the owners of FF were conducted. In this study we tested weather previously reported changes still had an effect on the business and activities participants practiced after the EU accession.

Data Analysis

Interviews for both studies were organised in the hotels and were tape recorded. Interviews conducted during first study lasted on average ninety minutes while the average length of the interviews during the second study was forty minutes. All interviews were recorded and transcribed. Interviews are referred to by the labels H1 through H17, where H stands for a 'hotel'.

Data analysis is guided by the principles of Framework Analysis (Ritchie & Spencer, 1994) which consists of six stages: familiarization; identifying a thematic framework; indexing or coding; charting; mapping and interpretation. The process is iterative as the analysis requires to go forwards and backwards between the stages. In order to become familiar with the data, it was necessary to listen the recordings and to read the interview transcripts which provided exposure to the text and the identified meanings. This stage of analysis led to identification of the main concepts according to which the data could be examined (the thematic framework). The thematic framework went through a number of iterations through the process of constant comparison to different data categories. Once the first and second order coding process was completed, 10 master codes were extracted.

The next stage of analysis required clustering of master codes into meta-codes, where the analysis identified meaningful connection. As a result, 10 master codes were synthesised to three meta-codes, 'environment for entrepreneurship'; 'business operations' and 'family resources'. In qualitative research, findings can be discussed and organised according to the main analytical themes (see for example Alsos, Carter, and Ljunggren, 2014; Drakopoulou- Dodd, 2014) or these themes can serve as 'anchors' around which the main story will be developed (see for example Dana, Calin, and Lasch, 2014). This research adopts the latter approach because entrepreneurial understandings of resilience and their responses are characterised by the complex interactions of various elements which cannot be discussed in isolation but rather call for a holistic view.

FINDINGS

Table 1 provides an overview of the investigated entrepreneurs. Almost all entrepreneurs graduated from university but only two of them hold a tourism and hospitality degree. Before starting up a hotel business, all entrepreneurs except the novice entrepreneur started and operated a business in a range of sectors such as real estate and accounting. Thus, only one entrepreneur is considered as novice (H2) while all others are either portfolio or serial entrepreneurs. With respect to gender, most of the entrepreneurs interviewed are men, however, four interviews involved a male and female couple. Returning migrants are those entrepreneurs who either emigrated themselves or their families emigrated while they were young and returned after the breakdown of Yugoslavia.

The Period of Communism

Although after 1974 people could run a small firm in the tourist industry, the reality was quite different. Private ownership was against the main principles of the dominant ideology and entrepreneurial endeavours were constantly hindered. Among 17 investigated entrepreneurs, 6 entrepreneurs or their families attempted to start a business when the former Yugoslavia was still in existence (Table 1). The communist party associated entrepreneurship with exploitation and dangerous Western movement. Individuals did not want to be associated with entrepreneurs, as this could cause problems, such as interrogations by state officers or job loss, illustrated by the following example:

You work against the country (Yugoslavia) and represent a danger to the state... you really had to be brave, no, you had to be crazy to start a private business during that time. That is the main reason why me and my family stopped with everything until Croatia declared independence. (H21)

Additionally, entrepreneurs did not have legal security as private ownership was mostly prohibited or nationalised:

They [the communist party] invented so-called property origin law which enabled them just to take the property from you. This is how they took a land from my father. (H2)

Entrepreneurs also reported that their initiatives to start a business in the T&H sector, although formally allowed, were stopped from the beginning:

My father's wish, he dreamt about it, was to open a nice family hotel. He had a land, the project was ready, but you know how it was, we were not perceived to be 'suitable' so the party blocked the attempts by nationalising our land. That was the turning point when my dad decided to immigrate. He set up his own restaurant in Germany at the beginning of the 1980s. (H2)

Table 1. Profile of the interviewed entrepreneurs

Participant	Sex	Age	Type	Owns/Was Owning Business in:	Number of Years in Hotel Business (at Second Study)	Business Growth	Family Firm Characteristics
H1	M/F	50s	Portfolio	Finance and Accounting	12	New hotel, hotel expansion	Husband and wife, children involved in a hotel business with clearly defined roles.
H2	M	30s	Novice	Real estate	11	Hotel expansion, real estate business	Family started a restaurant business abroad after being prohibited in former Yugoslavia (FY). Son took over the business and started a hotel (his wife runs daily operations).
H3	M/F	40s	Portfolio	Real estate	15	Construction, restaurants	Extended family (the entrepreneur, his wife and brother).
H4	M	50s	Portfolio	Finance and Accounting Restaurants	13	Hotel expansion, restaurant	Started a business in FY. Runs a business together with the wife and children.
H5	M	70s	Portfolio	Restaurant	50	Hotel expansion, food production, 'ethno' village	Started a restaurant business in FY. Runs a business with the wife and children with clearly defined succession plan.
H7	M	50s	Portfolio	Retail, Bakery	14	Hotel expansion, new hotel, retail	Wife runs a hotel business, the entrepreneur runs other businesses.
H8	M	60s	Serial (Returning migrant)	Restaurant	13	Hotel expansion	Started a restaurant business in FY. Involvement of the whole family (wife, kids, cousins).
H12	M/F	50s	Serial (Returning migrant)	Travel agency	15	Hotel expansion	Started a travel agency business abroad. Wife and husband run the business.
H16	F	60s	Portfolio	Transport Leasing (beach equipment)	14	Hotel expansion, retail	Husband and wife started a hotel business. Children joined when graduated. Retail business run by son.
H17	M	50s	Serial (Returning migrant)	Restaurant	14	Hotel expansion, new hotel, restaurants	Started a restaurant business abroad. Hotel is run by the family (wife and son).
H19	M	60s	Serial (Returning migrant)	Restaurant	13	Hotel expansion, retail	Started a restaurant business abroad. Hotel is run by the family (wife and daughter), son runs retail.
H21	M	70s	Portfolio	Manufacturing Trade	14	Restaurant	Involvement of the whole family (wife and two daughters). At the time of second study, the daughter took over the trade business, son in law manufacturing and younger daughter took over a restaurant.
H23	M	60s	Portfolio	Restaurant B&B	24	Hotel expansion, retail	Started a restaurant business in FY. Involvement of the whole family (wife, children).
H25	M	40s	Portfolio	Real estate Restaurant	13	New hotel	Family (grandfather and father) started a restaurant business in FY. Son took over at the time of second study.
H26	M	60s	Portfolio	Trade Restaurants	19	Hotel expansion, new hotel, real estate	Involvement of the whole family with clearly defined roles. A daughter to take over in the future.
H27	M/F	50s	Serial (Returning migrant)	Construction	17	Hotel expansion, restaurant, retail	Started a construction business abroad. Daughter runs a hotel, son runs retail.
H28	M	40s	Portfolio	Real estate Restaurant Travel agency	18	Hotel expansion	Family started a restaurant and a motel business in FY. Son took over and now runs the business with his own family.

Through this period, entrepreneurs and their families harboured strong ambitions to own businesses and attempted to pursue their dreams. However, those attempts were constantly obstructed by the state authorities. Some entrepreneurs and their families decided to emigrate and start their businesses abroad (for instance H2) while others managed to keep the businesses in the family (H4, H5, H8, H23, H25, H28). Those entrepreneurs reported that they purposively decided not to grow their businesses, but their main strategy was to keep the businesses in the family and await for the change to happen:

You see, we all knew that this [communism] cannot last forever... so this is why we decided to wait and to prepare for the future, our aim was to stay alive... like in the 80s, we had massive tourism in this area but we did not increase our seating capacity or anything like that, there was always a possibility that they will shut us down. (H23)

My father was hopeless, I was born in Germany, he was just waiting for the day to return back ...he was over the moon when we returned and I told him I want to pursue his idea and to start a hotel business. (H2)

The War and Transition Period

After the split from Yugoslavia, Croatia went through the war, transition and privatisation process. These changes were complex and have left considerable consequences for entrepreneurial activity, even today. The economy suffered from a lack of capital and entrepreneurs could not get any loans, except if they were willing to pay an interest rate which ranged between twenty to thirty percent. Racketeering was a normal activity and the system created a huge number of 'thieves' protected by the ruling party. Privatisation policy, instead of eliminating the category of 'socially-owned' enterprises and introducing new owners, masked them to enable frauds. Besides these issues, the major shortcomings of the transition process which affected entrepreneurs strongly were the treatment of ownership, which was not recognised in its full right by the state institutions and the speed of changes, which were not gradual and planned, but rapid and inadvertent. Despite this, all of the interviewed entrepreneurs (or their families in case of novice entrepreneur E2) started their business ventures immediately after the reforms in a range of sectors (Table 1). For these entrepreneurs, resilience is about change absorption. Therefore, their main mission is to keep the business unchanged while the change is occurring. When entrepreneurs practice resilience, the main focus is on the identity of the newly created business and to preserve it.

I felt all the transition changes. I opened a [manufacturing] firm in the craziest times, all those situations led me to do some things which I would not probably do, in normal circumstances. But me and my wife never let down any of our customers. We were among the rare private companies and our customers trusted us, but it was a hell of a job to gain that trust. (H21)

Four entrepreneurs also started their hotel businesses in addition to other businesses (H23, H26, H27, H28) during the transition period. These entrepreneurs perceive resilience as a proactive development of their business portfolios. However, all interviewed entrepreneurs dedicated a significant amount of time to discover changes on the global markets and were confident that the Croatian tourism industry would recover and experience significant growth in the coming years. As such, they wanted to develop businesses which would enable them to exploit this change. Some of the illustrative examples are shown below:

Every single time when we would walk around the town centre my husband would stop, he would say something to himself ...I did not have idea that he is thinking about the hotel in the most unpopular and problematic location in the town. But he read all the time about the small boutique family hotels located at a very distinct places... and he saw what will happen, he bought the old destroyed building with a vision to build a small hotel. (H1)

I travelled a lot and I saw Italy, I saw Austria, France... they do amazing stuff with small hotels, the offer they provide, how they change the local communities. I was always wondering why Croatia, with so many natural resources and tourism industry would not have something like that... I spent so much time reading, researching and more I knew, I was more confident that small hotels are the future of our tourism supply. (H4)

In addition, entrepreneurs stress all the changes which have happened in the tourism sector lately, like the end of the popular mass tourism model, very short tourist stays and evolving tourists' needs. The analysis reveals that entrepreneurs are constantly attempting to predict changes and to adapt business strategies and operations to meet all those needs, which witnesses their strong profit orientation. However, entrepreneurs have been faced with many challenges while attempting to provide a service more tailored to the tourists' needs. Nevertheless, entrepreneurs have been proactive in their endeavours and exhibited unconventional behaviour for entrepreneurs in the T&H sector.

When I started to construct a hotel my whole city thought that I went crazy. We never had a hotel in the city, former Yugoslavia was developing only mass tourism in the coastal parts of the country and people use to say (mimicking voice) 'the state did not build a hotel, how are you going to build it…today, almost twenty years after, nothing has changed…. We do not have a basic infrastructure, during the summer all islands have problems with the water, transport and our state is doing nothing. So my starting point was that if I want to have guests, I need to bring them into the destination... I am basically creating a destination, you know, although this is not the way it is supposed to be. I started to build a village. I also constructed a road, even a helideck; I am doing all those basic infrastructure things which I am not supposed to do in a normal world. (H5)

One of the most prohibiting elements reported by participants was an underdeveloped institutional framework in Croatia. It is important to stress that administrative and legal barriers caused many of the entrepreneurs for this study considerable delays in starting their hotel businesses. For instance, some entrepreneurs needed almost a decade to acquire a building permit. The entrepreneurs came up with various strategies to overcome this problem, as they wanted to make sure that the business would be ready once the tourism industry recovered. Only one entrepreneur decided to wait to obtain the permits, which has caused additional money losses, as they had to stop the construction works. A significant number of the entrepreneurs decided to work without all the permits risking significant penalties. The majority of the others were active in seeking strong connections and one proprietor even decided to enter into politics:

It took me 12 years to sort out all the paperwork… I became a vice mayor (laugh) but I could not do anything …, state officers are too powerful, and that is certainly such a huge damage that we will feel for a long time.. generations have to pass to change this, we are still dragging this from communism and my generation only cemented it.' (H5)

After 2000

Despite the turbulent environment, all of the interviewed entrepreneurs navigated through the transitional period and all of them created a portfolio of businesses, with the exception of the novice entrepreneur. However, even the novice entrepreneur reported that he started another business in 2015, in the real estate sector. All entrepreneurs at this stage were attempting to predict what the EU accession would bring and how they could prepare their businesses for new challenges. The main focus was on the operational side of the business.

Entrepreneurs considered that as small businesses they had limited power in the market, both nationally and globally. They were also thinking ahead to consider increasing competition which they perceived would come with Croatia's accession to the EU in 2013. To strengthen their position, they attempted to work closely not only with other hoteliers, but with all other complementary businesses, like food producers. They believed that this strategy could be extremely beneficial to both sides, as in the last couple of years there has been an incredible growth of small food and wine producers, offering authentic products. All interviewed entrepreneurs complained about the current practices of big, high quality hotels, which decrease prices during the winter below the average price of the three star small hotel. Small hotels cannot compete with this price trend because of two issues. Firstly, by decreasing prices significantly they cannot earn even enough to cover fixed costs, thus creating additional losses. Secondly, they do not want to send a signal to the market that they are becoming 'cheap doss houses' (H26). In other words, it is felt that lowering prices reduces customer perception of the quality of their businesses.

Another interesting reference made by participants was to tourist groups. Namely, most of the entrepreneurs complained that they did not have the capacity to accept an organised tour with a busload of guests. This implies that entrepreneurs cannot benefit from economies of scale and have limited possibility to sell their capacities during the pre- and post- season, when most tourists come as organised groups. In addition, they all expected a rise in tourist arrivals after the EU accession, especially during the pre-season:

Today, the number of rooms is our biggest shortcoming. We cannot accept travel agencies with buses in pre-season, which is a huge shortcoming in our business. (H3)

Besides preparing the business on the operational level by, for instance, increasing the quality of current offerings or cooperating with suppliers, all entrepreneurs also grew their hotel businesses. A vast majority of them either expanded the existing hotel or invested in a new one. Therefore, it was of crucial importance for the entrepreneurs to act in advance to prepare the business to be ready to adapt to the change. The change was perceived as positive as entrepreneurs did not identify any negative effects of the EU accession.

When entrepreneurs were asked about the coming recession in 2008, they did not consider it dangerous for their business, illustrated by the following quote:

Yeah, I know, everyone talks about it, but I think they [the Westerns] don't know what the real crisis is. We had three wars in 100 years, we live with the crisis all the time (laugh), Americans and the West, they don't know how to live with the crisis. (H8)

Entrepreneurs did not attempt to open up new markets by, for instance, attracting guests from the East, which is not their main market segment. Rather, they sought to preserve their position in already established markets such as Germany, Slovenia and Sweden. Interestingly, although the recession did hit Croatia in 2008 and there was a drop in tourist arrivals, the entrepreneurs did not experience any consequences as reported during the second study. All entrepreneurs at the time of second study were running successful businesses and were constantly growing, anticipating required activities before any disruptions occur. Entrepreneurs were optimistic about the future and instead of anticipating the negative aspects of disruption they were more focused on the available opportunities these events could bring. For instance, they were all identifying new market segments (the USA and South Korea) and were preparing their hotels' offerings to accommodate these guests. Also, almost all entrepreneurs were again looking to increase hotel capacities, either by increasing the number of rooms or opening up a new hotel. Besides preparing the business for increased supply, this expansion was also related to the business itself. Many entrepreneurs at this stage have children who are engaged with the hotel business and gradually are talking over specific roles. Entrepreneurs are thus confident that the business will stay in the family and that the children will bring new perspectives, illustrated by the following example:

It was always a family business, but now when my kids are in business it is much easier, I know they will continue the legacy I have built but also they are so forward looking, innovative and capable that I can see the business only prospering. My daughter is just like me, she likes numbers, so she is a revenue manager. My son was always keen to move around in the hotel, from the kitchen to the maintenance, so he is the head of operations. With such a team it was a no brainer to go for the third expansion (laugh). (H1)

DISCUSSION

During the period of communism, a large number of entrepreneurs or their families started a small business in the T&H sector. Despite being allowed, these entrepreneurs were in constant fear that they would lose everything since the communist party had a practice to nationalise private property (Sekelj and Lazic, 1997). Entrepreneurs during this period could not act in advance in terms of the business. However, they practiced resilience at the personal level, through monitoring closely all the trends happening in the former Yugoslavia and in Europe, so they could exploit new opportunities once they are allowed to. Following Conz et al.'s (2020) classification, they can be described as perpetuators, since their main goal was to keep the business unchanged due to the risk of property nationalisation, by absorbing the changes. During this time another group of entrepreneurs could be identified: those who emigrated and exploited the potential of the market economies. These entrepreneurs can be classified as developers, since they aimed to develop their businesses in a completely unfamiliar markets of Western economies acting ahead of the time, so their businesses would be ready to adapt to future change.

The study results reveal that the period of transition was the most challenging for the entrepreneurs. Entrepreneurs were faced with an underdeveloped institutional framework which made conducting a private business extremely difficult. Despite this, they all started businesses in a range of different sectors once the reforms took place in the 1990s. During this period, two groups of entrepreneurs were identified. The first group started a business but did not expand further until the institutional environment became more stable. These entrepreneurs wanted to keep the family business as it was despite all of the unexpected changes which a transitional environment could offer. Therefore, their understanding of resilience was about change absorption (perpetuators). Another group of entrepreneurs was more attracted with the possible opportunities and market gaps which could be exploited in the early years of transition. These entrepreneurs thought about their family business as a portfolio business and wanted to prepare for the changes that threatened the equilibrium of the business on a higher level (developers). Possible sources of variation (e.g. age, gender, previous experience) between these two groups could not be identified.

The period after 2000 was marked by entrepreneurial attempts to prepare hotel businesses to embrace all the positive effects of EU accession (developers). Entrepreneurs expected that the EU accession would bring a higher number of tourists looking for small authentic accommodation and were acting accordingly by improving the hotels' offerings and expanding their capacities to accommodate higher expected numbers of guests. It is interesting that entrepreneurs do not think of any possible side effects of the EU accession or even about the shocks which may happen due to the recession in 2008. Similar to Engeset's (2020) study, this may be explained by an entrepreneurial mindset, as entrepreneurs' response to unexpected challenges was related to the identification of new opportunities and the capacity to realise them. Another explanation may be related to the study context, which was experiencing crisis and disruption over a long period of time; resultantly, entrepreneurs learnt how to navigate successfully through this turbulence. Similarly, Dahles and Susilowati (2015) investigated tourism entrepreneurs in Indonesia during a decade of crisis and found that among survival, adaptation and innovation responses to crisis, most of the entrepreneurs oriented towards adaptation. This implies that resilience is related to the entrepreneurs and their own personal experiences (Conz et al., 2020).

Another point for discussion relates to the non-financial goals and preservation of SEW of family members. Although the literature, in particular T&H literature puts significant emphasis on the SEW and non-profit orientation of small T&H firms, this study has demonstrated that investigated entrepreneurs are strongly driven by profit. This does not imply that SEW is not one of their goals. For instance, there are strong binding ties between family members, who work closely with one another. All entrepreneurs who participated in this study had either already involved or were preparing the children for involvement. However, entrepreneurs are opportunistic and think of any possible change or shock as a potential opportunity for business development, something in contrast to the current portrayal of tourism entrepreneurs as lifestyle-driven.

CONCLUSION

This study contributes to the literature on family firm resilience and particularly the resilience of family firms in tourism, which so far have been largely overlooked in the research. Positioning itself in the middle of individual- and the firm-oriented research stream, this study offers a new perspective of resilience as perceived by entrepreneurs. In addition, this study encompassed three turbulent periods: the Communist Era; the War (1991-1995) and transition; and the EU accession, all of which have had significant effect on entrepreneurs themselves, their families and their businesses.

The study's findings contribute to the family business literature which so far has not paid significant attention to the resilience of FF. This study demonstrates that it is important to observe resilience from the point of entrepreneurial understandings (Conz et al., 2020). For instance, even though the study concentrated on turbulent periods, the entrepreneurs focused on emerging opportunities and understood resilience mainly through proactive business development. Additionally, entrepreneurs practiced resilience through business diversification; not to preserve family capital but to exploit opportunities and prepare the business for future change.

This study also contributes to the T&H entrepreneurship literature, which describes T&H entrepreneurs as 'lifestyle' (Thomas et al., 2011). This study demonstrates that a 'lifestyle' dichotomy may not provide adequate understandings of tourism entrepreneurs in different contexts and calls for a more holistic research approach.

Although entrepreneurship scholars call for more context-dependent research (Welter, 2011), resilience may be understood differently across different settings: a limitation of this study. Although qualitative studies are guided by transferability rather than generalisability (Pratt, Kaplan and Whittington, 2020), future research could address this challenge and study FF resilience across different industries, business types and countries. Another limitation of this study relates to the focus on entrepreneurs and not the whole family involved in the business. Therefore, it would be beneficial to investigate how other family members involved in the business understand resilience and how they manage possible conflicts if there is a disagreement between the entrepreneur and other family members.

REFERENCES

Acquaah, M., Amoako-Gyampah, K., & Jayaram, J. (2011). Resilience in family and nonfamily firms: An examination of the relationships between manufacturing strategy, competitive strategy and firm performance. *International Journal of Production Research*, *49*(18), 5527–5544. doi:10.1080/00207543.2011.563834

Ahmad, S. Z. (2015). Entrepreneurship Education in Tourism and Hospitality Programs. *Journal of Hospitality & Tourism Education*, *27*(1), 20–29. doi:10.1080/10963758.2014.998764

Aidis, R., Estrin, S., & Mickiewicz, T. (2008). Institutions and Entrepreneurship Development in Russia: A Comparative Perspective. *Journal of Business Venturing*, *23*(6), 656–672. doi:10.1016/j.jbusvent.2008.01.005

Alpeza, M., Oberman Peterka, S., & Has, M. (2018). Entrepreneurship in Croatia. In R. Palaić, L. P. Dana, & V. Ramadani (Eds.), *Entrepreneurship in former Yugoslavia: Diversity, Institutional Constraints and Prospects* (pp. 37–50). Springer. doi:10.1007/978-3-319-77634-7_3

Alsos, G. A., Carter, S., & Ljunggren, E. (2014). Kinship and business: How entrepreneurial households facilitate business growth. *Entrepreneurship and Regional Development*, *26*(1-2), 97–122. doi:10.1080/08985626.2013.870235

Ateljevic, I., & Doorne, S. (2000). 'Staying within the fence': Lifestyle entrepreneurship in tourism. *Journal of Sustainable Tourism*, *8*(5), 378–392. doi:10.1080/09669580008667374

Bamberger, P. (2008). From the editors beyond contextualization: Using context theories to narrow the micro-macro gap in management research. *Academy of Management Journal*, *51*(5), 839–846. doi:10.5465/amj.2008.34789630

Banki, M. B., & Ismail, H. N. (2015). Understanding the characteristics of family owned tourism micro businesses in mountain destinations in developing countries: Evidence from Nigeria. *Tourism Management Perspectives*, *13*, 18–32. doi:10.1016/j.tmp.2014.11.001

Berrone, P., Cruz, C., & Gomez-Mejia, L. R. (2012). Socioemotional wealth in family firms: Theoretical dimensions, assessment approaches, and agenda for future research. *Family Business Review*, *25*(3), 258–279. doi:10.1177/0894486511435355

Biggs, D., Hall, C. M., & Stoeckl, N. (2012). The resilience of formal and informal tourism enterprises to disasters: Reef tourism in Phuket, Thailand. *Journal of Sustainable Tourism*, *20*(5), 645–665. doi:10.1080/09669582.2011.630080

Brewton, K. E., Danes, S. M., Stafford, K., & Haynes, G. W. (2010). Determinants of rural and urban family firm resilience. *Journal of Family Business Strategy*, *1*(3), 155–166. doi:10.1016/j.jfbs.2010.08.003

Bui, H. T., Jones, T. E., Weaver, D. B., & Le, A. (2020). The adaptive resilience of living cultural heritage in a tourism destination. *Journal of Sustainable Tourism*, *28*(7), 1022–1040. doi:10.1080/09669582.2020.1717503

Butler, R. W. (2017). *Tourism and resilience*. CABI. doi:10.1079/9781780648330.0000

Cheung, K. S., & Li, L. H. (2019). Understanding visitor–resident relations in overtourism: Developing resilience for sustainable tourism. *Journal of Sustainable Tourism*, *27*(8), 1197–1216. doi:10.1080/09669582.2019.1606815

Chowdhury, M., Prayag, G., Orchiston, C., & Spector, S. (2019). Postdisaster social capital, adaptive resilience and business performance of tourism organizations in Christchurch, New Zealand. *Journal of Travel Research*, *58*(7), 1209–1226. doi:10.1177/0047287518794319

Chrisman, J. J., Chua, J. H., & Steier, L. P. (2011). Resilience of family firms: An introduction. *Entrepreneurship Theory and Practice*, *35*(6), 1107–1119. doi:10.1111/j.1540-6520.2011.00493.x

Chua, J. H., Chrisman, J. J., & Sharma, P. (1999). Defining the family business by behavior. *Entrepreneurship Theory and Practice*, *23*(4), 19–39. doi:10.1177/104225879902300402

Conz, E., Lamb, P. W., & De Massis, A. (2020). Practicing resilience in family firms: An investigation through phenomenography. *Journal of Family Business Strategy*, *11*(2), 2. doi:10.1016/j.jfbs.2020.100355

Conz, E., & Magnani, G. (2020). A dynamic perspective on the resilience of firms: A systematic literature review and a framework for future research. *European Management Journal*, *38*(3), 400–412. doi:10.1016/j.emj.2019.12.004

Dahles, H., & Susilowati, T. P. (2015). Business resilience in times of growth and crisis. *Annals of Tourism Research*, *51*, 34–50. doi:10.1016/j.annals.2015.01.002

Dana, L. P., Gurau, C., & Lasch, F. (2014). Entrepreneurship, tourism and regional development: A tale of two villages. *Entrepreneurship and Regional Development, 26*(3-4), 357–374. doi:10.1080/08985626.2014.918182

Dana, L. P., Palaić, R., & Ramadani, V. (2018). Entrepreneurship in former Yugoslavia: An Introduction. In R. Palaić, L. P. Dana, & V. Ramadani (Eds.), *Entrepreneurship in former Yugoslavia: Diversity, Institutional Constraints and Prospects* (pp. 1–8). Springer. doi:10.1007/978-3-319-77634-7_1

Danes, S. M., Lee, J., Amarapurkar, S., Stafford, K., Haynes, G., & Brewton, K. E. (2009). Determinants of family business resilience after a natural disaster by gender of business owner. *Journal of Developmental Entrepreneurship, 14*(04), 333–354. doi:10.1142/S1084946709001351

De Massis, A., Eddleston, K. A., & Rovelli, P. (2020). Entrepreneurial by Design: How Organizational Design Affects Family and Non-family Firms' Opportunity Exploitation. *Journal of Management Studies*. Advance online publication. doi:10.1111/joms.12568

DZS. (1991). *Statisticki ljetopis*. Drzavni Zavod za Statistiku.

DZS. (2020). *Statisticki ljetopis*. Drzavni Zavod za Statistiku.

Eisenhardt, K. (1989). Building theories from case study research. *Academy of Management Review, 14*(4), 532–550. doi:10.5465/amr.1989.4308385

Engeset, A. B. (2020). For better or for worse"–the role of family ownership in the resilience of rural hospitality firms. *Scandinavian Journal of Hospitality and Tourism, 20*(1), 68–84. doi:10.1080/15022250.2020.1717600

Fu, H., Okumus, F., Wu, K., & Köseoglu, M. A. (2019). The entrepreneurship research in hospitality and tourism. *International Journal of Hospitality Management, 78*, 1–12. doi:10.1016/j.ijhm.2018.10.005

Getz, D., & Carlsen, J. (2000). Characteristics and goals of family and owner-operated businesses in the rural tourism and hospitality sectors. *Tourism Management, 21*(6), 547–560. doi:10.1016/S0261-5177(00)00004-2

Getz, D., Carlsen, J., & Morrison, A. (2004). *The family business in tourism and hospitality*. CABI. doi:10.1079/9780851998084.0000

Ioannides, D., & Petersen, T. (2003). Tourism 'non-entrepreneurship' in peripheral destinations: A case study of small and medium tourism enterprises on Bornholm, Denmark. *Tourism Geographies, 5*(4), 408–435. doi:10.1080/1461668032000129146

James, H. (2006). *Family capitalism: Wendels, Haniels, Falcks, and the continental European model*. Harvard University Press.

Kobasic, A. (1987). *Turizam u Jugoslaviji: Razvoj, stanje, perspektive*. Informator.

Kornai, J. (2006). The great transformation of Central and Eastern Europe: Success and disappointment. *Economics of Transition*, *14*(2), 207–244. doi:10.1111/j.1468-0351.2006.00252.x

Kovac, B. (1990). Politicko-Ekonomska Uloga Poduzetnistva. In D. Njavro & V. Franicevic (Eds.), *Poduzetnistvo: Teorija, Politika, Praksa* (pp. 69–93). Privredni Vjesnik.

Kraiczy, N. D., Hack, A., & Kellermanns, F. W. (2015). What makes a family firm innovative? CEO risk-taking propensity and the organizational context of family firms. *Journal of Product Innovation Management*, *32*(3), 334–348. doi:10.1111/jpim.12203

Lengnick-Hall, C. A., & Beck, T. E. (2005). Adaptive fit versus robust transformation: How organizations respond to environmental change. *Journal of Management*, *31*(5), 738–757. doi:10.1177/0149206305279367

Marić, K., Samardžić, B., & Protrka, N. (2017). Analiza turističkih pokazatelja i sigurnost u turizmu Republike Hrvatske te mogućnosti jačanja konkurentnosti hrvatskog turizma nakon ulaska u Europsku uniju. *Obrazovanje za Poduzetništvo-E4E: Znanstveno Stručni Časopis o Obrazovanju za Poduzetništvo*, *7*(1), 301-318.

Memili, E., Patel, P. C., Koç, B., & Yazıcıoğlu, İ. (2020). The interplay between socioemotional wealth and family firm psychological capital in influencing firm performance in hospitality and tourism. *Tourism Management Perspectives*, *34*, 100651. doi:10.1016/j.tmp.2020.100651

Memili, E., Welsh, D. H., & Luthans, F. (2013). Going beyond research on goal setting: A proposed role for organizational psychological capital of family firms. *Entrepreneurship Theory and Practice*, *37*(6), 1289–1296. doi:10.1111/etap.12066

Miles, M. B., & Huberman, A. M. (1994). Qualitative data analysis: An expanded sourcebook. *Sage (Atlanta, Ga.)*.

Morrison, A., & Teixeira, R. (2004). Small business performance: A tourism sector focus. *Journal of Small Business and Enterprise Development*, *11*(2), 166–173. doi:10.1108/14626000410537100

Mzid, I., Khachlouf, N., & Soparnot, R. (2019). How does family capital influence the resilience of family firms? *Journal of International Entrepreneurship*, *17*(2), 249–277. doi:10.100710843-018-0226-7

Orchiston, C., Prayag, G., & Brown, C. (2016). Organizational resilience in the tourism sector. *Annals of Tourism Research*, *56*, 145–148. doi:10.1016/j.annals.2015.11.002

Pappas, N., & Papatheodorou, A. (2017). Tourism and the refugee crisis in Greece: Perceptions and decision-making of accommodation providers. *Tourism Management*, *63*, 31–41. doi:10.1016/j.tourman.2017.06.005

Paul, J. J., Winter, M., Miller, N. J., & Fitzgerald, M. A. (2003). Cross-institutional norms for timing and sequencing and the use of adjustment strategies in families affiliated with family-owned businesses. *Marriage & Family Review*, *35*(1-2), 167–191. doi:10.1300/J002v35n01_10

Peters, M., & Buhalis, D. (2013). SMEs in tourism destinations. In C. Costa, E. Panyik, & D. Buhalis (Eds.), Aspects of tourism: Vol. 60. Trends in European tourism planning and organisation (pp. 93–101). Bristol: Channel View Publications. doi:10.21832/9781845414122-011

Pratt, M. G., Kaplan, S., & Whittington, R. (2020). Editorial essay: The tumult over transparency: Decoupling transparency from replication in establishing trustworthy qualitative research. *Administrative Science Quarterly*, *65*(1), 1–19. doi:10.1177/0001839219887663

Prayag, G., Spector, S., Orchiston, C., & Chowdhury, M. (2020). Psychological resilience, organizational resilience and life satisfaction in tourism firms: Insights from the Canterbury earthquakes. *Current Issues in Tourism*, *23*(10), 1216–1233. doi:10.1080/13683500.2019.1607832

Ritchie, J., & Spencer, L. (1994). Qualitative data analysis for applied policy research. In A. Bryman & R. Burgees (Eds.), *Analyzing Qualitative Data* (pp. 173–194). Routledge. doi:10.4324/9780203413081_chapter_9

Russell, R., & Faulkner, B. (2004). Entrepreneurship, chaos and the tourism area lifecycle. *Annals of Tourism Research*, *31*(3), 556–579. doi:10.1016/j.annals.2004.01.008

Schönfelder, B. (2005). *The Impact of the War 1991: 1995 on the Croatian Economy: A Contribution to the Analysis of War Economies (No. 2005/14)*. Freiberger Arbeitspapiere.

Skokic, V., Lynch, P., & Morrison, A. (2016). Hotel entrepreneurship in a turbulent environment. *International Journal of Hospitality Management*, *53*, 1–11. doi:10.1016/j.ijhm.2015.11.008

Solvoll, S., Alsos, G. A., & Bulanova, O. (2015). Tourism entrepreneurship–review and future directions. *Scandinavian Journal of Hospitality and Tourism*, *15*(sup1), 120–137. doi:10.1080/15022250.2015.1065592

Stojanov, D. (2004). *Hungary and Bosnia and Herzegovina: A success and a failure of transition*. Institute for World Economics, Hungarian Academy of Sciences.

Thomas, R., Shaw, G., & Page, S. J. (2011). Understanding small firms in tourism: A perspective on research trends and challenges. *Tourism Management*, *32*(5), 963–976. doi:10.1016/j.tourman.2011.02.003

Tsao, C. Y., & Ni, C. C. (2016). Vulnerability, resilience, and the adaptive cycle in a crisis-prone tourism community. *Tourism Geographies*, *18*(1), 80–105. doi:10.1080/14616688.2015.1116600

Welter, F. (2011). Contextualizing entrepreneurship—Conceptual challenges and ways forward. *Entrepreneurship Theory and Practice*, *35*(1), 165–184. doi:10.1111/j.1540-6520.2010.00427.x

Williams, A. M., Shaw, G., & Greenwood, J. (1989). From tourist to tourism entrepreneur, from consumption to production: Evidence from Cornwall, England. *Environment & Planning A*, *21*(12), 1639–1653. doi:10.1068/a211639

Williams, C., You, J. J., & Joshua, K. (2020). Small-business resilience in a remote tourist destination: Exploring close relationship capabilities on the island of St Helena. *Journal of Sustainable Tourism*, *28*(7), 937–955. doi:10.1080/09669582.2020.1712409

Williams, N., & Vorley, T. (2014). Economic resilience and entrepreneurship: Lessons from the Sheffield City Region. *Entrepreneurship and Regional Development*, *26*(3-4), 257–281. doi:10.1080/08985626.2014.894129

WTTC. (2020). *Travel and Tourism Economic Impact: Croatia*. WTTC.

ENDNOTE

1 This research is a part of a larger study which investigated small tourism entrepreneurs, including family firms.

Section 2

Creating Opportunities Through Crisis

The second section is about 'Creating Opportunities Through Crisis' and consists of two chapters. In this section, both chapters apply a community-oriented model of the destination. First, Chapter 4 takes the destination and its stakeholders and focuses on opportunities for developing resilience strategies at a destination-wide level. Exploring such opportunities is vital, as there might be a trade-off between firm resilience and regional resilience. Secondly, the fifth chapter returns to the perspective of organizational resilience of family firms, while also following a community-oriented destination model.

Chapter 4
Resilience as an Opportunity Approach:
Challenges and Perspectives for Private Sector Participation on a Community Level

Daniel Zacher
Catholic University of Eichstaett-Ingolstadt, Germany

Harald Pechlaner
Catholic University of Eichstaett-Ingolstadt, Germany

ABSTRACT

Against the background of its conceptual vagueness, practicing resilience is a complex task and a particular challenge for management. Resilience initiatives on the local and regional levels pursue participation by using different methods of stakeholder integration. In this contribution, the private sector is presented as a particularly difficult stakeholder group to attract for cooperative resilience building. Due to limited resources, corporate development and responsibility often remain in the spheres of an operational context. Thus, involvement in regional resilience building is often seen as desirable, but not an integral part of strategic corporate management. This contribution gives recommendations on how coordinated resilience initiatives should act in order to achieve the systematic participation of the private sector. If resilience practice defines measurable outcomes for entrepreneurship and positive experiences are made in the regional network, a sustainable participation of firms in the resilience development should be manifested.

DOI: 10.4018/978-1-7998-7352-5.ch004

INTRODUCTION

The global tourism industry is in a challenging situation due to a global pandemic that is without comparison so far (Gössling et al., 2020). Tourists cannot go on holiday, or they travel in a severely restricted or altered form. It can be assumed that tourism as a whole will sooner or later find a way to deal with the longer-term changes caused by the pandemic, and some authors even speak of the crisis as an opportunity to reorientate and "reinvent" tourism (Brouder, 2020; Chang et al., 2020; Niewiadomski, 2020). Even though it is too early to make sound conclusions, tourism is an example of a sector in the process of change that has serious effects on a regional economic structure due to a massive external shock.

This chapter begins with the premise that these processes of change have to be actively and constantly shaped. This applies both to immediate crisis situations and to noncrisis times of success and economic growth. There is a need for mechanisms, structures, and processes that are designed to ensure the best possible mitigation and even prevention. The structural development of these mechanisms affects both the public and private sectors, as the necessary resources for change in times of cutbacks of public expenditures and because the manifold financial responsibilities of firms for their stakeholders cannot be solely provided by one side (Schmidt et al., 2017). Tourism destinations, with their interplay of private product development, public infrastructures, and community resources, are a particularly illustrative example of the need for integrative strategies of change (Saarinen and Gill, 2018). In pre-COVID-19 times, it was already clear that corresponding structures and processes had not been sufficiently developed. This can be seen in the following examples:

- In some destinations, the negative side effects of quantitatively successful tourism (overtourism) were only responded to very late (Cheer et al., 2019; Seraphin et al., 2020).
- In many economically strong regions, tourism businesses lack the personnel or competitiveness against other, more profitable industries (Joppe, 2012; Marcher et al., 2020; Zehrer, 2019).
- The climate crisis is a threat scenario for international tourism that remains unanswered in many respects (Dogru et al., 2019).
- Digitalisation is not always seen as an opportunity, but also as a threat (Dredge et al. 2018).
- Tourism companies often perceive themselves as being driven by change in social values (Cheer et al., 2017).

These and similar signals suggest that the necessary connections between individual tourism firms and destination networks in the joint assumption of responsibility are not defined (Pechlaner et al., 2019). This chapter looks at the challenges for proactive resilience building, and highlights ways and perspectives for greater private-sector participation and engagement in coordinated resilience activities. The framework is provided by the following research question:

Which challenges and perspectives do resilience initiatives face in integrating the private sector into a coordinated resilience-building process?

In order to answer this question, the first step is an examination of the existing resilience literature with a special focus on the resilience approach as an active design and planning tool in tourism related contexts. Then, various approaches to resilience on the organisational level are presented. Furthermore, practical initiatives related to resilience are presented in order to illustrate a possible framework of action for practical resilience agency.

Selected results from an empirical research project show concrete challenges in the cooperation of public resilience initiatives with the private sector, and present first approaches for the strategic integration of the private sector in coordinated processes on the regional level.

This leads to a presentation of different design options for resilience as an opportunity approach in the field of tension in public–private cooperation. These perspectives are placed in the context of current development issues with an outlook on the expected "career" of the resilience approach in this context. Lastly, research perspectives for empirical work on this young topic are derived.

BACKGROUND

The resilience approach has developed over the past few decades as a conceptual approach to the retrospective analysis and prospective formulation of personal, organisational, and regional characteristics and circumstances (Boschma, 2015; Pfefferbaum et al., 2007; Walker and Salt, 2012). The respective resilience characteristics are presented in different disciplinary research traditions, while a generally applicable definition of the resilience approach is hardly possible due to its diverse use (Caldarice et al., 2019; Davidson et al., 2016). Originally based in particular on psychology and socioecological research (e.g., Holling, 1973; Stainton et al., 2019), the resilience approach has become an established subject of research and application on the regional and organisational levels (Annarelli and Nonino, 2016; Fröhlich and Hassink, 2018).

In the course of this shift, resilience is not only regarded as a descriptive characteristic, but increasingly as an active field of action in which changes are not understood as a threat, but as necessary and actively shaped processes (Martin and Sunley, 2015). In this understanding, resilience finds its way into the organisational structures and policy processes of regional development strategies (Joseph, 2013; Linnenluecke and Griffiths, 2010).

In the context of private companies, organisational resilience is an established branch of resilience research (Annarelli and Nonino, 2016; Lengnick-Hall et al., 2011). A particular feature on the firm level is the systemic interconnection of individual, group-specific, and overall organisational characteristics of a resilient organisational system (Horne, 1997). Organisations can be resilient in different ways, for example, by consistently observing their environment and changing framework conditions, and being able to draw the right conclusions, or by learning from past experiences for the assessment of events that are taking or will take place (Hollnagel, 2011). The defined patterns of relationships within an organisation can contribute to the use of managed change processes (change management) to specifically drive resilience building (Ates and Bititci, 2011). However, the intended development of resilience is limited by two factors. On the one hand, the individual motives and motivation of individual organisational members to build organisational resilience can barely be identified and controlled (Lengnick-Hall et al. 2011). On the other hand, and particularly against the background of stakeholder constellations in tourist destinations, organisations are embedded in a higher-level systemic context, for example, a region, and interact with it by contributing to regional resilience as an organisation or, the other way round, by benefiting from the resilience of a region (Martin and Sunley, 2015; Williams and Vorley, 2014).

Accordingly, the approach of regional resilience is considered, and an economic–geographical perspective is adopted. The resilience of a region can be expressed in the fact that critical events such as an economic crisis or natural disaster affect that region less negatively than they do other regions (Martin, 2012). In the processual understanding of regional resilience research, this is complemented by the evolutionary perspective of resilience. The hallmark of a resilient region in the evolutionary understanding is a positive development perspective in which a region develops longer-term competences to pursue economic-growth paths (Boschma, 2015) or, in a more growth-critical perspective, the competence to successfully pursue alternative development paths beyond economic growth (Evenhuis, 2020; Hassink and Gong, 2020). In this context, it is important to promote the development and appreciation of decentrally distributed knowledge and competences in a regional network in order to be able to comprehensively display development potential.

The participation of local companies plays an important role. Entrepreneurial interests regarding their own development are reflected on the regional level, and there is a distributed assumption of tasks and responsibilities in the field of the tension of public–private organisation and the need for collaboration between sectors (Pinz et al., 2018). These arrangements can be formal and informal or ad hoc activities (Marana et al., 2018). This form of cooperation is partly regulated by contract and then known as the public–private partnership model (PPP). Shared objectives and a mutual agreement on the division of labour by committing resources, and sharing risks and benefits are core PPP characteristics (Buse and Walt 2000). In the context of the resilience approach, PPPs are an effective tool for critical infrastructure resilience (Dunn-Cavelty and Suter, 2009). In the reality of multilevel systems in public–private relationship contexts, the resilience level of one system depends on another system's resilience levels (Katina et al., 2014). Therefore, when it comes to multilevel systems and their resilience, a specific system's resilience level should be improved without losing a holistic perspective (Marana et al. 2018). This becomes clear when looking at urban systems (Vedeld et al., 2016). Corporate responsibility can contribute to increasing urban resilience (Twigg, 2015), but the need to design participatory processes is reflected in the stakeholder consideration of the inhabitants.

Companies that are tied to places and regions due to their specific service provision, as in tourism, also have a special responsibility towards the resilient development of this region out of self-interest (Lew, 2014; Pechlaner et al., 2019). From an entrepreneurial point of view, there might be a trade-off between the resilience of an individual enterprise and regional resilience, namely, when entrepreneurial resources are not used for the regional good, or even work against it (Aguirre, 2007; Cutter, 2016). Some authors thus emphasised that formalised cooperation between the public and private sectors contributes to regional resilience building, while others suggest that this cooperation faces multiple challenges that inhibit collaborative resilience building in practice. In relation to tourism and its multisystemic relationships on the destination level, these challenging inter-relationships are commonly presented.

Research on tourism resilience is usually considered in the regional context. A fundamental distinction can be made between studies that (1) describe the handling of specific catastrophic events (Orchiston, 2013), (2) consider ecosystems that are out of balance due to high tourism volume (Biggs, 2011) and (3) focus on long-term processual changes such as climate change and the adequate adaptation of destinations (Luthe and Wyss, 2014). The increasingly evolutionary understanding of regional resilience is also reflected in the tourism literature and leads to a growing importance of the opportunity-orientated view of resilience. Here, tourism is seen as a key regional sector for ensuring regional resilience, in that it is seen as a comparatively fast-developing opportunity for value creation, especially in rural areas or in areas of major economic structural disruption (Cheer and Lew, 2018; Steiner and Atterton, 2015). Other case studies show that this relationship cannot be generalised, and that the positive contribution of tourism to regional resilience depends on several factors, e.g., the dynamics of the labour market or the demographic composition of the population (Ibanescu et al., 2020). Tourism enterprises in the destinations undoubtedly play a central role in this and family businesses have a special motivation to also ensure sustainable continuity in rural areas (Engeset, 2020).

In order to understand the specific characteristics that private-sector companies can have in regional resilience building, typical resources of companies in resilience building are shown before strategies are presented for valorising these resources on the regional or national level.

PRIVATE SECTOR PROCESSES AND ORGANISATIONAL RESILIENCE BUILDING

The literature on organisational resilience has a decades-long tradition (Annarelli and Nonino, 2016). The private sector develops its own resilience-building mechanisms as part of its strategy, receiving impacts from outside and in turn impacting the environment. Different focal points have emerged that are presented below as an overview based on Linnenluecke (2017).

Resilience as Answer to Internal and External Threats

Organisational resilience began to be scientifically studied as early as the beginning of the 1980s (Linnenluecke, 2017; Meyer, 1982). The focus of these studies was on internal threats to operations. These referred to the negative aspects of a corporate culture that conceals risks and considers emerging crises too late in operational management practice (Staw et al., 1981). Meyer (1982) examined sudden internal crisis events, such as employee strikes, and the organisation's respective handling of these events. Resilience is seen in these contexts as a positive aspect of coping with these events. From today's perspective, this predominantly reactive understanding of resilience seems outdated. First, because it is too limited to the internal processes in the emergence and management of crises; and second, because the organisational social potential for actively building resilience within the company is largely disregarded (Brewton et al., 2010; Conz and Magnani, 2020).

With the increasing discussion of organisational resilience in the context of possible and actual disaster events (e.g., natural hazards, artificial disasters, and terrorism), the role of organisations in a superordinate system is being more strongly addressed. So-called high-reliability organisations (HROs) are critical to the functioning of societal systems and therefore have a special responsibility in ensuring resilient operations (Haavik et al., 2019). Their organisational structure is designed to prevent errors as far as possible, and also to identify and avoid potential crises in advance (Weick and Quinn, 1999).

Employees as a Source of Resilience

With its adaptation in psychology, individual resilience is also gaining increasing importance in its role within the organisation (Shin et al., 2012). Studies showed that a good organisational climate or job satisfaction has an impact on both individual performance and organisational resilience (Avey et al., 2008). Furthermore, employee resilience can be promoted and trained within an organisational context, making resilience building a part of human-resource management and development (Bardoel et al., 2014).

Adaptation of Business Models in Response to Changing Circumstances

In contrast to the initially more resistance-orientated management practices of organisational resilience, and in the course of the more competence-orientated view of employees as a strategic corporate resource, the evolutionary resilience approach has become increasingly important in questions of organisational development. Sutcliffe and Vogus (2003) emphasise the adaptive nature of resilient organisations. These adaptations can be small-scale and implemented on an ongoing basis. They can also be fundamental and involve a change in the business model or business migration (Buliga et al., 2016). Here, the organisational environment as a framework for the resilient organisation is an important determinant.

Designing Resilient Supply Chains

An explicitly cross-organisational approach to resilience building can be identified when looking at supply chains (Ponomarov and Holcomb, 2009). Certain supply chains are more affected by crises than others are, which characterises them in their varying degrees of resilience. Principles of resilient supply chains are flexibility in the choice of network partners or redundancy in the use of alternative transport routes (Sheffi and Rice, 2005). Commonly, very efficiently organised supply chains contradict their resilient characteristics (Ponomarov and Holcomb, 2009). Supply chains are international and global in many aspects of the global economy. Therefore, local regional economic systems and the role of companies in these systems are only indirectly addressed. However, regionally confined supply chains are said to be more resilient (Sabahi and Parast, 2020). Recently, the ethical evaluation of supply chains and proposals for supply-chain legislation in various European parliaments has become a focus of scientific and public discussion (von Westphalen, 2020).

On the basis of the literature presented above, it is clear that studies on organisational resilience are situated in the area of tension between organisational development, national interests in key industries, and international trade and supply relationships. The central goal of organisational resilience remains within the oragnizational boundaries and is rarely considered in relation to local and regional economic and social systems. In the following, the regional level is considered, and the potential and actual interplay between organisation and region in the course of resilience building is examined on the basis of existing practical initiatives.

STRATEGIES TO IMPLEMENT RESILIENCE BUILDING IN REGIONAL SYSTEMS

Here, selected strategies for building resilience are presented, and their different approaches are discussed. These instruments are targeted interventions to promote resilience building within a spatially or thematically defined framework (Taylor and Derudder, 2016). Development trajectories shown in Figure 1 systematise strategies for explicitly and actively building resilience that were observed in practice. These approaches refer, on the one hand, to an individual and organisational build-up of information on resilience development, and on the other hand, to framework laws and procedures in political systems. This field of tension is an important basis for entrepreneurial decisions and their consequences for entrepreneurial participation in resilience building on the regional-network level (Williams and Vorley, 2014). A distinction can be made between (1) time-limited programmes that aim to develop guidelines or toolkits, (2) more or less formalised organisational structures on the local and network levels that promote the development of resilience through active process support, and (3) governmental policies through national strategy papers and legislation (Figure 1).

Figure 1. Development trajectories of resilience building.
Source: own elaboration.

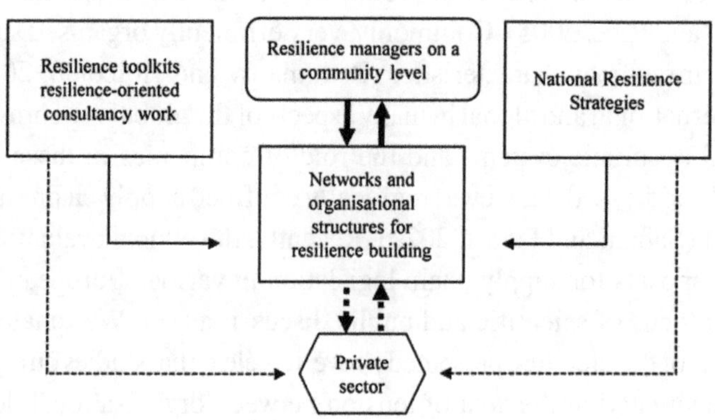

Community Resilience Toolkits and Resilience-Orientated Consultancy

The development of guidelines and toolkits over the course of regional development is a popular funding objective on the federal or national level. With high-quality tools that are graphically presented, a high degree of outreach to target groups is intended to be achieved with a comparatively low investment of resources. Especially in resilience building, a structured approach is often required that systematises engagement with potential crises and challenges.

The Communities Advancing Resilience Toolkit (CART) is an exemplary resilience-building toolkit that contains various methodological building blocks such as interview guides, questionnaires, and focus-group road maps. It focuses on its application in communities that want to develop strategies for coping with catastrophic crisis events, addressing both pre- and postcrisis perspectives. CART aims to bring stakeholders together to address common challenges through actions and group processes (Pfefferbaum et al., 2013). Intervention through CART was designed to direct the engagement of public actors towards the identification of problems, and to stimulate the collective use of resources to address them. The awareness-raising and participation process is carried out by organisations called CART community sponsors. These mostly public organisations have an interest in building resilience in a community under their responsibility and guiding the process. The process begins with status quo analysis that is often conducted by academic institutions or consultancies (Pfefferbaum et al., 2013). As a second step, participation processes are carried out with different methodological approaches in order to identify deficits, knowledge gaps, and development perspectives in deeper analysis. As a third step, the result of the participation process is transferred into a goal and strategy formulation. As a fourth and final step, the strategy is adopted by the participants, and prioritised actions in the action plan are implemented. Here, intensified interaction with as many community members as possible occurs by disseminating the strategy. The core working group remains in place and defines tasks in the process of implementation. The defined strategy may require changes through interaction with a broader stakeholder group due to changing conditions, where a steering group can control the integration of new ideas into the existing strategy.

Often, the CART model remains a purely analytical tool that includes quantitative and qualitative descriptions of the status quo (Binder et al., 2015). This is also due to the lack of comprehensive financial resources for the long-term implementation of such a process. Nevertheless, such toolkits were developed within the framework of different projects, and they contribute to the communicative dissemination of the resilience approach on the regional level (Jones, 2019).

A directly tourism-related project is the Hotel Resilient initiative, a spin-off of the Karlsruhe Institute of Technology. Researchers developed a certification tool for disaster risk management and disaster change adaptation for hotels and resorts on the basis of scientific criteria. A corresponding certification procedure allows for hotels to prove their resilience to different risks and challenges, which is intended to create a favourable basis for decision making in cooperation with customers, tour operators, insurance companies, and financiers. The certification thus directly aims at understanding demonstrated resilience as a competitive advantage. The business model of Hotel Resilient includes, in addition to the certification process, a range of workshops and trainings for hotel management to build an understanding of disaster risk reduction and climate adaptation management. Education for hotel staff with the same goal is also offered through e-learning and webinars. Hotel Resilient also responds to current challenges. In the wake of the COVID-19 crisis in 2020, a special certification programme was offered (Bentfeld et al., 2018; HotelResilient, 2020).

Establishing Networks and Organisational Structures for Resilience Building

A second strategy for implementing resilience building is to promote the development of a resilience network among network partners with similar challenges. The European Union launched the ResilientEurope network under the umbrella of URBACT—a Europe-wide initiative to promote sustainable urban development. ResilientEurope consisted of a network of 11 cities in 9 countries and pursued the exchange of cocreative methods of future-orientated urban development based on experimental governance approaches. The network had project character and largely ceased its publicly visible activity at the end of the project period (ResilientEurope, 2020).

A much larger initiative was launched in 2013 with the Rockefeller Foundation's initiative 100 Resilient Cities. In a three-stage process, a total of 100 cities were nominated to develop an independent resilience strategy on the local level over a period of 2 years. These cities were financed by the Rockefeller Foundation for this task and accompanied by free consulting services in the process. Cities around the world could apply for participation in this network by agreeing to the framework conditions of the funding foundation. These included (Rockefeller Foundation 2015):

- Outlining the most significant vulnerabilities on the local and regional levels, and making a coherent case for the need for action.
- Presenting a range of relevant stakeholders in the urban environment with whom a strategy would be developed, and the strategy process would be implemented (with particular reference to a cross-sectoral coalition to address different vulnerabilities).

- A statement of intent to develop a resilience strategy for the long term, and a continuity plan for implementation.
- A willingness to engage in regular exchange in the global network of 100 Resilient Cities.
- Establishing a new type of position within the city administration, a Chief Resilience Officer.

The initiative was designed to develop organisational structures within the municipality that decidedly related to resilience building. The stakeholder coalition on the local level should explicitly involve the private sector, particularly to answer questions of follow-up funding.

Objectives of National Resilience Strategies

The promotion of local and regional resilience initiatives can also be achieved through the adoption of national or federal policy frameworks. Different focal points can be identified that are aligned with prevailing resilience definitions (see also the beginning of this section). Back around 2010, countries such as Australia, the Netherlands, New Zealand, the United Kingdom, and the United States adopted resilience strategies that focused on national security issues related to maintaining critical infrastructure and disaster management (Fjäder, 2014). Critical infrastructure such as energy and water supply, health services, transport, and food supply chains is often privately owned or operates as public–private partnerships (Brown et al., 2017). A key objective of national resilience strategies designed to address security issues is, therefore, to integrate private infrastructure into the national resilience strategy (Fjäder, 2014). Therefore, the attention of national executives is either on high-reliability organisations (Dalgaard-Nielsen, 2017) or on organisations of which the decline was assessed as posing a systemic risk (e.g., large banks; Ioannou et al., 2019).

More recent national resilience strategies place less emphasis on disaster protection and functional preservation, and more on the proactive developmental nature of the resilience approach. However, despite the impression of an increasingly complex world, the diversity of stakeholders to be addressed in a national resilience strategy is only slightly changing. The Austrian government, as part of the development of a national resilience strategy, is promoting projects that include the innovative capacity of smaller companies, in addition to usual organisations and networks for the preservation of critical infrastructure (Pfosser and Seper, 2020). France is developing a resilience strategy through a network of public authorities and businesses brought together as an association to exchange expertise, and this network is supported by the state. The focus of members here is on local authorities, nongovernmental organisations (NGOs), and large and infrastructure companies. Small companies and start-ups are not part of this network (HCFCD, 2020).

Overall, the general impression is that national resilience strategies have so far displayed a rather defensive and protective understanding of resilience. This follows lines of argumentation in political theory that the role of the state in liberal democracies should concentrate on immediate security issues (Joseph, 2013). In this respect, the observed emphases are no surprise, but nevertheless make it apparent that entrepreneurial capacity as a potential resilience resource is structurally disregarded. This reluctance is ultimately also expressed in the concentration on the role of the citizen and predominantly indirect addressing of private-sector responsibility in the context of community resilience toolkits. In the following, insight is given into an empirical study that looked at the issue of private-sector involvement in the context of local resilience initiatives.

EMPIRICAL RESULTS: CHALLENGES AND PERSPECTIVES ON PRIVATE SECTOR PARTICIPATION

For a closer look on the practical building of resilience on the regional level, interviews were held with leaders of practical resilience initiatives that were launched as part of the 100 Resilient Cities network and other local community initiatives that use the term resilience in their name. These organised structures were established on the local level within local-government frameworks. The central task of these organisational units is the development and implementation of a local resilience strategy with the participation of civil society and the collaboration of the local private sector providing the framework. A total of 8 in-depth interviews with the responsible directors of these initiatives were conducted online or by telephone, in English, and they lasted an average of 40 minutes. This approach allowed for insight into different city contexts and their approach to actively building resilience. The studied initiatives saw tourism as a relevant part of their regional economy, although the focus of the resilience strategy did not explicitly refer to tourism. However, cross-sectoral opportunities and challenges of resilience building, and the challenges of private-sector participation in such initiatives were discussed. The interviews were transcribed and analysed using the GABEK® method and the associated WinRelan® software (Zelger and Oberprantacher, 2002). In this process, sense units of key terms were aggregated across single interviews. The figures resulting from this process represent the keywords of the interview conversations, and show relations to other keywords by means of a connecting line. The thickness of the connecting line can be related to the frequency of the mapped relations. In the present figures 2 and 3, the connections are usually supported by 2 or more sense units. Occasionally, connections that were only mentioned once are also shown. Due to the explorative character of the study, the aim was to present the different aspects in relation to the research question as comprehensively and diversely as possible, so that even individually presented statements would find their way into the study. As a result, so-called association-network graphs are presented and interpreted below. Direct quotations are used to underline central aspects of the results and terms shown in italics provide a direct link to the respective network graphics.

Figure 2. Association-network graph: challenges of integrating private sector in resilience building.
Source: own elaboration based on WinRelan®.

Figure 2 shows the challenges addressed in the interviews with regard to private-sector participation. At the centre of the figure is the *business community*, which is seen by the interviewed resilience experts as a key actor in contributing to the *continuous funding* of local or regional resilience initiatives through financial means. This is seen as a challenge, as resilience initiatives are currently located in a *fuzzy*

area or structurally new ground where network relationships between public and private bodies are not yet sufficiently differentiated and established. The resilience initiatives under consideration do not yet see themselves in a position to map out a sustainable *business model* from their work. Although many *projects* are attempting to actively involve the private sector in resilience building, they are *difficult to convince*:

"If you talk to other cities in our network, the private sector is often the most difficult to engage with because people need a good reason why they are coming to a meeting." (Resilience Manager A)

In basic terms, it is considered necessary for the private sector to become involved in resilience building not only financially, but also by actively contributing specific skills and experience. Hence, in times of scarce resources, many private companies prefer to engage for their immediate business survival, as the following statement underlines:

"Very few people will understand in the private sector. They like to give funding to certain projects which gives something tangible. They want immediate outcomes and something that people can see and touch." (Resilience Manager B)

Currently, resilience initiatives are rarely aimed at short-term and *immediately perceptible outcomes*. In the assessment of the interviewed resilience experts, there is currently a lack of concepts that can present the results of the work of resilience initiatives as direct added economic value for individual companies. Where projects are implemented that integrate both visible short- and long-term development goals, there is a lack of communication strategies capable of establishing complex connections in resilience building between strategic thinking on the regional level and the immediate profit thinking of companies.

With the challenges outlined above, resilience initiatives find themselves in uncertain times in terms of their future funding. They are either dependent on *money* from large funding bodies on the national or supranational level, or on continued funding from municipal budgets. Thus, the demand for the structural development of resilience organisations is articulated, expressed in a consistent diversification strategy with regard to their funding. It is examined in more detail on the basis of the statements provided by the associations shown in Figure 3.

In the resilience approach, often only the downsides of organisational and regional development are illuminated, which is partly explained by the research tradition in engineering and psychology. The notion of "*opportunity*" is, therefore, at the heart of the reorganisation of resilience initiatives, and serves as the narrative for this process:

Figure 3. Association-network graph: perspectives and possible tactics to integrate private sector.
Source: own elaboration based on WinRelan®.

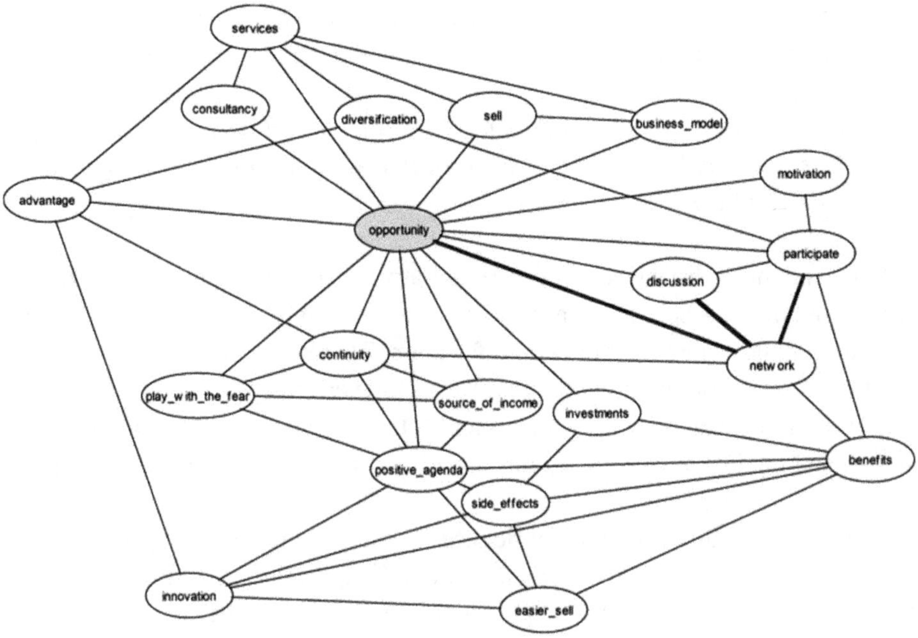

" I can easily see in our example that we could be helping the city having a more economically dynamic innovative ecosystem and prevent risk areas to be occupied. (...) This is what we call in our network of cities the 'Values of Resilience' (...) Mainly focusing the risk as an opportunity." (Resilience Manager C)

From this narrative, concrete development options arise on the organisational level in the view of the interviewees. Resilience organisations could structurally develop and, in terms of their personal competencies, in such a way that they provide concrete *consultancy* services for their stakeholders, presenting these as *saleable* offers. These consultancy services have a dual function. On the one hand, they can finance a certain part of the work of resilience initiatives; on the other hand, the exchange that takes place in this framework serves to build networks. Positive experiences in cooperation increase the willingness of the private sector to *participate* in resilience building in their region, according to the interviewees. In the process, a broader understanding of resilience develops, which makes connections between organisational and regional resilience more obvious. Recognising these connections has a *motivating* effect.

Resilience managers are aware of their central role in this process. They are asked to actively steer the desired development, whereby they have two different approaches at their disposal. They can market resilience building predominantly as a *positive agenda* by establishing links to innovation-orientated development policies and by actively communicating the synergy effects of integrated development strategies:

"The idea is to start to look at planning for resilience not only on the aspect if it raises the resilience, but if it also gives you day-to-day benefits in the community." (Resilience Manager D).

Some of the interviewed resilience initiatives come from the organisational background of disaster management. The perspective of regional development and innovation policy is a rather new field for them. Therefore, addressing immediate threat scenarios is often within their area of competence, wherein they have a high credibility. In certain actor constellations, it may likewise be appropriate to play on fear, and to clearly address negative development scenarios regarding potential crises and disaster events.

The overall picture is one of confidence in actively engaging the private sector in overarching resilience development, although pathways need to be differently shaped in each local system. This applies to the tourism industry as part of the overall regional economy to a similar extent, whereby in the opinion of the interviewees, the local tourism industry should take a special role, as it can positively influence resilience building on a local level through its products and outcomes on both the economic and the societal level.

DISCUSSION

Challenges in Participatory Resilience Building

Resilience found its way into practical issues of urban and regional development which makes it relevant for tourism destinations. The discussed resilience initiatives are developing a cross-sectoral resilience strategy where objectives and priorities are developed in a participatory process. Since the outbreak of the COVID-19 pandemic, tourism destinations need integrative development strategies that consciously and actively address the necessary change processes. In the experience of the interviewed resilience managers, a participatory strategy process on resilience can be launched where, in pioneering resilience networks, the substantial participation of private-sector actors is still in its infancy. The locally and regionally based private sector still lacks a self-understanding as part of a collaborative resilience-building process, and can rarely be persuaded to actively participate. The participation of tourism enterprises in activities on the network level of the destination is known in principle through new forms of collaboration (e.g., productive coalitions, marketing cooperation) (d'Angella and Go, 2009; Hartman et al., 2020). However, the missing or methodologically and procedurally not yet developed direct utilisation context of regional resilience initiatives for tourism firm performance suggests a motivation deficit for actual participation, as do correlations in the present empirical study.

Community resilience toolkits aim at the active participation of the private sector. However, these practice guides often remain unapplied. There is a lack of commissioned leadership and management personnel that could further develop such toolkits for the specific contextual conditions in each case, centrally accelerating resilience-related network building. Raising awareness, and explaining the resilience approach and its perspectives are necessary first steps to increase active participation. Tourism destinations have gained experience in this regard, e.g., in brand development processes (Saraniemi and Komppula, 2019). Future participation processes on the destination level can integrate the experiences of dedicated resilience projects.

National resilience strategies are only just emerging in many countries. Where they have already been formulated for several years, they focus on security policy issues, and on preparing communities in relation to natural and artificial disasters. Resilience as a regional development strategy is hardly present in national regulatory frameworks so far. Tourism development issues on the destination level represent an illustrative application area for national resilience guidelines, as they address diversity and capacity for change in regional economic structures, particularly in rural areas (Watson and Deller, 2021). Furthermore, there is little concrete guidance on issues of resilient spatial planning and development that could provide important orientation for organisational resilience on the regional level (White and O'Hare, 2014). While the tourism industry, severely affected by the COVID-19 pandemic, is facing financial constraints, a restart also offers an opportunity to rethink collaborative tasks of resilient destination development and its financing. In doing so, concepts of existing resilience initiatives can be linked to the specific needs of a rapidly changing tourism industry. One focus can be on alternative and innovative organisational and ownership structures that are centrally coordinated on the destination level (Cave and Dredge, 2020).

Institutionalising Resilience as a Service

Existing resilience initiatives are on the threshold of institutionalisation in their respective urban societies, seeking a sustainable funding mix for their activities. The COVID-19 pandemic is accelerating profound changes in regional governance, providing a window of opportunity for the implementation of resilience management.

This also applies in a special way to regional tourism in which network relationships with businesses within a region can be renegotiated. This requires a conceptual development of the resilience approach as an opportunity approach in resilience practice. Both the literature on regional economic resilience and on post-COVID-19 destination strategies favour the evolutionary resilience approach (Boschma, 2015; Brouder, 2020; Ioannides and Gyimothy, 2020), whereas developments in practical resilience management are still to be proven.

For this to succeed, two perspectives for change in local and regional resilience management need to be more closely considered. First, existing or emerging resilience initiatives are in an organisational and collective learning process that includes the degree of visibility and measurement of their respective contribution to building organisational resilience in the region. Appropriate scales and indicators have not yet been developed, but provide a justification for organisational funding in general, and the direct involvement of the private sector in particular (Lee et al., 2013). Second, companies in a region are also called upon to understand the resilience of their organisation more in the context of regional development. Existing priorities of organisational resilience in the area of capacity building of employees or in relation to resilient supply chains still carry weight. A next big step in building organisational resilience could be immediate local and regional contextualisation, with large and small companies looking at their respective scope of action in changing existing business models.

FUTURE RESEARCH DIRECTIONS

The concretisation of the fuzzy resilience concept (Caldarice et al., 2019) in the framework of destination and regional development issues requires scientific debate. Academically developed criteria and indicators can be helpful in the institutionalisation of resilience initiatives. Appropriate impact research is, therefore, an important perspective for future research in the field of destination resilience. The role of the private tourism sector should be given weight.

Governance research can examine suitable actor constellations on the network level for sustainable resilience building. In general, it is important to analyse emerging and existing resilience management in its development process, and generalise development paths that are favourable for building resilience and shape them into being transferable to other contexts of action. The tourism network should be analysed both in itself, and in relation to the interactions with other sectoral and social systems.

Interdisciplinary research related to organisational resilience needs stronger economic–geographical and regional–political collaboration members, which have so far been under-represented compared to the individual psychological perspective. In addition to research on the regional-network level, projects in the field of resilience research should be undertaken in the organisational context. This should focus on how to increase the motivation to participate in supraorganisational resilience projects, and how to measure and evaluate the organisational contribution to the regional economy in terms of indicators of resilience in a new way.

Research activities in the field of resilience gain importance, and new directions often occur in the wake of major disaster and crisis events – as seen in the wake of the experiences of the September 11 terrorist attacks (Linnenluecke, 2017). In this respect, the COVID-19 pandemic can be a development framework for research with a stronger regional focus, especially for the tourism industry. This goes hand in hand with changing social values in the area of regional sustainability, and regional value creation ultimately becomes more attractive (Stankov et al., 2020). In this respect, research in the field of developing regional business models can enrich firm resilience research. Tourism in its destination system is particularly suitable as a direct object of research due to its local reference, the special importance of local resources, and the heterogeneous organisational structure of small and medium-sized enterprises.

CONCLUSION

The discussion of resilience as an opportunity approach suggests that resilience can be achieved through economic growth. However, major change processes can also lead to a decline in certain sectors, with a large proportion of the individual companies involved facing economically challenging times. In these times, threat situations for companies become concrete, which should increase the motivation to actively build resilience. At the same time, the financial scope for involvement in regional resilience initiatives is particularly narrow.

It can be asked quite fundamentally to what extent the commercialisation of resilience management can be a viable basis for regional resilience. Even though the evolutionary approach to resilience has gained importance, the conservative preservation characteristics of resilience are still relevant. Private-sector business models have rarely been established for areas of disaster management and immediate crisis management on the regional level (Tingsanchali, 2012). Against this background, future resilience management on the regional level should strike a balance between private-sector financing and public services, which should also be reflected in its financing models. In any case, however, a network of entrepreneurs in a region could help to jointly and actively address potential future crisis scenarios. The crystallisation of this cooperative engagement through a resilience initiative could be an organisationally viable solution.

REFERENCES

Aguirre, B. E. (2007). Dialectics of vulnerability and resilience. *Georgetown Journal on Poverty Law & Policy*, *14*(1), 39.

Annarelli, A., & Nonino, F. (2016). Strategic and operational management of organizational resilience: Current state of research and future directions. *Omega*, *62*, 1–18. doi:10.1016/j.omega.2015.08.004

Ates, A., & Bititci, U. (2011). Change process: A key enabler for building resilient SMEs. *International Journal of Production Research*, *49*(18), 5601–5618. doi:10.1080/00207543.2011.563825

Avey, J. B., Wernsing, T. S., & Luthans, F. (2008). Can positive employees help positive organizational change? Impact of psychological capital and emotions on relevant attitudes and behaviors. *The Journal of Applied Behavioral Science*, *44*(1), 48–70. doi:10.1177/0021886307311470

Bardoel, E. A., Pettit, T. M., De Cieri, H., & McMillan, L. (2014). Employee resilience: An emerging challenge for HRM. *Asia Pacific Journal of Human Resources*, *52*(3), 279–297. doi:10.1111/1744-7941.12033

Bentfeld, M., Huppertz, S., Khazai, B., & Girard, T. (2018). Hotel resilient – plan ahead, stay ahead. Promoting the business case for disaster and climate resilience in the tourism sector. In E. Innerhofer, H. Pechlaner, & M. Fontanari, M. (Eds.), Destination Resilience (pp. 153-163). Routledge.

Biggs, D. (2011). Understanding resilience in a vulnerable industry: The case of reef tourism in Australia. *Ecology and Society*, *16*(1), 30. doi:10.5751/ES-03948-160130

Binder, S. B., Baker, C. K., & Barile, J. P. (2015). Rebuild or relocate? Resilience and postdisaster decision-making after Hurricane Sandy. *American Journal of Community Psychology*, *56*(1-2), 180–196. doi:10.100710464-015-9727-x PMID:25903679

Boschma, R. (2015). Towards an evolutionary perspective on regional resilience. *Regional Studies*, *49*(5), 733–751. doi:10.1080/00343404.2014.959481

Brewton, K. E., Danes, S. M., Stafford, K., & Haynes, G. W. (2010). Determinants of rural and urban family firm resilience. *Journal of Family Business Strategy*, *1*(3), 155–166. doi:10.1016/j.jfbs.2010.08.003

Brouder, P. (2020). Reset redux: Possible evolutionary pathways towards the transformation of tourism in a COVID-19 world. *Tourism Geographies*, *22*(3), 484–490. doi:10.1080/14616688.2020.1760928

Brown, C., Seville, E., & Vargo, J. (2017). Measuring the organizational resilience of critical infrastructure providers: A New Zealand case study. *International Journal of Critical Infrastructure Protection*, *18*, 37–49. doi:10.1016/j.ijcip.2017.05.002

Buliga, O., Scheiner, C. W., & Voigt, K. I. (2016). Business model innovation and organizational resilience: Towards an integrated conceptual framework. *Journal of Business Economics*, *86*(6), 647–670. doi:10.100711573-015-0796-y

Buse, K., & Walt, G. (2000). Global public-private partnerships: Part I - a new development in health? *Bulletin of the World Health Organization*, *78*, 549–561. PMID:10885184

Caldarice, O., Brunetta, G., & Tollin, N. (2019). The challenge of urban resilience: Operationalization. In G. Brunetta, O. Caldarice, N. Tollin, M. Rosas-Casals, & J. Morato (Eds.), *Urban Resilience for Risk and Adaptation Governance* (pp. 1–6). Springer.

Cave, J., & Dredge, D. (2020). Regenerative tourism needs diverse economic practices. *Tourism Geographies*, *22*(3), 503–513. doi:10.1080/14616688.2020.1768434

Chang, C. L., McAleer, M., & Ramos, V. (2020). A charter for sustainable tourism after COVID-19. *Sustainability*, *12*(9), 3671. doi:10.3390u12093671

Cheer, J. M., & Lew, A. A. (Eds.). (2017). *Tourism, resilience and sustainability: Adapting to social, political and economic change*. Routledge. doi:10.4324/9781315464053

Cheer, J. M., & Lew, A. A. (2018). Understanding tourism resilience: Adapting to social, political, and economic change. In J. M. Cheer & A. A. Lew (Eds.), *Tourism, resilience, and sustainability: Adapting to social, political and economic change* (pp. 3–17). Routledge.

Cheer, J. M., Milano, C., & Novelli, M. (2019). Tourism and community resilience in the Anthropocene: Accentuating temporal overtourism. *Journal of Sustainable Tourism*, *27*(4), 554–572. doi:10.1080/09669582.2019.1578363

Conz, E., & Magnani, G. (2020). A dynamic perspective on the resilience of firms: A systematic literature review and a framework for future research. *European Management Journal*, *38*(3), 400–412. doi:10.1016/j.emj.2019.12.004

Cutter, S. L. (2016). Resilience to what? Resilience for whom? *The Geographical Journal*, *182*(2), 110–113. doi:10.1111/geoj.12174 d'Angella, F., & Go, F. M. (2009). Tale of two cities' collaborative tourism marketing: Towards a theory of destination stakeholder assessment. *Tourism Management, 30*(3), 429–440. doi:10.1016/j.tourman.2008.07.012

Dalgaard-Nielsen, A. (2017). Organizational resilience in national security bureaucracies: Realistic and practicable? *Journal of Contingencies and Crisis Management, 25*(4), 341–349. doi:10.1111/1468-5973.12164

Davidson, J. L., Jacobson, C., Lyth, A., Dedekorkut-Howes, A., Baldwin, C. L., Ellison, J. C., Holbrook, N. J., Howes, M. J., Serrao-Neumann, S., Singh-Peterson, L., & Smith, T. F. (2016). Interrogating resilience: Toward a typology to improve its operationalization. *Ecology and Society, 21*(2), 27. doi:10.5751/ES-08450-210227

Dogru, T., Marchio, E. A., Bulut, U., & Suess, C. (2019). Climate change: Vulnerability and resilience of tourism and the entire economy. *Tourism Management, 72*, 292–305. doi:10.1016/j.tourman.2018.12.010

Dredge, D., Phi, G. T. L., Mahadevan, R., Meehan, E., & Popescu, E. (2018). *Digitalisation in Tourism: In-depth analysis of challenges and opportunities. Low Value procedure GRO-SME-17-C-091-A for Executive Agency for Small and Medium-sized Enterprises (EASME) Virtual Tourism Observatory*. Aalborg University.

Dunn-Cavelty, M., & Suter, M. (2009). Public–Private Partnerships are no silver bullet: An expanded governance model for Critical Infrastructure Protection. *International Journal of Critical Infrastructure Protection, 2*(4), 179–187. doi:10.1016/j.ijcip.2009.08.006

Engeset, A. B. (2020). "For better or for worse"–the role of family ownership in the resilience of rural hospitality firms. *Scandinavian Journal of Hospitality and Tourism, 20*(1), 68–84. doi:10.1080/15022250.2020.1717600

Evenhuis, E. (2020). New directions in researching regional economic resilience and adaptation. In G. Bristow & A. Healy (Eds.), *Handbook on Regional Economic Resilience*. Edward Elgar Publishing. doi:10.4337/9781785360862.00010

Fjäder, C. (2014). The nation-state, national security and resilience in the age of globalisation. *Resilience, 2*(2), 114–129. doi:10.1080/21693293.2014.914771

Fröhlich, K., & Hassink, R. (2018). Regional resilience: A stretched concept? *European Planning Studies, 26*(9), 1763–1778. doi:10.1080/09654313.2018.1494137

Gössling, S., Scott, D., & Hall, C. M. (2020). Pandemics, tourism and global change: A rapid assessment of COVID-19. *Journal of Sustainable Tourism, 29*(1), 1–20. doi:10.1080/09669582.2020.1758708

Haavik, T. K., Antonsen, S., Rosness, R., & Hale, A. (2019). HRO and RE: A pragmatic perspective. *Safety Science*, *117*, 479–489. doi:10.1016/j.ssci.2016.08.010

Hartman, S., Wielenga, B., & Heslinga, J. H. (2020). The future of tourism destination management: building productive coalitions of actor networks for complex destination development. *Journal of Tourism Futures*.

Hassink, R., & Gong, H. (2020). Regional Resilience. In A. Kobayashi, M. Boyle, N. Castree, F. Collins, J. Crampton, S. de Leeuw, E. Ho, I. Luginah, B. Maharaj, & J. Tyner (Eds.), *International Encyclopedia of Human Geography* (pp. 351–355). Elsevier. doi:10.1016/B978-0-08-102295-5.10126-X

HCFCD. (2020). *Haut Comité Français pour la Résilience Nationale*. https://www.hcfdc.org/fr/guest/nosmembres

Holling, C. S. (1973). Resilience and stability of ecological systems. *Annual Review of Ecology and Systematics*, *4*(1), 1–23. doi:10.1146/annurev.es.04.110173.000245

Hollnagel, E. (2011). RAG – The resilience analysis grid. In E. Hollnagel, J. Pariès, D. Woods, & J. Wreathall (Eds.), *Resilience Engineering in Practice. A Guidebook* (pp. 275–296). Ashgate.

Horne, J. F. (1997). The coming age of organizational resilience. In *Business Forum* (Vol. 22, No. 2/3, pp. 24 - 28). California State University, Los Angeles, School of Business and Economics.

HotelResilient. (2020). *About us. Hotel Resilient*. https://hotelresilient.org/about-us

Ibanescu, B. C., Eva, M., & Gheorghiu, A. (2020). Questioning the role of tourism as an engine for resilience: The role of accessibility and economic performance. *Sustainability*, *12*(14), 5527. doi:10.3390u12145527

Ioannides, D., & Gyimothy, S. (2020). The COVID-19 crisis as an opportunity for escaping the unsustainable global tourism path. *Tourism Geographies*, *22*(3), 624–632. doi:10.1080/14616688.2020.1763445

Ioannou, S., Wójcik, D., & Dymski, G. (2019). Too-Big-To-Fail: Why Megabanks Have Not Become Smaller Since the Global Financial Crisis? *Review of Political Economy*, *31*(3), 356–381. doi:10.1080/09538259.2019.1674001

Jones, L. (2019). Resilience isn't the same for all: Comparing subjective and objective approaches to resilience measurement. *Wiley Interdisciplinary Reviews: Climate Change*, *10*(1), 552. doi:10.1002/wcc.552

Joppe, M. (2012). Migrant workers: Challenges and opportunities in addressing tourism labour shortages. *Tourism Management*, *33*(3), 662–671. doi:10.1016/j.tourman.2011.07.009

Joseph, J. (2013). Resilience as embedded neoliberalism: A governmentality approach. *Resilience*, *1*(1), 38–52. doi:10.1080/21693293.2013.765741

Katina, P. F., Pinto, C. A., Bradley, J. M., & Hester, P. T. (2014). Interdependency-induced risk with applications to healthcare. *International Journal of Critical Infrastructure Protection*, *7*(1), 12–26. doi:10.1016/j.ijcip.2014.01.005

Lee, A. V., Vargo, J., & Seville, E. (2013). Developing a tool to measure and compare organizations' resilience. *Natural Hazards Review*, *14*(1), 29–41. doi:10.1061/(ASCE)NH.1527-6996.0000075

Lengnick-Hall, C. A., Beck, T. E., & Lengnick-Hall, M. L. (2011). Developing a capacity for organizational resilience through strategic human resource management. *Human Resource Management Review*, *21*(3), 243–255. doi:10.1016/j.hrmr.2010.07.001

Lew, A. A. (2014). Scale, change and resilience in community tourism planning. *Tourism Geographies*, *16*(1), 14–22. doi:10.1080/14616688.2013.864325

Linnenluecke, M. K. (2017). Resilience in business and management research: A review of influential publications and a research agenda. *International Journal of Management Reviews*, *19*(1), 4–30. doi:10.1111/ijmr.12076

Linnenluecke, M. K., & Griffiths, A. (2010). Corporate sustainability and organizational culture. *Journal of World Business*, *45*(4), 357–366. doi:10.1016/j.jwb.2009.08.006

Luthe, T., & Wyss, R. (2014). Assessing and planning resilience in tourism. *Tourism Management*, *44*, 161–163. doi:10.1016/j.tourman.2014.03.011

Marana, P., Labaka, L., & Sarriegi, J. M. (2018). A framework for public-private-people partnerships in the city resilience-building process. *Safety Science*, *110*, 39–50. doi:10.1016/j.ssci.2017.12.011

Marcher, A., Kofler, I., Innerhofer, E., & Pechlaner, H. (2020). Perceptions of and interactions between locals, migrants, and tourists in South Tyrol. *Tourism Geographies*, 1–17. Advance online publication. doi:10.1080/14616688.2020.1795709

Martin, R. (2012). Regional economic resilience, hysteresis and recessionary shocks. *Journal of Economic Geography*, *12*(1), 1–32. doi:10.1093/jeg/lbr019

Martin, R., & Sunley, P. (2015). On the notion of regional economic resilience: Conceptualization and explanation. *Journal of Economic Geography*, *15*(1), 1–42. doi:10.1093/jeg/lbu015

Meyer, A. D. (1982). Adapting to Environmental Jolts. *Administrative Science Quarterly*, *27*(4), 515–537. doi:10.2307/2392528 PMID:10257768

Niewiadomski, P. (2020). COVID-19: From temporary de-globalisation to a re-discovery of tourism? *Tourism Geographies*, *22*(3), 651–656. doi:10.1080/14616688.2020.1757749

Orchiston, C. (2013). Tourism business preparedness, resilience and disaster planning in a region of high seismic risk: The case of the Southern Alps, New Zealand. *Current Issues in Tourism*, *16*(5), 477–494. doi:10.1080/13683500.2012.741115

Pechlaner, H., Zacher, D., Eckert, C., & Petersik, L. (2019). Joint responsibility and understanding of resilience from a DMO perspective–an analysis of different situations in Bavarian tourism destinations. *International Journal of Tourism Cities*, *5*(2), 146-168.

Pfefferbaum, B. J., Reissman, D. B., Pfefferbaum, R. L., Klomp, R. W., & Gurwitch, R. H. (2007). Building resilience to mass trauma events. In L. S. Doll, S. E. Bonzo, J. A. Mercy, & D. A. Sleet (Eds.), *Handbook of Injury and Violence Prevention* (pp. 347–358). Springer. doi:10.1007/978-0-387-29457-5_19

Pfefferbaum, R. L., Pfefferbaum, B., Van Horn, R. L., Klomp, R. W., Norris, F. H., & Reissman, D. B. (2013). The communities advancing resilience toolkit (CART): An intervention to build community resilience to disasters. *Journal of Public Health Management and Practice*, *19*(3), 250–258. doi:10.1097/PHH.0b013e318268aed8 PMID:23524306

Pfosser, R., & Seper, L. (2020) *Die österreichische Resilienzstrategie kommt.* http://resilienzstrategie.at/

Pinz, A., Roudyani, N., & Thaler, J. (2018). Public–private partnerships as instruments to achieve sustainability-related objectives: The state of the art and a research agenda. *Public Management Review*, *20*(1), 1–22. doi:10.1080/14719037.2017.1293143

Ponomarov, S. Y., & Holcomb, M. C. (2009). Understanding the concept of supply chain resilience. *International Journal of Logistics Management*, *20*(1), 124–143. doi:10.1108/09574090910954873

Resilient Europe. (2020). *Improving city resilience*. https://urbact.eu/resilient-europe

Rockefeller Foundation. (2015). *Challenge Eligibility and Criteria. Who can submit an application?* http://www.100resilientcities.org/challenge-eligibility-criteria

Saarinen, J., & Gill, A. M. (Eds.). (2018). *Resilient destinations and tourism: Governance strategies in the transition towards sustainability in tourism*. Routledge. doi:10.4324/9781315162157

Sabahi, S., & Parast, M. M. (2020). Firm innovation and supply chain resilience: A dynamic capability perspective. *International Journal of Logistics Research and Applications*, *23*(3), 254–269. doi:10.1080/13675567.2019.1683522

Saraniemi, S., & Komppula, R. (2019). The development of a destination brand identity: A story of stakeholder collaboration. *Current Issues in Tourism*, *22*(9), 1116–1132. doi:10.1080/13683500.2017.1369496

Schmidt, E., Groeneveld, S., & Van de Walle, S. (2017). A change management perspective on public sector cutback management: Towards a framework for analysis. *Public Management Review*, *19*(10), 1538–1555. doi:10.1080/14719037.2017.1296488

Seraphin, H., Ivanov, S., Dosquet, F., & Bourliataux-Lajoinie, S. (2020). Archetypes of locals in destinations victim of overtourism. *Journal of Hospitality and Tourism Management*, *43*, 283–288. doi:10.1016/j.jhtm.2019.12.001

Sheffi, Y., & Rice, J. B. Jr. (2005). A supply chain view of the resilient enterprise. *MIT Sloan Management Review*, *47*(1), 41–48.

Shin, J., Taylor, M. S., & Seo, M. G. (2012). Resources for change: The relationships of organizational inducements and psychological resilience to employees' attitudes and behaviors toward organizational change. *Academy of Management Journal*, *55*(3), 727–748. doi:10.5465/amj.2010.0325

Stainton, A., Chisholm, K., Kaiser, N., Rosen, M., Upthegrove, R., Ruhrmann, S., & Wood, S. J. (2019). Resilience as a multimodal dynamic process. *Early Intervention in Psychiatry*, *13*(4), 725–732. doi:10.1111/eip.12726 PMID:30126047

Stankov, U., Filimonau, V., & Vujičić, M. D. (2020). A mindful shift: An opportunity for mindfulness-driven tourism in a post-pandemic world. *Tourism Geographies*, *22*(3), 703–712. doi:10.1080/14616688.2020.1768432

Staw, B. M., Sandelands, L. E., & Dutton, J. E. (1981). Threat Rigidity Effects in Organizational Behavior: A Multilevel Analysis. *Administrative Science Quarterly*, *26*(4), 501–524. doi:10.2307/2392337

Steiner, A., & Atterton, J. (2015). Exploring the contribution of rural enterprises to local resilience. *Journal of Rural Studies*, *40*, 30–45. doi:10.1016/j.jrurstud.2015.05.004

Sutcliffe, K. M., & Vogus, T. J. (2003). Organizing for resilience. In K. S. Cameron, J. E. Dutton, & R. E. Quinn (Eds.), *Positive organizational scholarship: Foundations of a new discipline* (pp. 94–110). Berret-Koehler.

Taylor, P. J., & Derudder, B. (2016). *World city network: a global urban analysis*. Routledge.

Tingsanchali, T. (2012). Urban flood disaster management. *Procedia Engineering*, *32*, 25–37. doi:10.1016/j.proeng.2012.01.1233

Twigg, J. (2015). *Disaster risk reduction*. Overseas Development Institute, Humanitarian Policy Group.

Vedeld, T., Coly, A., Ndour, N. M., & Hellevik, S. (2016). Climate adaptation at what scale? Multi-level governance, resilience, and coproduction in Saint Louis, Senegal. *Natural Hazards*, *82*(2), 173–199. doi:10.100711069-015-1875-7

von Westphalen, F. G. (2020). Einige Vorüberlegungen zum bevorstehenden Lieferkettengesetz. *ZIP: Zeitschrift für Wirtschaftsrecht*, *41*(49), 2421–2422.

Walker, B., & Salt, D. (2012). *Resilience thinking: sustaining ecosystems and people in a changing world*. Island Press.

Watson, P., & Deller, S. (2021). Tourism and economic resilience. *Tourism Economics*. Advance online publication. doi:10.1177/1354816621990943

Weick, K. E., & Quinn, R. E. (1999). Organizational change and development. *Annual Review of Psychology*, *50*(1), 361–386. doi:10.1146/annurev.psych.50.1.361 PMID:15012461

White, I., & O'Hare, P. (2014). From rhetoric to reality: Which resilience, why resilience, and whose resilience in spatial planning? *Environment and Planning. C, Government & Policy*, *32*(5), 934–950. doi:10.1068/c12117

Williams, N., & Vorley, T. (2014). Economic resilience and entrepreneurship: Lessons from the Sheffield City Region. *Entrepreneurship and Regional Development*, *26*(3-4), 257–281. doi:10.1080/08985626.2014.894129

Zehrer, A. (2019). Structure, significance and challenges of family firms in community-type Alpine tourism destinations. In H. Pechlaner (Ed.), *Destination und Lebensraum* (pp. 25–38). Springer Gabler. doi:10.1007/978-3-658-28110-6_2

Zelger, J., & Oberprantacher, A. (2002, May). Processing of verbal data and knowledge representation by GABEK®-WinRelan®. *Forum Qualitative Social Research*, *3*(2).

Chapter 5
Family Firm Management in Turbulent Times:
Opportunities for Responsible Tourism

Sarah Eichelberger
University of Innsbruck, Austria

Mike Peters
University of Innsbruck, Austria

ABSTRACT

Family firms in tourism and hospitality (FFTH) are often characterized by their long-term orientation and strong "familiness" and embeddedness in tourism destinations. Families however are often confronted with and have to adapt to global change and can sometimes struggle with drastic changes in the tourism industry. And while the current COVID-19 pandemic will mean even more challenges for FFTH, it may also lead to opportunities for responsible tourism development. This chapter attempts to shed light on FFTH's ability to manage change, discussing their specific characteristics through the lens of sustainability and responsibility. The following chapters will analyze FFTH characteristics to develop recommendations for FFTH in supporting responsible tourism, and present ideas for a tourism policy that promotes FFTH.

DOI: 10.4018/978-1-7998-7352-5.ch005

INTRODUCTION

Family firms continually face increasing global competition, handle technological changes, and adapt their strategies to maintain their competitive advantage (De Massis et al., 2018). Most notably in European countries (Astrachan and Shanker, 2003), they are strong contributors to economic prosperity due to their often long-established presence and importance in their respective regions. Family businesses are essential not only because they create jobs in remote and rural areas (Peters and Kallmuenzer, 2015), but also because they provide services and products, strengthen local communities, contribute to social change and economic stability, and generate opportunities for innovation (Della Piana et al., 2019). For instance, in Austria, nearly half a million companies are family-run, with over 90% of them being SMEs that are often long-established key players (Austrian Federal Economic Chamber, 2019). This holds true for the tourism and hospitality industry, which is dominated by family firms. Along the tourism value chain, family businesses are especially prominent in the accommodation and restaurant industries, as well as in the provision of additional services such as ski rental or local transport. In European community-oriented tourism destinations, these firms tend to be small, with the majority operating with less than ten employees (Peters and Buhalis, 2013).

The COVID-19 crisis hit the tourism and hospitality industries particularly hard. Travel bans and long-term lockdowns have seen many tourism firms suffer from a dramatic drop in demand. They have responded in various ways, especially during the first wave of the crisis during the spring of 2020. While some firms placed their entire effort in doing what they could to take care of their employees, others started to renovate, and still others began innovating. The longer the crisis lasted, the lower the motivation of firms to plan for the future became. The following attempts to elaborate on whether family firms in tourism and hospitality (FFTH) display characteristics which promote or hinder effective and efficient crisis management. After discussing both the nature of crises and the characteristics of family firms, we discuss the option of further developing paths of responsible tourism management. The paper concludes by highlighting both FFTH actions to support responsible tourism, along with implications for destination and tourism policy to develop the ideal "responsible tourism" framework.

CRISES IN TOURISM

All tourism systems face numerous vulnerabilities (Lew, 2014). The internationalization of tourism has made this sector more susceptible to global risks (Jessop, 1999). Tourism is argued as being impacted by the global change which forms contemporary destinations and influences them (Cheer and Lew, 2018). In other words: global change determines the current state of tourism and leads to policy, governance, and management challenges (Dogru et al., 2019; Luthe and Wyss, 2014). In this vein, tourism is extremely dependent on external factors (Ritchie, 2004). A globalized, accelerated world with its respective change processes requires tourism to respond and adapt to changing environmental, social, and economic conditions (Luthe and Wyss, 2014). Moreover, the increasing number of crises and disasters affecting the tourism industry makes it necessary to react to these changes as effectively as possible (Prayag, 2020).

The current COVID-19 crisis is however regarded as the "deadliest epidemy" (Fotiadis et al., 2021) compared to events such as the SARS, Ebola, and H1N1 epidemics, or the 2008 financial crisis. Although the tourism industry has proven itself resilient, the COVID-19 crisis with its global, multidimensional, and interconnected impacts is intensively challenging and questioning current tourism systems and values (Sigala, 2020). Moreover, the prevailing COVID-19 pandemic is viewed as a larger-scale crisis compared to previous ones, particularly due to its global scope and widespread shutdown of tourism, business activities, and general activity (Higgins-Desbiolles, 2020). One year after the COVID-19 outbreak, uncertainty remains about when normal tourism activity can resume. In this regard, the pandemic is still impacting socio-cultural and political systems as well as the global economy (Sigala, 2020). But despite the absence of tourists, there is a need to develop recovery strategies that will set the pathway for future tourism (Assaf and Scuderi, 2020). As the tourism and hospitality industries prepare for recovery (Fotiadis et al., 2021), recommendations for their upcoming strategies are needed.

Tourism research understands crisis management as occurring before, during, and after a crisis to respond to, recover, and restart within different phases of crises (Sigala, 2020). Blake and Sinclair (2003) refer to the issues in crisis management in accordance with the severity of the decrease in tourism activity and the duration of a crisis, which steers policymakers into adaptation to the situation or minimization of costs along a path towards a new equilibrium, assessing the possible responses from stakeholders to implement proper tourism policies. For Ritchie (2004), proactive strategic crisis management approaches in tourism are required, i.e. proactive planning and as well as an implementation of strategies when crises occur, along with an evaluation of these strategies' effectiveness. With the COVID-19 crisis' increasing complexity, operations that can decrease its impact are crucial (Beirman, 2003; Wang, 2009); these should be comprised of quick, proactive approaches and effective decisions (Okumus et al., 2005; Okumus and Karamustafa, 2005; Ritchie, 2004).

According to Faghfouri et al. (2015), crisis preparedness also plays a crucial role in family firms' long-term survival. It is therefore argued in favor of formalized crisis procedures which decrease organizational decline probability (Faghfouri et al., 2015).

However, as family firms are often labelled as less professional in comparison with non-family firms (Chua et al., 2009), Faghfouri et al. (2015) propose that family firms implement less formalized and standardized crisis management.

Moreover, the present COVID-19 crisis will result in exceptional challenges for family businesses due to its economic and social reverberations (De Massis and Rondi, 2020). Kraus et al. (2020) demonstrated in a recent study on the COVID-19 pandemic and family businesses that family firms apply divergent crisis approaches that specifically relate to family firm characteristics.

CHARACTERISTICS OF FAMILY FIRMS IN TOURISM AND HOSPITALITY (FFTH)

Family firms are defined as firms where 50% of the ownership lies in the hands of one or more families. These families are furthermore active in the company's management with at least two family members (Chua et al., 1999). The tourism and hospitality industries have a large number of family firms. By nature, these are often small- and medium-sized enterprises employing about 80% of the workforce in this industry. For instance, in the province of Tyrol (Austria), about 17,500 family firms in tourism and hospitality employ more than 150,000 people (Haushofer, 2013).

Although many FFHT characteristics can be explained by their size, their peculiarities tend to be family firm-driven:

1. **FFHTs are long-term oriented:** Family management and leadership are long-term oriented. Several studies have identified the fact that family firms usually foster continuity and family traditions (e.g. Peters and Kallmuenzer, 2018; Veider and Kallmuenzer, 2016). Founder-centered family firms are much more interested in keeping the "old," while successor-centered family firms are aware of the fact that managing the change resulting from shifting environmental conditions is crucial for their survival (Veider and Kallmuenzer, 2016). Furthermore, succession is a driver of long-term prosperity. Recent research shows that the quality of this period has a strong effect on the future development and success of FFTH. So the multidimensional succession process represents a very critical issue family firms are faced with, as succession constitutes the central issue for survival (Ibrahim et al., 2001), the subsequent performance of the firm, the ultimate viability of the business, and the satisfaction of FFTHs' stakeholders (Cabrera-Suárez et al., 2001; Sharma et al., 2001).
2. **FFTH are regionally embedded:** Earlier studies underline that FFTH are very much embedded in the areas they operate within. This holds especially

true for rural family firms in hospitality and tourism, with solid community networks developed over many generations (Kallmuenzer and Peters, 2017).

3. **FFTH are main stakeholders in community-oriented tourism destinations:** Nurtured by the fact that FFTH are mainly SMEs and that tourism in many European countries is dominated by SMEs, we find a large number of FFTH SMEs at European tourism destinations (Pikkemaat et al., 2018). Families run the businesses at many of these destinations, creating so-called "dynasties" that have accumulated various SMEs in tourism and hospitality operations. Over time, complex business and family stakeholder networks have emerged which are characterized by multiple roles, duties, and reciprocal relationships (Beritelli et al., 2007).
4. **Coopetition instead of competition for FFTH:** Analyzed in the context of tourism and hospitality, it has become obvious that entrepreneurial orientation (EO) needs to be redefined in the context of family firms. The dimension of "competitive aggressiveness" in particular needs to be reconsidered for FFTH. Here, Kallmuenzer and Peters (2017) have introduced the term "competitive agreeableness" to signalize firm coopetition among community-oriented destinations.
5. **Ecology and social acceptance are of major importance for FFTH:** Previous studies analyzing FFTH actions and attitudes underline the fact that ecological and social attributes can be of higher priority than economic factors such as financial outcome. This fact often helps justify and explain family firms' long-term orientation (Gómez-Mejía et al., 2007; Huang et al., 2015; Kallmuenzer et al., 2018).
6. **Decision making in FFTH is characterized by flexibility:** FFTH mostly employ family firm members in their management. Strong motivation to overcome crises can therefore be expected. Furthermore, family firms usually have a non-hierarchical structure, allowing rapid information exchange and decision making. On a strategic level, family firms' strategic response to crises is shaped by the family's commitment to the company (Zahra et al., 2008). However, due to the smallness of FFTH, we usually find a strong amalgamation of operative and strategic decision making (Peters and Buhalis, 2004). De Massis et al. (2013) point out that family owners often influence and control core decision making processes, allowing them to remain flexible during a crisis.
7. **Familiness differentiates FFTH from other non-family firms:** Several studies underline the role and importance of familiness (Martín-Santana et al., 2020; Presas et al., 2011). This social component plays an important role for a family running a business, and it persists over time, giving the family business a unique character and culture (Chrisman et al., 2005; Peters et al., 2012).

Family members feel strongly and emotionally attached to their firm, meaning that familiness can increase competitive advantages for FFTH (Habbershon and Williams, 1999).

These characteristics have the potential to support an effective restart of tourism following the COVID-19 pandemic. But responsible tourism also demands a strong reorientation of tourism destinations, from mass consumption to resource protection policies, all the way to mass market promotion and target segmentation that are based on the re-evaluation of a destination's socio-cultural, ecological, and economic resources.

TOWARDS RESPONSIBLE TOURISM – THE NEED FOR CHANGE IN TIMES OF CRISIS

The COVID-19 crisis according to Higgins-Desbiolles (2020) has "the potential to transform the tourism industry as well as the context in which it operates" (p. 610). Governmental intervention will help strengthen tourism's resilience in the post- pandemic world. Furthermore, the COVID-19 pandemic offers opportunities to consolidate and design transition (Ioannides and Gyimóthy, 2020). Put differently, tourism will soon experience the opportunity to reset and rethink itself by exposing the possibilities found within this transformative, historic moment (Higgins-Desbiolles, 2020). The COVID-19 crisis can thus be seen as a chance to react to the "overtourism" debate that prevailed prior to the crisis (Eichelberger and Peters, 2020). COVID-19 therefore might be the perfect chance to embark on a more sustainable path and new direction (Ioannides and Gyimóthy, 2020). This is why Higgins-Desbiolles (2021) argue for a more responsible, ethical, as well as sustainable tourism industry.

Sustainable tourism as a popular field of research (Liu, 2003) demonstrates the relevance of sustainability in the tourism industry. This is also apparent in practice, as sustainable tourism is used in most strategic tourism planning around the globe (Hardy et al., 2002). Although there is a wide agreement in research about the relevance (Fang, 2020), importance, and subsequent implementation of sustainability, in tourism it still varies based on the respective arguments being used (Hardy et al., 2002). Prior research on sustainable tourism focused heavily on tourism development models pursuing more environmentally friendly, sustainable, and social tourism (Ritchie and Crouch, 2011). With this in mind, Liu (2003) criticized the sustainable tourism debate as flawed, and making the wrong arguments and assumptions. Sustainable tourism has in fact often been condemned for its inadequacy and difficulty in application (Dávid, 2011; Higgins-Desbiolles, 2010).

This is why research on sustainable tourism has been accompanied by the concept of responsible tourism (Mihalic, 2016).

Responsible tourism was introduced by Krippendorf (1982) in the 1980s in an attempt to demonstrate the importance of sociocultural and environmental factors in the industry. In this vein, responsible tourism suggests taking responsibility for action, with suppliers, tourists, service providers, and governments also getting involved (Mathew and Sreejesh, 2017). Fang (2020) recently referred to responsible tourism as "one way of maximizing the possibilities of positive impacts and minimizing the negative impacts of tourism" (p. 132). Responsible tourism thus pursues the goals of providing economic and social benefits to the local community while taking into account its culture, traditions, and religions; protecting the natural environment and resources; and reducing pollution (Fang, 2020). Responsible tourism consists of the dimensions of economic, social, cultural, and environmental responsibility (Fang, 2020; Frey and George, 2010; Gong et al., 2019; Mathew and Sreejesh, 2017). Economic responsibility refers to the support of local products, services, and facilities. Culture, heritage, and traditions, as well as safety, human rights, and community engagement relate to social and cultural responsibility. Environmental responsibility concentrates on environmental conservation, as well as pollution and waste management (Gong et al., 2019; Mathew and Sreejesh, 2017). Responsible tourism as a whole consists of the interests of the people involved in it while protecting the natural and cultural environments of the hosts (Smith, 1992).

For Gössling et al. (2020), tourism faces the challenge to "collectively learn from this global tragedy to accelerate the transformation of sustainable tourism" (p. 15), in order to move towards sustainable post-COVID-19 pandemic tourism (Gössling et al., 2020). Tourism here is regarded as something to be reshaped into a more inclusive and sustainable model, which furthermore cares about the many stakeholders who rely on it (Cheer, 2020). Sustainable tourism as a model (Cheer, 2020) building on proper sustainable strategies and policies (Mathew and Sreejesh, 2017) needs to be designed to ensure responsible tourism (Blackstock et al., 2008). Subsequently, responsible tourism signifies the major facet of sustainable tourism, representing the right interaction between individuals and organizations (Blackstock et al., 2008). Responsible tourism is thus defined as engaging processes to accomplish the action in line with the pillars and requirements of sustainability: stakeholders' participation, cooperation, leadership, consensus, action, and visitor satisfaction (Mihalic, 2016).

On the business level, the COVID-19 pandemic offers the opportunity to develop corporate social responsibility (CSR) and address environmental as well as social challenges (He and Harris, 2020). Responsible tourism entrepreneurship can contribute to a more balanced, social, and environmentally friendly tourism by taking responsibility for its economic, social, cultural, ecological, and governmental dimensions (Eichelberger and Peters, 2020). According to Dodds and Joppe (2009), sustainable tourism signifies the multi-stakeholder perspective of sustainability, while CSR indicates a corporation's role in pursuing sustainable development. CSR refers to the responsibility of a business or organization to contribute to the society within which it operates as a legal entity (Tran et al., 2018). Although early CSR focused on maximum financial return for shareholders, a consideration of national and international laws and interests developed the concept much more broadly (Carroll, 1991). CSR these days means more than obeying the law and fulfilling firm interests; it now involves the generation of social good as well (McWilliams and Siegel, 2001).

RESPONSIBLE FAMILY FIRM MANAGEMENT IN TOURISM AND HOSPITALITY

According to Le Breton-Miller and Miller (2016), family firms' sustainability practices are "the tendency to behave in economically, socially, and environmentally responsible ways in a manner that benefits all stakeholders and the community at large" (p. 26). Here, a focus on sustainability is described as critical for family firms, as sustainability practices guide the relationships with both stakeholders such as the local community, as well as sustainability related to business continuity (Memili et al., 2018). For Kallmuenzer et al. (2018), family firms are likely to pursue environmental and social sustainability while being influenced by their specific family firm's characteristics.

Zehrer and Raich (2020) argue in this regard that the specific characteristics of family firms serve as triggers for CSR. On the one hand, the regional embeddedness of family firms contributes to CSR by enabling cooperation with regional stakeholders and resources, as well as through a focus on environmentally friendly activities. Kallmuenzer et al. (2018) demonstrated the importance of regional embeddedness in the sustainable decision making of family firms, with local networks and interchange being emphasized. This becomes evident in family firms' tendencies towards coopetition (Gast et al., 2019) which according to Manzhynski and Figge (2020) has the potential to contribute to sustainable development when organizations attempt to positively contribute on the societal level. On the other hand, familiness contributes to an increased socially responsible perception and supports the corresponding decisions of family firms. Also, the identification of family members with the business positively contributes to CSR. In this regard, Memili et al. (2018) describe family involvement in business as being able to facilitate various strategic behavior and practices such as sustainability. Moreover, the long-term orientation of family firms in tourism fosters attachment to the local community and motivates family businesses to protect local residents and the environment. Entrepreneurial flexibility of family businesses can lead to an increased awareness of the need for social and environmental trade through a quick and straightforward response from family businesses (Zehrer and Raich, 2020). Additionally, the importance of ecology and social acceptance for FFTH contributes to the higher priority given to CSR in family businesses (Gómez-Mejía et al., 2007; Huang et al., 2015; Kallmuenzer et al., 2018).

These FFTH characteristics correspond to responsible tourism elements (Mihalic, 2016). In this regard, the specific characteristics of family firms guide strategic actions and planning in family businesses by making them receptive to social concerns, motivating them to embrace responsibility (Zehrer and Raich, 2020). Figure 1 illustrates the contribution of FFTH to responsible tourism development. Here, the family firm follows sustainability attempts to further improve CSR action and develop local social and environmental resources supporting the long-term development of the firm in the respective region (Kallmuenzer et al., 2018), contributing to responsible tourism as a result (Mihalic, 2016).

Figure 1. Responsible family firm management in tourism and hospitality
Source: own illustration

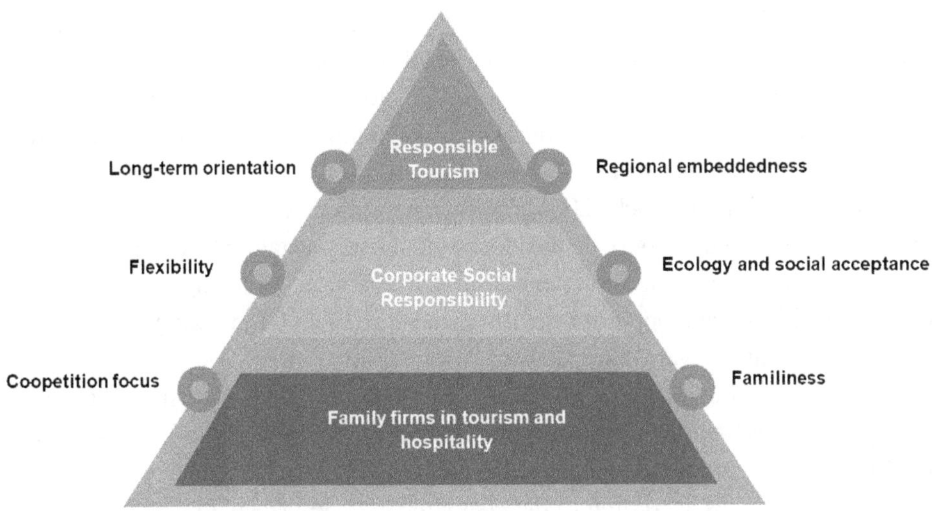

This pyramid illustrates family firms' attempts to responsibly act economically, socially, and environmentally (Le Breton-Miller and Miller, 2016). Because responsible tourism is furthermore focused on resource exchange with e.g. knowledge (Manente et al., 2014) or ecological resources, it tends towards incorporating the local community and natural environment (Fang, 2020). The pyramid of responsible

FFTH management is as a result even more pronounced in community-focused destinations consisting of many diverse players in the tourism value chain.

FFTH IN COMMUNITY-ORIENTED DESTINATIONS TRIGGERING RESPONSIBLE TOURISM

According to Higgins-Desbiolles (2020) responsible tourism alone will not be able to provide sufficient capacity for tourism reset following the COVID-19 pandemic. Indeed, community-centered tourism frameworks are necessary that define and orient tourism based on the local communities' interests (Higgins-Desbiolles, 2020).

In this regard, community-oriented destinations, particularly those focusing on the relationships between the destination management organization (DMO) and its stakeholders (e.g. providers of tourism services, leader networks, as well as cooperation between stakeholders) are emphasized (Beritelli et al., 2007; Komppula, 2016; Zach, 2012). According to Beritelli et al. (2007), community-oriented destinations are characterized by personal and transactional relationships within various networks, as compared to corporate-oriented destinations where hierarchical relationships are highlighted. Moreover, the interdependence in community-oriented destinations achieves a diversity based on the market portfolio and its services; trust and control are also described here as the basis of solid decision making. Community-oriented destinations furthermore have a number of recognizable personal and informal connections through networks (Beritelli et al., 2007).

According to Blackstock (2005), community-oriented destinations contribute to a more sustainable tourism development. Here, sustainable tourism largely focuses on community development approaches (Joppe, 1996) by engaging with the responsibilities and opportunities of citizenship (Li, 2004), making community-oriented destinations valuable arenas of sustainability (Li, 2004).

According to Pikkemaat et al. (2018), the tourism sector consists of complex industries and stakeholder networks dominated by SMEs and family businesses. In this context, community-oriented destinations can also consider the specifics of family businesses according to the characteristics described above, particularly when focusing on the regional embeddedness of family firms. However, FFTH often rely on destination management organizations or other institutions to develop a communication platform for forming common strategic community goals. And FFTH tend to have an inward orientation; due to their predominantly family-based goals, they hardly ever monitor the competitive environment surrounding them (Cater and Justis, 2010; Gudmundson et al., 1999). As opposed to corporate-oriented destinations, community-oriented destinations need a strong DMO that supports product- and service development processes and options to gather and share knowledge and know-how (Bichler and Lösch, 2019).

IMPLICATIONS AND ACTIONS FOR FFTH: TOWARDS RESPONSIBLE TOURISM

Tourism is continuously changing due to global developments and the external factors that impact it (Cheer and Lew, 2018; Ritchie, 2004). Although tourism has survived several crises, COVID-19 is unique because of its global and multidimensional impact (Fotiadis et al., 2021; Higgins-Desbiolles, 2020; Sigala, 2020). The far-reaching consequences of the crisis could have never been mapped or predicted by existing crisis management systems. This is also reflected in family businesses in tourism, where crisis arrangements contribute to long-term survival (Faghfouri et al., 2015). Although the current economic and social impacts of the COVID-19 pandemic have resulted in exceptional challenges (De Massis and Rondi, 2020), COVID-19's impact will without question be an important element of "overtourism" debates (Eichelberger and Peters, 2020). It's an opportunity to rethink tourism developments in particular (Ioannides and Gyimóthy, 2020) with research arguing that this crisis presents the best opportunity to make tourism sustainable in the future (Higgins-Desbiolles, 2021).

Family businesses are characterized by peculiarities containing the potential to influence their crisis management (Kraus et al., 2020) as well as their sustainability (Kallmuenzer et al., 2018; Zehrer and Raich, 2020). Their regional embeddedness, familiness, long-term orientation, and entrepreneurial flexibility are described as contributors to CSR (Zehrer and Raich, 2020). The coopetition focus (Manzhynski and Figge, 2020) and emphasis on ecology and social acceptance contribute to sustainability attempts by FFTH (Gómez-Mejía et al., 2007; Huang et al., 2015; Kallmuenzer et al., 2018). CSR efforts then promote responsible tourism by taking responsibility for economic, environmental, and social impacts (Le Breton-Miller and Miller, 2016; Mihalic, 2016).

FFTH will have to work to apply their specific family business characteristics to facilitate CSR and contribute to responsible tourism. We propose the following action:

- **Regional embeddedness can be promoted through increased collaboration with regional stakeholders**

FFTH can leverage their ties to their region to collaborate with various regional stakeholders (Zehrer and Raich, 2020). This could strengthen the other stakeholders' attachment to the region via e.g. local networks (Kallmuenzer et al., 2018).

Subsequently, regional resources and environmental-friendly activities should be emphasized to contribute to responsible tourism (Mihalic, 2016; Zehrer and Raich, 2020), which furthermore corresponds to the possibility of family firms' sustainability through coopetition (Manzhynski and Figge, 2020). Regional embeddedness and long-term collaboration can here help to overcome crisis situations. For instance, firms in a rural destination can support each other in managing seasonal employees (providing various career options or additional positions throughout the year) or by creating long-term regional supplier networks supporting local producers (e.g. farm and restaurant cooperatives).

- **Focus on family values to increase the overall sense of responsibility**

As with family orientation, familiness can be applied in FFTH to increase the perception of social and community responsibility in family businesses (Zehrer and Raich, 2020). These family values furthermore have the potential to make strategic behavior and actions more responsibly- and sustainably focused (Memili et al., 2018). As FFTH contribute to authentic products and services, they also need to take advantage of family firm branding strategies. Family firm values should therefore play a main role in both internal and external marketing strategies to strengthen the competitive advantage of FFTH and make them less vulnerable to crises.

- **Long-term responsibility orientation towards environmental and social tourism impacts**

Continuity and tradition are key to the long-term orientation of FFTH (Peters and Kallmuenzer, 2018; Veider and Kallmuenzer, 2016). This is of course the case for responsible tourism when considering the local community and protecting the environment (Zehrer and Raich, 2020). Tourism research has made clear that family firm owners plan long-term and want to hand over their business to the next generation. But although succession planning is crucial for the survival of FFTH, they often start planning it far too late, causing succession processes to fail. Long-term responsibility in this case means the current generation needs to involve the next one as early on as possible in the business to jointly shape the sustainable future of their FFTH (Heimerl and Peters, 2019).

In addition, destinations and tourism policies can support FFTH to collectively contribute to responsible tourism.

Community-oriented destinations in particular emphasize the relationships between DMOs and all other stakeholders (Beritelli et al., 2007; Komppula, 2016; Zach, 2012). These personal relationships can be supported by DMOs and tourism policy by focusing on and promoting the formation of networks (Beritelli et al., 2007). These networks then contribute to increased informal and personal relationships that can promote the regional embeddedness of FFTH (Kallmuenzer et al., 2018; Zehrer and Raich, 2020), as well as FFTH coopetition (Manzhynski and Figge, 2020). In this context, DMOs are furthermore able to take on the role as initiators of knowledge transfer. Accordingly, community-oriented destinations as well as their regional embeddedness and coopetition can contribute to sustainable and responsible tourism development.

Moreover, as succession represents a driver of long-term success in FFTH (Ibrahim et al., 2001), policymakers should support succession processes. This critical issue in many family businesses impacts their future development, while the succession process determines FFTH future performance and viability (Cabrera-Suárez et al., 2001; Sharma et al., 2001).

Crisis management systems are needed for FFTH that emphasize continuous crisis reporting. Since crisis preparations also play a role in family businesses, especially with regard to long-term survivability (Faghfouri et al., 2015), more advanced crisis management approaches are needed to handle the specifics of the COVID-19 crisis (De Massis and Rondi, 2020).

Because family businesses play a major role in the tourism and hospitality industry (Haushofer, 2013), FFTH need a voice in destination governance. Since the particularities of family businesses need to be considered (among other things) in terms of responsible future development (Kallmuenzer et al., 2018; Memili et al., 2018), the inclusion of family businesses in destination governance is recommended. However, due to the fact that FFTH are often SMEs (making their tourism tax contributions quite low as a result) their influence on destination governance might suffer from a low representation in destination management organizational (supervisory) boards. And as outlined in Heimerl and Peters (2019), the succeeding generation is often not able to take part in shaping the future of their tourism destination. If the young generation cannot realistically work to create the ideal destination, they may very well refrain from taking over the family firm if they see no perspectives for future entrepreneurship.

Crises such as COVID-19 reveal how vulnerable SMEs are (Basnyat and Sharma, 2021). Although this holds true for FFTH, family firms might have certain advantages helping them to survive crisis situations. On the one hand, FFTH have an advantage during crisis phases because they "benefit from their controlling families' financial support to secure investments and employment" (Kraus et al., 2020, p. 1083). On the other hand, family values serving as the basis of the family culture provide clarity regarding how the business should be managed in turbulent times. Furthermore, it can be expected that family bonds influence the way FFTH manage the COVID-19 crisis: strong connections among family members as well as to their loyal customer base (Breier et al., 2021), suppliers, and local stakeholders can positively influence FFTH management practices during any crisis.

REFERENCES

Assaf, A., & Scuderi, R. (2020). COVID-19 and the recovery of the tourism industry. *Tourism Economics*, *26*(5), 731–733. doi:10.1177/1354816620933712

Astrachan, J. H., & Shanker, M. C. (2003). Family Businesses' Contribution to the U.S. Economy: A Closer Look. *Family Business Review*, *16*(3), 211–219. doi:10.1177/08944865030160030601

Basnyat, S., & Sharma, S. (2021). Effects of COVID-19 crisis on small and medium-sized hotel operators: Insights from Nepal. *Anatolia*, *4*(1), 1–15. doi:10.1080/13032917.2021.1879184

Beirman, D. (2003). *Restoring tourism destinations in crisis: A strategic marketing approach*. CABI Publishing.

Beritelli, P., Bieger, T., & Laesser, C. (2007). Destination Governance: Using Corporate Governance Theories as a Foundation for Effective Destination Management. *Journal of Travel Research*, *46*(1), 96–107. doi:10.1177/0047287507302385

Bichler, B. F., & Lösch, M. (2019). Collaborative Governance in Tourism: Empirical Insights into a Community-Oriented Destination. *Sustainability*, *11*(23), 6673. doi:10.3390u11236673

Blackstock, K. (2005). A critical look at community based tourism. *Community Development Journal: An International Forum*, *40*(1), 39–49. doi:10.1093/cdj/bsi005

Blackstock, K. L., White, V., McCrum, G., Scott, A., & Hunter, C. (2008). Measuring Responsibility: An Appraisal of a Scottish National Park's Sustainable Tourism Indicators. *Journal of Sustainable Tourism*, *16*(3), 276–297. doi:10.1080/09669580802154090

Blake, A., & Sinclair, M. T. (2003). Tourism crisis management. *Annals of Tourism Research*, *30*(4), 813–832. doi:10.1016/S0160-7383(03)00056-2

Breier, M., Kallmuenzer, A., Clauss, T., Gast, J., Kraus, S., & Tiberius, V. (2021). The role of business model innovation in the hospitality industry during the COVID-19 crisis. *International Journal of Hospitality Management*, *92*, 102723. doi:10.1016/j.ijhm.2020.102723

Cabrera-Suárez, K., de Saá-Pérez, P., & García-Almeida, D. (2001). The Succession Process from a Resource- and Knowledge-Based View of the Family Firm. *Family Business Review*, *14*(1), 37–46. doi:10.1111/j.1741-6248.2001.00037.x

Carroll, A. B. (1991). The pyramid of corporate social responsibility: Toward the moral management of organizational stakeholders. *Business Horizons*, *34*(4), 39–48. doi:10.1016/0007-6813(91)90005-G

Cater, J. J. III, & Justis, R. T. (2010). The development and implementation of shared leadership in multi-generational family firms. *Management Research Review*, *33*(6), 563–585. doi:10.1108/01409171011050190

Cheer, J. M. (2020). Human flourishing, tourism transformation and COVID-19: A conceptual touchstone. *Tourism Geographies*, *22*(3), 514–524. doi:10.1080/14616688.2020.1765016

Cheer, J. M., & Lew, A. A. (2018). Understanding tourism resilience: Adapting to social, political, and economic change. In J. M. Cheer & A. A. Lew (Eds.), *Routledge advances in tourism. Tourism, resilience and sustainability: Adapting to social, political and economic change* (pp. 3–17). Routledge.

Chrisman, J. J., Chua, J. H., & Sharma, P. (2005). Trends and Directions in the Development of a Strategic Management Theory of the Family Firm. *Entrepreneurship Theory and Practice*, *29*(5), 555–575. doi:10.1111/j.1540-6520.2005.00098.x

Chua, J. H., Chrisman, J. J., & Bergiel, E. B. (2009). An Agency Theoretic Analysis of the Professionalized Family Firm. *Entrepreneurship Theory and Practice*, *33*(2), 355–372. doi:10.1111/j.1540-6520.2009.00294.x

Chua, J. H., Chrisman, J. J., & Sharma, P. (1999). Defining the Family Business by Behavior. *Entrepreneurship Theory and Practice*, *23*(4), 19–39. doi:10.1177/104225879902300402

Dávid, L. (2011). Tourism ecology: Towards the responsible, sustainable tourism future. *Worldwide Hospitality and Tourism Themes*, *3*(3), 210–216. doi:10.1108/17554211111142176

De Massis, A., Frattini, F., Majocchi, A., & Piscitello, L. (2018). Family firms in the global economy: Toward a deeper understanding of internationalization determinants, processes, and outcomes. *Global Strategy Journal*, *8*(1), 3–21. doi:10.1002/gsj.1199

De Massis, A., Kotlar, J., Campopiano, G., & Cassia, L. (2013). Dispersion of family ownership and the performance of small-to-medium size private family firms. *Journal of Family Business Strategy*, *4*(3), 166–175. doi:10.1016/j.jfbs.2013.05.001

De Massis, A., & Rondi, E. (2020). Covid-19 and the Future of Family Business Research. *Journal of Management Studies, 57*(8), 1727–1731. doi:10.1111/joms.12632

Della Piana, B., Vecchi, A., Marino, V., & Arco, D. (2019). How do family firms grow? The strategic goals of innovation and internationalization. *International Journal of Business and Management, 14*(2), 1. doi:10.5539/ijbm.v14n2p1

Dodds, R., & Joppe, M. (2009). The demand for, and participation in corporate social responsibility and sustainable tourism: Implications for the Caribbean. *Ara: Journal of Tourism Research, 2*(1), 1–11.

Dogru, T., Marchio, E. A., Bulut, U., & Suess, C. (2019). Climate change: Vulnerability and resilience of tourism and the entire economy. *Tourism Management, 72*(3), 292–305. doi:10.1016/j.tourman.2018.12.010

Eichelberger, S., & Peters, M. (2020). Responsible Tourism Entrepreneurship nach COVID-19: Neue Wege im Tourismus. In H. Pechlaner & S. Speer (Eds.), *Responsible Entrepreneurship* (pp. 237–260). Springer Fachmedien Wiesbaden., doi:10.1007/978-3-658-31616-7_11

Faghfouri, P., Kraiczy, N. D., Hack, A., & Kellermanns, F. W. (2015). Ready for a crisis? How supervisory boards affect the formalized crisis procedures of small and medium-sized family firms in Germany. *Review of Managerial Science, 9*(2), 317–338. doi:10.100711846-014-0158-0

Fang, W.-T. (2020). Responsible Tourism. In W.-T. Fang (Ed.), *Tourism in Emerging Economies* (Vol. 50, pp. 131–151). Springer Singapore., doi:10.1007/978-981-15-2463-9_6

Fotiadis, A., Polyzos, S., & Huan, T.-C. T. C. (2021). The good, the bad and the ugly on COVID-19 tourism recovery. *Annals of Tourism Research, 87*, 103117. doi:10.1016/j.annals.2020.103117 PMID:33518847

Frey, N., & George, R. (2010). Responsible tourism management: The missing link between business owners' attitudes and behaviour in the Cape Town tourism industry. *Tourism Management, 31*(5), 621–628. doi:10.1016/j.tourman.2009.06.017

Gast, J., Kallmünzer, A., Kraus, S., Gundolf, K., & Arnold, J. (2019). Coopetition of small- and medium-sized family enterprises: Insights from an IT business network. *International Journal of Entrepreneurship and Small Business, 38*(1/2). doi:10.1504/IJESB.2019.102493

Gómez-Mejía, L. R., Haynes, K. T., Núñez-Nickel, M., Jacobson, K. J. L., & Moyano-Fuentes, J. (2007). Socioemotional Wealth and Business Risks in Family-controlled Firms: Evidence from Spanish Olive Oil Mills. *Administrative Science Quarterly, 52*(1), 106–137. doi:10.2189/asqu.52.1.106

Gong, J., Detchkhajornjaroensri, P., & Knight, D. W. (2019). Responsible tourism in Bangkok, Thailand: Resident perceptions of Chinese tourist behaviour. *International Journal of Tourism Research, 21*(2), 221–233. doi:10.1002/jtr.2256

Gössling, S., Scott, D., & Hall, C. M. (2020). Pandemics, tourism and global change: A rapid assessment of COVID-19. *Journal of Sustainable Tourism, 12*(5), 1–20. doi:10.1080/09669582.2020.1758708

Gudmundson, D., Hartman, E. A., & Tower, C. B. (1999). Strategic Orientation: Differences between Family and Nonfamily Firms. *Family Business Review, 12*(1), 27–39. doi:10.1111/j.1741-6248.1999.00027.x

Habbershon, T. G., & Williams, M. L. (1999). A Resource-Based Framework for Assessing the Strategic Advantages of Family Firms. *Family Business Review, 12*(1), 1–25. doi:10.1111/j.1741-6248.1999.00001.x

Hardy, A., Beeton, R. J. S., & Pearson, L. (2002). Sustainable Tourism: An Overview of the Concept and its Position in Relation to Conceptualisations of Tourism. *Journal of Sustainable Tourism, 10*(6), 475–496. doi:10.1080/09669580208667183

Haushofer, C. (2013). *Familienunternehmen in Österreich*. Academic Press.

He, H., & Harris, L. (2020). The impact of Covid-19 pandemic on corporate social responsibility and marketing philosophy. *Journal of Business Research, 116*, 176–182. doi:10.1016/j.jbusres.2020.05.030 PMID:32457556

Heimerl, P., & Peters, M. (2019). Shaping the future of Alpine tourism destinations' next generation: An action research approach. *Tourism: An International Interdisciplinary Journal, 67*(3), 281–298.

Higgins-Desbiolles, F. (2010). The elusiveness of sustainability in tourism: The culture–ideology of consumerism and its implications. *Tourism and Hospitality Research, 10*(2), 116–129. doi:10.1057/thr.2009.31

Higgins-Desbiolles, F. (2020). Socialising tourism for social and ecological justice after COVID-19. *Tourism Geographies, 22*(3), 610–623. doi:10.1080/14616688.2020.1757748

Higgins-Desbiolles, F. (2021). The "war over tourism": Challenges to sustainable tourism in the tourism academy after COVID-19. *Journal of Sustainable Tourism, 29*(4), 551–569. doi:10.1080/09669582.2020.1803334

Huang, M., Li, P., Meschke, F., & Guthrie, J. P. (2015). Family firms, employee satisfaction, and corporate performance. *Journal of Corporate Finance, 34*, 108–127. doi:10.1016/j.jcorpfin.2015.08.002

Ibrahim, A. B., Soufani, K., & Lam, J. (2001). A study of succession in a family firm. *Family Business Review, 14*(3), 245–258. doi:10.1111/j.1741-6248.2001.00245.x

Ioannides, D., & Gyimóthy, S. (2020). The COVID-19 crisis as an opportunity for escaping the unsustainable global tourism path. *Tourism Geographies, 22*(3), 624–632. doi:10.1080/14616688.2020.1763445

Jessop, B. (1999). Reflections on globalisation and its (il)logic(s). In K. Olds, P. Dicken, P. F. Kelly, L. Kong, & H. W. Yeung (Eds.), *Globalisation and the Asia-Pacific: Contested territorie* (pp. 19–38). Routledge.

Joppe, M. (1996). Sustainable community tourism development revisited. *Tourism Management, 17*(7), 475–479. doi:10.1016/S0261-5177(96)00065-9

Kallmuenzer, A., Nikolakis, W., Peters, M., & Zanon, J. (2018). Trade-offs between dimensions of sustainability: Exploratory evidence from family firms in rural tourism regions. *Journal of Sustainable Tourism, 26*(7), 1204–1221. doi:10.1080/09669582.2017.1374962

Kallmuenzer, A., & Peters, M. (2017). Exploring entrepreneurial orientation in family firms: The relevance of social embeddedness in competition. *International Journal of Entrepreneurship and Small Business, 30*(2). doi:10.1504/IJESB.2017.081436

Komppula, R. (2016). The role of different stakeholders in destination development. *Tourism Review, 71*(1), 67–76. doi:10.1108/TR-06-2015-0030

Kraus, S., Clauss, T., Breier, M., Gast, J., Zardini, A., & Tiberius, V. (2020). The economics of COVID-19: Initial empirical evidence on how family firms in five European countries cope with the corona crisis. *International Journal of Entrepreneurial Behaviour & Research, 26*(5), 1067–1092. doi:10.1108/IJEBR-04-2020-0214

Krippendorf, J. (1982). Towards new tourism policies. *Tourism Management, 3*(3), 135–148. doi:10.1016/0261-5177(82)90063-2

Le Breton-Miller, I., & Miller, D. (2016). Family firms and practices of sustainability: A contingency view. *Journal of Family Business Strategy*, *7*(1), 26–33. doi:10.1016/j.jfbs.2015.09.001

Lew, A. A. (2014). Scale, change and resilience in community tourism planning. *Tourism Geographies*, *16*(1), 14–22. doi:10.1080/14616688.2013.864325

Li, Y. (2004). Exploring Community Tourism in China: The Case of Nanshan Cultural Tourism Zone. *Journal of Sustainable Tourism*, *12*(3), 175–193. doi:10.1080/09669580408667232

Liu, Z. (2003). Sustainable Tourism Development: A Critique. *Journal of Sustainable Tourism*, *11*(6), 459–475. doi:10.1080/09669580308667216

Luthe, T., & Wyss, R. (2014). Assessing and planning resilience in tourism. *Tourism Management*, *44*(3), 161–163. doi:10.1016/j.tourman.2014.03.011

Manente, M., Minghetti, V., & Mingotto, E. (2014). *Responsible tourism and CSR: Assessment systems for sustainable development of SMEs in tourism. CSR, Sustainability, Ethics & Governance*. Springer International Publishing., doi:10.1007/978-3-319-06308-9

Manzhynski, S., & Figge, F. (2020). Coopetition for sustainability: Between organizational benefit and societal good. *Business Strategy and the Environment*, *29*(3), 827–837. doi:10.1002/bse.2400

Martín-Santana, J. D., Cabrera-Suárez, M. K., & La Déniz-Déniz, M. C. (2020). Familiness, social capital and market orientation in the family firm. *European Journal of Marketing*, *54*(7), 1731–1760. doi:10.1108/EJM-04-2018-0274

Mathew, P. V., & Sreejesh, S. (2017). Impact of responsible tourism on destination sustainability and quality of life of community in tourism destinations. *Journal of Hospitality and Tourism Management*, *31*, 83–89. doi:10.1016/j.jhtm.2016.10.001

McWilliams, A., & Siegel, D. (2001). Corporate Social Responsibility: A Theory of the Firm Perspective. *Academy of Management Review*, *26*(1), 117–127. doi:10.5465/amr.2001.4011987

Memili, E., Fang, H., Koç, B., Yildirim-Öktem, Ö., & Sonmez, S. (2018). Sustainability practices of family firms: The interplay between family ownership and long-term orientation. *Journal of Sustainable Tourism*, *26*(1), 9–28. doi:10.1080/09669582.2017.1308371

Mihalic, T. (2016). Sustainable-responsible tourism discourse – Towards 'responsustable' tourism. *Journal of Cleaner Production*, *111*, 461–470. doi:10.1016/j.jclepro.2014.12.062

Okumus, F., Altinay, M., & Arasli, H. (2005). The impact of Turkey's economic crisis of February 2001 on the tourism industry in Northern Cyprus. *Tourism Management*, *26*(1), 95–104. doi:10.1016/j.tourman.2003.08.013

Okumus, F., & Karamustafa, K. (2005). Impact of an economic crisis: Evidence from Turkey. *Annals of Tourism Research*, *32*(4), 942–961. doi:10.1016/j.annals.2005.04.001 PMID:32572284

Peters, M., & Buhalis, D. (2004). Family hotel businesses: Strategic planning and the need for education and training. *Education + Training*, *46*(8/9), 406–415. doi:10.1108/00400910410569524

Peters, M., & Buhalis, D. (2013). SMEs in Tourism Destinations. In C. Costa, E. Panyik, & D. Buhalis (Eds.), *Aspects of tourism. Trends in European tourism planning and organisation* (pp. 92–101). Channel View Publications. doi:10.21832/9781845414122-011

Peters, M., & Kallmuenzer, A. (2015). Entrepreneurial orientation in family firms: The case of the hospitality industry. *Current Issues in Tourism*, *21*(1), 21–40. doi:10.1080/13683500.2015.1053849

Peters, M., & Kallmuenzer, A. (2018). Entrepreneurial orientation in family firms: The case of the hospitality industry. *Current Issues in Tourism*, *21*(1), 21–40. doi:10.1080/13683500.2015.1053849

Peters, M., Raich, M., Märk, S., & Pichler, S. (2012). The role of commitment in the succession of hospitality businesses. *Tourism Review*, *67*(2), 45–60. doi:10.1108/16605371211236169

Pikkemaat, B., Peters, M., & Chan, C.-S. (2018). Needs, drivers and barriers of innovation: The case of an alpine community-model destination. *Tourism Management Perspectives*, *25*, 53–63. doi:10.1016/j.tmp.2017.11.004

Prayag, G. (2020). Time for Reset? Covid-19 and Tourism Resilience. *Tourism Review International*, *24*(2), 179–184. doi:10.3727/154427220X15926147793595

Presas, P., Muñoz, D., & Guia, J. (2011). Branding familiness in tourism family firms. *Journal of Brand Management*, *18*(4-5), 274–284. doi:10.1057/bm.2010.41

Ritchie, B. W. (2004). Chaos, crises and disasters: A strategic approach to crisis management in the tourism industry. *Tourism Management, 25*(6), 669–683. doi:10.1016/j.tourman.2003.09.004

Ritchie, J. R. B., & Crouch, G. I. (2011). *The competitive destination: A sustainable tourism perspective*. CABI Pub.

Sharma, P., Chrisman, J. J., Pablo, A. L., & Chua, J. H. (2001). Determinants of Initial Satisfaction with the Succession Process in Family Firms: A Conceptual Model. *Entrepreneurship Theory and Practice*, *25*(3), 17–36. doi:10.1177/104225870102500302

Sigala, M. (2020). Tourism and COVID-19: Impacts and implications for advancing and resetting industry and research. *Journal of Business Research, 117*, 312–321. doi:10.1016/j.jbusres.2020.06.015 PMID:32546875

Smith, R. A. (1992). Beach resort evolution. *Annals of Tourism Research, 19*(2), 304–322. doi:10.1016/0160-7383(92)90083-2

Su, L., Swanson, S. R., & He, X. (2020). A scale to measure residents perceptions of destination social responsibility. *Journal of Sustainable Tourism*, *28*(6), 873–897. doi:10.1080/09669582.2019.1708372

The Austrian Federal Economic Chamber (Ed.). (2019). *Das Serviceangebot der Wirtschaftskammer*. https://www.wko.at/service/Startseite.html

Tran, H., Hwang, Y., Yu, C., & Yoo, S. (2018). The Effect of Destination Social Responsibility on Tourists' Satisfaction: The Mediating Role of Emotions. *Sustainability*, *10*(9), 3044. doi:10.3390u10093044

Veider, V., & Kallmuenzer, A. (2016). Assessing long-term orientation among founder- and descendant-led firms. *Journal of Family Business Management*, *6*(1). Advance online publication. doi:10.1108/JFBM-07-2015-0027

Wang, Y.-S. (2009). The impact of crisis events and macroeconomic activity on Taiwan's international inbound tourism demand. *Tourism Management, 30*(1), 75–82. doi:10.1016/j.tourman.2008.04.010 PMID:32287727

Zach, F. (2012). Partners and Innovation in American Destination Marketing Organizations. *Journal of Travel Research*, *51*(4), 412–425. doi:10.1177/0047287511426340

Zahra, S. A., Hayton, J. C., Neubaum, D. O., Dibrell, C., & Craig, J. (2008). Culture of Family Commitment and Strategic Flexibility: The Moderating Effect of Stewardship. *Entrepreneurship Theory and Practice*, *32*(6), 1035–1054. doi:10.1111/j.1540-6520.2008.00271.x

Zehrer, A., & Raich, F. (2020). Corporate Social Responsibility im Tourismus - der Einsatz von CSR-Aktivitäten in familiengeführten touristischen Unternehmen. In H. Pechlaner & S. Speer (Eds.), *Responsible Entrepreneurship* (pp. 215–236). Springer Fachmedien Wiesbaden. doi:10.1007/978-3-658-31616-7_10

Section 3
Reacting in Times of Change

The third section, 'Reacting in Times of Change', consists of three chapters. All chapters focus on family firm performance during turbulent times. Chapters 6 and 7 therefore conduct interviews with family firm members of different generations working in the family business. Thus, the studies represent the perception of the owner-family towards the own family firm. Chapter 7 in contrast interviews non-family member employee to obtain a perception of the family firms' ambidexterity.

Chapter 6
The Risk Perception of Family Business Owner-Manager in the Tourism Industry:
A Qualitative Comparison of the Intra-Firm Senior and Junior Generation

Gundula Glowka
https://orcid.org/0000-0002-2152-1644
MCI Management Center Innsbruck, Austria

Martin Tusch
MCI Management Center Innsbruck, Austria

Anita Zehrer
https://orcid.org/0000-0003-0325-7170
MCI Management Center Innsbruck, Austria

ABSTRACT

Successfully dealing with risk has become a major challenge for firms to stay competitive in the long run. The tourism industry is vulnerable to any crisis and therefore has to deal with a high degree of uncertainty including many risks. The Alpine tourism industry is shaped by small and medium-structured family businesses. Such firms have more informal decision-making processes and more generations involved, while risk taking depends on the decisions of an owner-manager. Understanding the risk perception owner-managers is thus a complex task. This study conducts n=12 qualitative interviews with senior and junior owner-managers of family SME in tourism. Understanding the risk perception for tourism family SME might therefore contribute to build resilience and long-term competitiveness.

DOI: 10.4018/978-1-7998-7352-5.ch006

INTRODUCTION

Tourism heavily shapes community-type destinations of the Alpine mountain regions (Glowka & Zehrer, 2019; Zehrer & Hallmann, 2015). In these communities, small and medium-structured family enterprises (family SME), play a central role by providing employment and contributing to the economic development of the region (Pechlaner et al., 2014). In such businesses, the family is heavily involved in daily business operations and often led by an owner-manager who shapes the vision and goals of the business (Chua, Chrisman, & Sharma, 1999; Feltham, Feltham, & Barnett, 2005). Regarding risk management, the risk behaviour of the owner-manager typically forms the basis for management decisions and, thus, the amount of risk taken by the business (Glowka, Kallmünzer, & Zehrer, 2020). Since the family SME is more informally structured, the individual risk perception of the owner-manager plays a central role in the assessment of risks (Rohrmann, 2008), which influences the business performance (Child, 1972; Das & Teng, 2001).

In addition, the requirements of a family business and its family members as well as their risk attitude and family orientation change with the participation of new generations (Molly, Laveren, & Jorissen, 2012). Typically, various generations participate in the family SME (Gersick, 1997), which complicates the risk management process of family SME even more. It might appear obvious that the senior and the junior generation have a divergent risk behaviour due to the generational differences (Hiebl, 2012). Casillas, Moreno, and Barbero (2011) find that more than one generation in the business increases risk-taking. On the other hand, Hall, Melin, and Nordqvist (2001) propose that the junior generation is more risk averse than the founding generation (Cassia, Massis, & Pizzurno, 2012; Hall et al., 2001). Falkner and Hiebl (2015) stress that future research should relate risk behaviour in family businesses to generational succession.

Following the call for more evidence on generational effects in family business research (Falkner & Hiebl, 2015; Muñoz-Bullon, Sanchez-Bueno, & Suárez-González, 2018; Nordqvist, Sharma, & Chirico, 2014), this study examines the risk perceptions of different generations. We therefore investigate the following research question: Which risks do senior and junior generations in tourism family SME perceive, and, are there any differences in their perception? We conduct n=12 qualitative, semi-structured interviews, with the senior and junior generations from touristic family SME in the Alpine tourism regions of the Tirol, Austria. Six interviews are conducted with senior owners, while the other six interviews are conducted with the successors of the same family SME that are already involved in daily management, which allows a comprehensive analysis on the risk perception in the owner family.

We choose the tourism industry for conducting this research due to two reasons. First, the tourism industry is vulnerable to any kind of economic crisis – not only the Covid-19 pandemic – which spill over into tourism, and therefore has to deal with a high degree of uncertainty (Ritchie, 2004, 2009). Secondly, Alpine tourism destinations are shaped by family SME, which all face similar risks they have to manage (Kallmuenzer, Nikolakis, Peters, & Zanon, 2018). For example, the competitive market pressure on investments and marketing performance as well as staff shortage are identified as main risks by almost all family SME in this sector (Glowka & Zehrer, 2019).

In the following, the different risk perceptions are structured into different areas of tasks, namely financing and investment management, marketing management, human resources management, cost management, eco-management and succession risks. The results show that both generations similarly perceive risks in the business areas of financing and investment management and human resources management. The junior generation does not perceive many risks in the field of marketing, whereas the senior generation does. Regarding the corporate sector of ecological management, only discussions with the senior generations take place. Cost management, as well as risks on the level of the entrepreneurial family are not perceived strongly by both generations.

This study contributes to the existing body of literature in various ways. It adds to our knowledge about the complex dynamics of risk management in family SME, where subjective risk perceptions of the owner-manager as well as family involvement of different generations and family members play a role. We also take the heterogeneity of family firms (Nordqvist et al., 2014) into account by examining only family SME in tourism, which all face a similar environment. Further, this study is relevant for the tourism industry, as is discusses different risks in the industry and how they are perceived.

In the following, the literature review discusses the concept of risk and its perception, including the risk taking behaviour of family firms and a discussion of types of tourism risks. Then, the method sections gives an overview in the data collection and analysis. Finally, the results are presented and the findings are related to the literature in the discussion section, before we outline some limitations and suggestions for future research and discuss implications for family business owners. In the end, we present our conclusions.

BACKGROUND

In accordance with the ISO 31000 where risk is the "effect of uncertainty" (Hopkin, 2018, p. 17), we define risk as "an event with the ability to impact (inhibit, enhance or cause doubt about) the effectiveness and efficiency of the core processes of an organization" (Hopkin, 2018, p. 17). Entrepreneurial risk is conceived as a psychological construct that comprises two elements: Risk as an opportunity and risk as a threat (Barbosa, Kickul, & Liao-Troth, 2007; Dickson & Giglierano, 1986; Mullins & Forlani, 2005). Perceiving risk as an opportunity is important, as profit can be increased, if a substantial risk is taken and properly managed. If, on the other hand, the risk is perceived as a threat, its interpretation focusses on the potential losses for the person viewing the risk (Dickson & Giglierano, 1986; Mullins & Forlani, 2005; Venkatarman, 2002).

For Lowrance (1976), risk is associated with a conscious human decision, e.g. of the family SME owner-manager. The resulting damage can, therefore, always be attributed to a specific person such as the owner-manager and traced back to his or her choices (Japp, 2010). The evaluation and assessment of hazards to which owner-managers are or may be exposed is called 'risk perception' (Rohrmann, 2008). Risk perceptions are interpretations of the world based on experience and beliefs (Sjoberg, 2000). They are embedded in the norms, value systems and cultural characteristics of societies (Finucane & Holup, 2006; French, Sutton, Kinmonth, & Marteau, 2006; Slovic & Peters, 2006). Risk perceptions are the basic background for decisions regarding the acceptance of risks and influence behaviour before, during and after the concrete manifestation of a threat (Rohrmann, 2008). In family SME, owner-managers typically decide how much risk is taken, which makes the decision process fast, however, dependent on the owner-managers ability to identify and assess risks (Feltham et al., 2005). How the extent of risk is assessed, and the extent to which owner-managers are willing to take risk, depend on the type of risk, personal experiences, beliefs and attitudes, and various social influences (Sjoberg, 2000; Slovic & Peters, 2006). These factors are embedded in a broader cultural and political context; therefore, social (sub) groups might differ significantly in their acceptance of risk (Rohrmann, 2008).

Conventional attitude theory assumes that attitude is a function of beliefs and values (Fishbein & Ajzen, 1975). Regarding the expected benefits derived from owner-managers' decisions, risk attitudes play a central role in the discussion of risk perception (Weber & Milliman, 1997). Risk attitudes are used to explain the intentions of owner-managers. They assess a risky situation as either favourable or unfavourable and act accordingly (Vasvári, 2015). The underlying characteristic of the manager is his or her risk propensity (Sjoberg, 2000).

Family SME are characterized by following non-economic goals, taking social responsibility and perusing the vision of handing over the business to the next generation (Dyer & Whetten, 2006). Due to the complexity of family businesses, the construct of socio-emotional wealth (SEW) is suitable as an explanatory model for studying risk taking in family SME (Berrone, Cruz, Gomez-Mejia, & Larraza-Kintana, 2010; Gomez-Mejia, Cruz, Berrone, & Castro, 2011; Gomez-Mejia, Makri, & Kintana, 2010; Gómez-Mejía, Haynes, Núñez-Nickel, Jacobson, & Moyano-

Fuentes, 2007). Gómez-Mejía et al. (2007) describe the non-economic benefit that family entrepreneurs derive from their business as the socio-emotional value. SEW explains how emotions, social capital and altruism influence the management of the company, especially in situations where an attempt is made to maintain SEW in the family business (Cleary, Quinn, & Moreno, 2019). La Porta, Lopez-De-Silanes, and Shleifer (1999) point out that the concentration of assets in a single company leads to greater risk aversion in family businesses as family entrepreneurs do not only focus on financial returns, but also the socio-emotional value (Gómez-Mejía et al., 2007).

The SEW construct is based on Behavioural Agency (BA) Theory (Gomez-Mejia et al., 2011). Therefore, the occasionally opportunistic behaviour of family members is also included in SEW (Berrone, Cruz, & Gomez-Mejia, 2012). Gómez-Mejía et al. (2007) argue that in the context of BA theory, the primary reference point for family businesses is the loss of their socio-emotional value. To avoid this loss, family businesses are willing to take a significant risk for their performance. Thus, family members in family businesses make decisions not to maximize profits, but consider the specific interests of the owner family, which might include profits among others. The SEW model also helps to explain results that do not match the predictions of agency theory by allowing for different risk preferences for family members (Berrone et al., 2010). The decision on whether family businesses act risk-affine or risk-averse depends on the distinction between two types of risk: performance hazard risk and venturing risk (Gómez-Mejía et al., 2007). Family businesses generally try to avoid risks. However, to protect their socio-emotional value, they may be willing to take a performance hazard risk (Gomez-Mejia et al., 2010).

Additionally, various generations are involved in the family SME (Chua et al., 1999). In the context of families, generation refers to the other members of the line of descent in the family group, for example, parents, children and grandchildren (Szydlik & Künemund, 2009). The position in the lineage of descent remains constant, which means being the child of one's parents throughout one's life. Family constellations are in a continuous state of change, e.g. the parenthood of children results in a grandparenthood. This will eventually lead to a change in roles at both the family and social level (Szydlik & Künemund, 2009). The junior generation often has to contend with problems arising from the strong dependence of the company on the skills and expertise of the transferors and the unwillingness of them to reduce their control (Handler, 1990, 1994). Family entrepreneurs of later generations must react to changing environmental conditions during an internal realignment and develop their own approach to risk (Beck, Janssens, Debruyne, & Lommelen, 2011). Studies have examined generational differences, e.g. when it comes to diversification or internationalization strategies (Fang, Kotlar, Memili, Chrisman, & Massis, 2018). For example, succeeding generation family firms are more willing to diversify, than founding generations with SEW preservation across generations as theoretical explanation (Muñoz-Bullon et al., 2018). Fang et al. (2018) suggests that family firms, which are managed by a later generation, are more likely to internationalize, than family firms in the first generation are. This shows that the willingness to take certain risks differs among generations.

Thus, the risk perception of family members and their risk behaviour is much more complicated than previously described due to the complexity of interests of the different parties involved (Brustbauer & Peters, 2013).

METHOD

As is obvious from the discussion above, investigating the research question is not feasible to be investigated by a standardized survey as this may miss important differences in the risk perception of different generations. To understand the different interpretations of risk it is important to look at underlying differences of the meaning risks have across generations, which in turn is conditional on the type of risks perceived and one's experiences. Therefore, qualitative data are a more promising method. Choosing a qualitative research approach, data was collected using semi-standardized, guideline-based interviews (Wittkowski, 1994, 1994). Here, the researcher personality can deepen a relevant topic area through spontaneous questions (Wittkowski, 1994). In this way, statements on all topic areas are generated in the individual interview, although they differ in scope and depth (Brosius, Haas, & Koschel, 2012). Bortz and Döring (2015) also recommend this data collection method for difficult and unpleasant topics, such as risk decisions within generational conflicts (Birley, 2002; Chen, 2017; Chirico, Sirmon, Sciascia, & Mazzola, 2011). The interviewee first addressed the risk perception of investments, marketing, human resource management, costs, eco-management, and succession. This is followed up by discussing specific risks that the interviewee addressed just before, e.g. the risk of staff shortage. We ask the interviewee to tell us about the experiences and strategies of dealing with this risk. Then, we ask them about their expertise and risk tolerance for some of the risks, before addressing generational differences and decision taking under risk in the family SME.

For the selection of a family business, it is crucial whether the family business perceives itself as such (Sieger, Zellweger, & Aquino, 2013). The purpose was to interview owner-managers from the senior as well as from the junior generation. All interviewees are thus actively involved in the family business at the moment when the interviews were conducted. The interviewees from the junior generation have already taken on initial responsibilities and are expected to assume full control of the family business in the near future. Finally, 12 interviews were conducted, six interviews with the senior generation and six interviews with the junior generation.

In most cases, the predecessor and successor both come from the same family business. In two family businesses, we could only interview one generation. In the results, F stands for family firm and is followed by a number, which is the number of interviews as given in the first column of Table 1. Additionally, HO (handing over) is used for the senior generation and TO (taking over) for quotes from the junior generation. Table 1 and 2 summarize descriptive statistics of the family SME.

The data collection took place in the period from mid-March to the end of April 2020 i.e. at the start of the Covid-19 pandemic in Austria. The interviews were

Table 1. Family SME descriptive statistics

Family Firm	Type of Family Firm	Year of Foundation	Number of Beds	Number of Employees	Annual Turnover
1	3 Stars Hotel	1581	65	< 50	£ 10m
2	Alpine Hut	1926	25	< 50	£ 10m
3	4 Stars Superior Hotel	1985	147	<250	£ 10m
4	4 Stars Hotel	1995	133	< 50	£ 10m
5	3 Stars Hotel	1955	70	< 50	£ 10m
6	4 Stars Hotel	1976	139	< 50	£ 10m
7	4 Stars Superior Hotel	1920	148	< 50	£ 10m

Table 2. Family SME descriptive statistics regarding family firm characteristics

Family Firm	Involved Generations Since Foundation	Involved Family Members	Management by Owner Family	Majority of Ownership by Owner Family	Interviewees (Age, Gender)
1	22nd	1	Yes	Yes	Senior (47, male)
2	1st	4	No	No	Senior (54, female) Junior (25, female)
3	1st	5	Yes	Yes	Senior (61, male) Junior (29, female)
4	1st	4	Yes	Yes	Junior (24, female)
5	3rd	4	Yes	Yes	Senior (53, female) Junior (27, female)
6	2nd	4	Yes	Yes	Senior (58, female) Junior (28, female)
7	4th	5	Yes	Yes	Senior (64, male) Junior (30, male)

conducted in German. The original German quotes are indicated in the footmarks and the English translations are written in the main text. The interview duration was 30 to 50 minutes. Both generations were interviewed separately to ensure that the interview partners do not influence each other. The family businesses are all operating in the Alpine tourism industry of the Tirol, Austria.

The data was then coded using maxQDA20 software using codes and sub-codes in a first and second coding cycle procedure (Saldaña, 2015). Finally, we present the results in the risk perceptions of senior and junior generation family business owners along six risk areas: financing and investment management, marketing management, human resources management, cost management, eco-management and succession risks. This separation is commonly used in the hospitality industry (Andersson et al., 2002; Getz & Carlsen, 2000; Henschel et al., 2013; Memili et al., 2020) and helps to structure the result section. However, the emphasis of the operational tasks can vary in the hotel management of a family business (Getz & Carlsen, 2000).

RESULTS AND DISCUSSION

The results are structured along the six operational task areas often used in the hospitality industry (Andersson et al., 2002; Getz & Carlsen, 2000; Henschel et al., 2013; Memili et al., 2020) and presented below.

Risk Perception in Financing and Investment Management

The financing and investment behaviour of the hotel industry differs from other sectors (Serrasqueiro & Nunes, 2014). The capital-intensive character of companies in the hotel industry can increase corporate risk. This is particularly evident in the fluctuations in revenues caused by high fixed costs (Devesa & Esteban, 2011; Elgonemy, 2002; Lee & Xiao, 2011). In this research, the senior generation perceived such risks for the financing and investment management as industry-specific. Examples of this are the investment intensity of the family business or competition with others.

Two interviewees of the junior generation (F6_TO; F7_TO) perceive the investment intensity that the hotel industry entails as a risk: "There is almost no standstill in investments because we are a well-established company and have been for over sixty years"[1] (F7_HO). The Alpine hotel trade possesses a high-quality standard, which needs constant investment to keep up with. This stimulates a kind of investment obligation according to the senior generation: "Many take a risk and say, their hotel must make a certain amount of turnover, because they invested so much for having the most beautiful, the largest and the best"[2] (F6_HO). Ongoing investment is perceived as a competition: "Maybe people will start to think about what this fierce competition does. The willingness to take part in this form of a contest has prompted entire tourist regions to make large investments"[3] (F6_HO). Consequently, financial dependence on debt capital is often seen as a risk (F7_HO).

Consequently, the senior generation sees only limited scope for their actions and reacts mainly reactive. The junior generation is less concerned with external risks, probably because a large part of them is still lacking knowledge with this area of the company. Generally, we find a lack of confidence in their own skills and experiences of the junior generation regarding investment decisions. This finding coincides with research by Sjoberg (2000), who found that an individual's subjective risk perception is based on personal experience and convictions. The senior generation thus showed itself to be more affine to risk; the same insight is found in the research results of Cassia et al. (2012).

Risk averseness was evident in the senior generation based on the experience gained in the areas of financing and investment (F6_HO; F7_HO). One interviewee of the senior generation stated: "In the past, I was quite willing to take risks, but that has changed over time. I am willing to take risks, but they must be within a manageable range, and a certain amount of flexibility must remain"[4] (F6_HO). Perceived failures by competitors led to increased risk aversion among some respondents (F5_HO; F7_HO). However, the findings of this study show that as soon as the junior generation had first responsibilities in the investment area, the generation handing over the company was more willing to invest. This could be related to the fact that the involvement of the junior generation brings in new knowledge and insights into the company. This seems to have a positive effect on the willingness of the senior generation to innovate and grow the company. Zahra (2005) comes to similar conclusions in this context.

The influence of the SARS-CoV-2 pandemic led to a change in investment behaviour: "At the moment I am very cautious, although I have to say that we are investing. We are still investing one million or something. However, we had planned to invest five million and cancelled this project. So I would already say that I have become more careful"[5] (F7_HO). However, uncertainties can be identified, caused by the pandemic in family businesses. The effects of the crisis are difficult to assess; family businesses are proving to be economically resilient to the crisis (F2; F3; F5; F6; F7).

In all family businesses, the senior generation has control over strategic investment decisions. However, in some areas the control was handed over to the junior generation (F3; F5; F6; F7) probably to smoothen the transition. One predecessor stated: "If I want to hand over my business, I will have to relinquish control sooner or later"[6] (F3_HO). In finance and investment management, a complete transfer of control rights could not be established: "The decisions are now largely made by my son. So my son deals with the architects and other construction companies, but if things get a bit more special, I still take care"[7] (F7_HO). Another interviewee states: "That may depend on the situation, because I leave the small decisions to my children, but in case of big ones I like to have the last word"[8] (F3_HO). The involvement of the junior generation in investment decisions was welcomed by the senior generation (F2; F3; F5, F6; F7): "The younger generation brings a breath of fresh air into the hotel. This has helped to make us look better"[9] (F5_HO).

Even if the junior generation has already been given control, the senior generation had difficulties to pass on full decision-making power in major investment decisions. It became clear that both generations want to prevent financial misconduct and take the specific interests of the entrepreneurial family into account. This behaviour corresponds to the research of Gómez-Mejía et al. (2007). They argue that not considering the specific interests of members of the entrepreneurial family leads to a loss of their socio-emotional value (Gomez-Mejia et al., 2010). To avoid failure, some family businesses are willing to take a significant economic risk to protect the interests of the entrepreneurial family. However, the junior generation recognizes the connection that the risk-taking investment behaviour of the transferring generation contributes to the continued existence of the family business.

One respondent from the junior generation (F7) stated that the own risk affinity is reduced due to the reluctant handover of control by the predecessor, although company growth plays a role for the junior generation (F7_TO). The hesitant transfer of authority is seen as disturbing by two predecessors (F6_TO; F7_TO), but two others show understanding for it (F3_TO; F4_TO). A junior generation interviewee explains that "the positive development of a hotel has to do with the willingness to invest"[10] (F7_TO3). This attitude was shared equally by other take-overs (F3_TO; F6_TO). However, there were situations in which the junior generation (F3; F4; F5) got responsibility for investment decisions and behaved in a risk-averse manner. This risk-averse behaviour is associated with a lack of experience in this area. One female transferee explained: "I have little experience so far because my parents hold the investment sovereignty and power. And therefore, I'm insecure, because I haven't made any big investments by myself yet and because I lack experience"[11] (F3_TO).

In all interviews, the unsuccessful investment decisions are considered a personal failure rather than risky bets. A senior stated: "It is essential to me, and I mean very important, that we take the correct decisions, because the enterprise is very close to my heart"[12] (F5_HO). A transferee explained: "I could not stand to be responsible, if the company were to go bankrupt, because of my [unsuccessful] investment"[13] (F5_TO). This identification with the family business is linked to emotional values (F3_TO; F4_TO; F5_HO; F5_TO).

There is clear communication between family members in strategic investment decisions (F2; F3; F4; F5; F6; F7). To minimize risk in investment decisions, decisions are made within the family (F2; F3; F4; F5; F6; F7). Here, the family members look for practical solutions to implement projects as efficiently and effectively as possible (F6; F7). Harmonious cooperation (F5; F6; F7) and mutual trust (F4; F5; F6; F7) promote the implementation of projects and reduce uncertainties for both the senior and the junior generation (F3; F4; F5; F6; F7). Regarding investment decisions, the junior generation is afraid of having too much debt (F4_TO) or even of losing the company (F5_TO). In the same vain, the junior generation stated that they have great trust in the transferring generation (F3_TO; F4_TO; F6_TO, F7_TO). To further minimize risks, the junior generation plans to include external information in their decision (F6_TO; F7_TO).

The fact that the junior generation expresses its interest in taking over the family business makes it easier for the senior generation to act in a risk-oriented manner in the investment area (F3_TO; F7_TO). One interviewee of the senior generation declared: "Above all, because I know that my children will continue to run this company, this also makes me more willing to invest, than I would otherwise be"[14] (F7_HO, 62). Among those handing over the family businesses, concerns about the transfer of debts could be identified (F2_ HO; F7_HO). In this context, the persons taking over the family businesses declared themselves willing to take on financial debts (F3_TO F7_TO). One interviewee taking over, stated: "We are a family there, and I am his son. He does not hand over the hotel to a stranger, where everything must fit. I will continue to run the business and not him. I think he worries too much, which he doesn't need, but I think it's completely natural"[15] (F7_ TO).

Summing up, there are no great differences in the risk perception of the different generations in the area of investment and financial decisions. Some frictions, however, seem to arise from an interaction with succession dynamics and the question of control.

Risk Perception in Human Resources Management

In this study, staff shortage was perceived as a significant risk by the senior and the junior generation. A central issue for both, the transferring generation (F1_HO; F2_HO; F3_HO; F6_HO) and the junior generation (F2_TO; F6_TO; F7_TO) was the difficulty in finding qualified staff. This was described as an industry-specific risk (F3_TO). One predecessor stated: "To find a chef is currently impossible and it doesn't really matter what you pay, because the market has no one to offer"[16] (F1_HO). The cause of staff shortage is seen in poor payment and the image of the hospitality industry and is considered a structural problem in the tourism industry. To counteract the staff shortage, many entrepreneurial families are trying to establish social relationships with the staff in their family businesses. The interviewed persons of both generations feel equally responsible for the development of social relationships. The behaviour of those surveyed is congruent with Cruz, Justo, and Castro (2012), who consider it essential to build a sense of belonging and identity with non-family employees, as this can promote a sense of stability and commitment to the company. However, building and maintaining relationships with the workforce tends to be perceived as a challenge by the junior generation.

One of the owners spoke of a shift in the balance of power in the staff market: "Only now is it the case that for every five vacancies there is one employee, and he or she selects. This means that in the meantime, it is no longer we who select, but the employee who chooses"[17] (F6_HO). Both generations addressed the image of the tourism industry as an inhibiting factor for staff acquisition (F1_HO; F2_HO_TO; F3_HO; F6_HO; F7_TO).

The employment relationships of family SME owner-managers are based on personal relationships as well as employment contracts (Andersson, Carlsen, & Getz, 2002). In family businesses, building personal relationships with staff can be a significant source of competitive advantages (Das & Teng, 2001). Since tourism has an above-average staffing level, the relationship between family members and employees plays an important role (Getz & Carlsen, 2000).

Uncertainties are identified by both, the senior generation (F2_HO; F6_HO) and the junior generation (F2_TO; F6_TO; F7_TO) in the recruitment interviews of potential staff. Although the senior generation hires the staff (F2_HO; F3_HO; F6_HO), the right of the junior generation to have a say is of central importance for the senior generation, since this can prevent future conflicts with the staff (F2_TO; F4_TO; F6_TO; F7_TO). One predecessor explained that her daughter is present at every job interview so that "she can get to know the person, and perhaps assess and help with the decision, because a decision made together is one that will last longer, because we all have to work together"[18] (F6_HO).

Managing employees, the predecessor and the successor must focus on using human capital that can sustain the existence of the company (Sirmon & Hitt, 2003). In other words, the future generation must motivate the staff to implement the company's strategies and achieve the company's goals (Sirmon & Hitt, 2003). Nevertheless, risk can arise from conflicts in the relationship between staff and family members, which can unsettle family businesses. Uncertainty, in turn, can lead to managers increasingly using lower-risk strategies and thus contribute to passivity (Tripsas & Gavetti, 2000).

For the senior generation (F1; F2; F3; F6), the relationship of trust with the employees is of great importance. The junior generation shares this attitude (F6; F7). Only in one family business (F2) conflicts arise because the generations involved have different views on the relationship of trust with the employees.

Other risks mentioned by junior generations are conflicts with employees (F6_TO; F7_TO) and disputes between employees (F2_TO; F6_TO; F7_TO). One successor (F6) stated that she is not perceived as a person of authority by individual employees: "It depends on the employees. We have many long-standing employees, and some employees, although they have been working here for a long time, do not yet see and accept me as their boss. In this area, there is often a conflict"[19] (F6_TO). Conflicts of this kind require the support of the senior generation (F6_HO). When family members of the next generation take over, they have the responsibility to manage both, family and business relationships, to win the support of non-family employees and other key stakeholders (DeNoble, Ehrlich, & Singh, 2007; Tabor, Chrisman, Madison, & Vardaman, 2018).

All in all, it seems that also in this area there seem to be not a lot of differences in the risk perception of different generations with the possible exception of problems in relation to authority because of a minor junior position.

Risk Perception in Marketing Management

The term market orientation refers to the implementation of the marketing concept, i.e. listening to customers and responding to their interests and needs (Beck et al., 2011). If predecessor and successor have different views on market orientation, this can lead to conflicts, which in turn can threaten the existence of the family business (Beck et al., 2011). As a result, later generations tend to tread new paths (Kepner, 1991) and therefore have a more robust market orientation than e.g. the first generation (Beck et al., 2011).

In marketing management, the senior generation (F2; F5; F6) perceived deficits and lack of knowledge in dealing with digital marketing. Two persons of the senior generation reported that they felt insecure in dealing with traditional marketing methods (F2; F6). One senior generation interviewee stated: "This is the hardware, I have huge deficits with the software"[20] (F6_HO). The junior generation (F5; F6) introduced modern sales instruments into the family business. "Now, marketing and management have changed in such a way, we work using OTA's and can enter markets, where one did not even know they existed before"[21] (F6_HO).

To reduce uncertainties in dealing with digital marketing, control is handed over to the junior generation (F5; F6). This transfer of control is associated with trust and with the perception of competencies in this area. A predecessor (F6) reported: "Yes, in marketing. That is what my daughter is taking over. I trust her 100%. I took over the business at a time when a fax was the ultimate"[22] (F6_HO). The generation handing over the business was sometimes hesitant in acquiring additional knowledge in marketing management (F5; F6). The respondents argue that they do not have the time resources (F6) or want to save financial resources (F3).

The interviewees of the junior generation in this study felt very confident in the field of marketing. This security is based on the experience gained in education and trainings, as well as their general competencies in this area. The senior generation stated that they lacked knowledge in marketing management and perceived this as a personal deficit. Cruz et al. (2012) come to similar conclusions by observing that later generations in family businesses place greater emphasis on the external environment, which leads to better insights into opportunities such as the development of new customer segments.

In the area of marketing, one finds differences in risk perception, which are based on different levels of information. Interestingly, the firms are aware of this and solve the discrepancy by shifting control to the younger generation in an effort to manage the changes efficiently.

Risk Perception in Cost Management

Small family businesses often have informal business practices and processes (Andersson et al., 2002). Although this can be used to their advantage through flexibility and rapid response, they often lack a systematic management approach (Peters & Buhalis, 2004), which means a lack of procedures that leads to improvisation by family members, affecting product standardization and quality control (Beck et al., 2011). Besides, accounting and financial management are often inaccurate, leading to a potential loss of revenue, lack of statistics for rational decision-making and miscalculation of critical performance indicators such as return on investment, variable costs and pricing structures (Peters & Buhalis, 2004). Conflicts of interest in cost management can arise between generations, as individual attitudes to collaboration can change with each generation (Chen, 2017).

The senior generation (F3; F5; F7) stated that their cost structure was efficiently structured and that they felt no uncertainty in this respect. Some companies transferred cost management to external service providers (F3; F7). A transferor (F5) would like the junior generation taking a closer look at the cost structure: "My son always has so many ideas, sometimes I don't know where to put them. But the fact is that in the end we have to bear the costs and I have the feeling that my son sometimes forgets this"[23] (F5_HO).

The combination of the ideas from the succeeding generation and a realistic assessment of the predecessor proves to be productive. The junior generation interviewee (F5) stated that she felt more uncertainty in cost management compared to other divisions of the company. She explained that she has little experience and therefore feels insecure, but that she has the will to accumulate knowledge in this area of the company to increase her feeling of security.

In this study, the senior generation stated that they felt secure about managing costs, because they had an efficient cost structure in the family business. In some cases, they outsourced accounting and financial management to external service providers. Although business areas are outsourced, and this may result in a loss of control, the senior generation does not seem to perceive this as a risk. This is probably related to the professionalism of external service providers, which contributes to a sense of security. However, the outsourcing of corporate divisions to external service providers can also give rise to the risk that future generations may lack an understanding of particular areas due to an absence of discussion (Bertrand & Schoar, 2006).

It seems that in this area again differences in knowledge and role drive differences in risk perceptions.

Risk Perception in Eco-Management

According to Henschel, Gruner, and Freyberg (2013) the orientation of the entrepreneurs concerning the Eco-Management is to limit the damage to the environment by the business to a minimum. However, it generates many negative environmental impacts, which are perceived by both, the family business and the public, as it uses local and imported consumer goods, energy and water and emits a large amount of carbon dioxide (Bohdanowicz, 2005). The contribution of proactive environmental management to achieving a competitive advantage lies in minimizing costs and differentiating from competitors (Galdeano-Gómez, Céspedes-Lorente, & Martínez-Del-Río, 2008). However, decisions such as the purchase of 'green' technology, the consideration of environmentally-friendly distribution and transport systems, or the fair product design are associated with an entrepreneurial risk assessment by family members (Fraj-Andrés, Martínez-Salinas, & Matute-Vallejo, 2009).

Two persons of the senior generation felt uncertainty in ecological management compared to the other divisions. Among some family businesses (F1; F3), there is an awareness of eco-management in their own hotel business. In this context, one of the seniors declared: "This trend is unstoppable, and we must go with it. Global warming and the associated constraints will force us to act"[24] (F3_HO).

The other senior explained: "I hope it is not only a trend, but a sustainable development that is happening now. I think the market offers other alternatives"[25] (F1_HO). Two companies mentioned the cost-intensive integration of sustainable infrastructures as a risk (F1; F3). Investments in sustainable infrastructure are only made once the technology meets the expectations of the junior generation. To minimize the risk of a wrong decision in eco-management, one senior owner-manager (F3) considers the gathering of expert knowledge to be indispensable.

The predecessors (F1; F3) intend to integrate their ecological objectives into the corporate policy of the family business. One stated here: "I also believe that this should be incorporated into our DNA, so we should work on it and communicate it more"[26] (F3_HO). One senior generation interviewee (F1) explained: "There are many chances that can be used at present, also as a marketing instrument"[27] (F1_HO, 60).

Concerning eco-management in this study, the senior generation recognizes opportunities to integrate sustainable technologies. Despite the willingness to invest in this area, the integration of sustainable technologies into the family business seems to be associated with high costs and thus high risks.

In the junior generation, the topic of eco-management remained primarily uncommented. The answers of the junior generation in the field of eco-management were comparatively undifferentiated. Possible reasons for this are that this division is mainly in the control area of the senior generation and the junior generation could hardly gain any experience. Therefore, an evaluation of these risks was difficult and there was a lack of awareness.

Risk Perception of the Succession Process

Handler (1990) describes the process of succession as a reciprocal adaptation of roles between members of the current leading generation and members of the following generations. In the process of succession, several uncertainties arise, such as the early departure of the person handing over the business due to illness (Chrisman, Chua, & Litz, 2004). Furthermore, poor communication creates a potential for conflict between family members (Chen, 2017). The lack of trust in the entrepreneurial skills of both generations can be differently pronounced and thus endanger the continued existence of the company (Memili, Patel, Koç, & Yazıcıoğlu, 2020). Parental altruism, which has positive connotations in principle (Schulze & Lubatkin, Michael, Dino, Richard, 2003), can pose a danger in the emotional relationship construct of family members (Berrone et al., 2010).

Regarding risks affecting the entrepreneurial family, the senior generation showed less uncertainty than the junior generation. This was particularly evident in the high level of trust the senior generations put into the professional competence of their successors. In addition to the professional qualities and training of the junior generation, the collaboration between the generations was also positively emphasized (F2; F3; F4; F5; F7). Among the risks cited by the junior generation is the voluntary exit of a family member from the family business (F2; F3), the involuntary loss due to illness (F3; F4; F5; F7) and concerns of unexpected developments which may cause difficulties in finding a successor within the family (F2; F3; F4; F5; F7). The junior generation (F3; F4; F5; F6) explained that the topic of internal family succession is communicated and that it is of great importance within the family business. However, different perceptions of the junior generation could be identified. Moral obligations and moral pressure can be observed among the interviewees taking over the family business (F3; F4; F5; F7). The whole succession process is perceived as a major risk by the junior generation (F2; F3; F4; F5; F7). Another successor explained: "Above all, I don't want to have to say that I am the generation that 'blew it'. I could not bear that and I do not want to experience that. Not at all!"[28] (F5_ TO, 58).

The internal succession of the company caused most of the uncertainties among the junior generation. The successors often deal with their strengths and weaknesses and questions the personal skills that are crucial for the continuation of the family business. To reduce the risk of succession failure, the junior generation gained experience through internships and by completing a degree in tourism, which increases their skills. Both the predecessor and the successor admitted that there are occasional conflicts, but these are harmless and are usually resolved quickly. This indicates a desire to avoid conflict-ridden relationships and to live together in harmony (Berrone et al., 2012).

CONCLUSION

The different generational risk perception was examined in the following tasks: Financing and investment management, human resources management, marketing management, cost management and ecological management. Besides, we examined the risk perception regarding the entrepreneurial family and generational relations. In the area of financing and investment management, it was found that the risk propensity of the senior generation is associated with past experiences. Although it is difficult for the senior generation to completely relinquish control due to their attachment to the family business, it leads to a greater willingness to take risks.

For human resources management, the shortage of staff is identified by both generations as a significant structural risk in the tourism industry. Due to deficits of knowledge of the senior generation, there is broad participation of the junior generation in marketing management. They introduce new marketing methods into the companies and feel secure in the knowledge that they have gained in this area. In cost management, the senior generation feels little uncertainty, since the task was frequently outsourced to external service enterprises. However, there might be a risk of loss of control due to a lack of competence among the next generation. At the level of the entrepreneurial families and their generational relationships, it was found that conflict-ridden relationships within the families are avoided as far as possible. The generations involved describe an emancipated relationship to the outside world, but moral pressure is evident in the junior generation. The succeeding generation sometimes feels pushed into their role, which can lead to various risks.

This study faces some limitations. First, there is a limited number of interviewed family businesses. While the small number of interviews allows a detailed description of the different perspectives, future research should address the topic in depth, e.g. by conducting case studies. For generalizability, future research should also focus on different industries. In the present study, only family businesses operating in the Alpine tourism industry of the Tirol were interviewed. The findings are thus set in a very specific tourism context, making generalizability of the results difficult. For reasons of clarity, the study is limited to five selected business areas. The selection of the business units was guided by the frequency with which the business units are mentioned in the literature and by results from Glowka and Zehrer (2019). Furthermore, the format of the video phone call due to the pandemic restrictions might represent an unfamiliar situation for some interviewees, which might be seen as a constraint. The line between risk perception and risk behaviour is fluid in this work. Any findings in the research of risk perceptions must be considered in their context. For future research, long-term studies also lend themselves to this area of research, since the risk perception of individuals is a dynamic construct. Various survey methods are available to investigate the change in risk attitudes. Not only narrative interviews, but also ethnographic research methods could be considered for this purpose. Future research could also focus on gender-specific risk perception.

The recommendations for action should help family members to feel more secure in decision-making processes. This should reduce possible unsuccessful decisions and prevent conflicts with family members. To facilitate the transfer of control rights in the family business, it is beneficial to strengthen trust between family members. The results show that belief in the predecessor provides security to the successor, which makes strategic decisions in different areas of the company easier. Mutual trust is essential in the area of financing and investment management, which involves many risky decisions.

Succession process decisions are made jointly by both generations within a certain period, since usually the junior generation of the company is given a period of familiarization with the different business areas. During this phase, the risk attitude of both generations may differ. Our research results illustrate the need for open communication of motives and emotions regarding individual decisions. Since open communication helps to avoid conflicts, it is recommended in situations of disagreement to speak openly about one's own feelings and risk attitudes and to put them in relation to the underlying economic decisions.

Due to their small-scale structure, family-owned tourism businesses have good starting conditions for creating lasting ties with the staff, which can be derived from the family atmosphere. By investing in social relationships with the team, the entrepreneurial family can prevent various risks. The research results show that a healthy working atmosphere facilitates the entry of the successor into the family business and increases his/her acceptance. Besides, social relationships serve to bind the staff, which, because of the tight staff market, reduces the risk of not finding qualified staff.

The knowledge of the senior generation turned out to be a valuable resource in the study. Therefore, this knowledge should be consciously conveyed during the internal succession process. In this way, the junior generation can benefit from the wealth of experience of their parents. This allows risk decisions to be assessed more adequately. The openness of the junior generation is just as necessary as the willingness of the person handing over their knowledge. Also, the results show that the preservation of the company for future generations is generally regarded as a central goal for family members in family businesses.

Companies can professionalize the internal assessment of risks by implementing a risk management system. A risk management system is used to create an overview of opportunities and threats. Based on this, company-specific guidelines can be developed. By integrating a risk management system, companies can create a better starting position for decisions taken on identifying, assessing and controlling financing and investment risks.

REFERENCES

Andersson, T., Carlsen, J., & Getz, D. (2002). Family Business Goals in the Tourism and Hospitality Sector: Case Studies and Cross-Case Analysis from Australia, Canada, and Sweden. *Family Business Review*, *15*(2), 89–106. doi:10.1111/j.1741-6248.2002.00089.x

Barbosa, S. D., Kickul, J., & Liao-Troth, M. (2007). Development and Validation of a Multidimensional Scale of Entrepreneurial Risk Perception. *Annual Meeting of the Academy of Management*, 2007(1), 1–6. 10.5465/ambpp.2007.26508257

Beck, L., Janssens, W., Debruyne, M., & Lommelen, T. (2011). A Study of the Relationships Between Generation, Market Orientation, and Innovation in Family Firms. *Family Business Review*, *24*(3), 252–272. doi:10.1177/0894486511409210

Berrone, P., Cruz, C., & Gomez-Mejia, L. R. (2012). Socioemotional Wealth in Family Firms. *Family Business Review*, *25*(3), 258–279. doi:10.1177/0894486511435355

Berrone, P., Cruz, C., Gomez-Mejia, L. R., & Larraza-Kintana, M. (2010). Socioemotional Wealth and Corporate Responses to Institutional Pressures: Do Family-Controlled Firms Pollute Less? *Administrative Science Quarterly*, *55*(1), 82–113. doi:10.2189/asqu.2010.55.1.82

Bertrand, M., & Schoar, A. (2006). The Role of Family in Family Firms. *The Journal of Economic Perspectives*, *20*(2), 73–96. doi:10.1257/jep.20.2.73

Birley, S. (2002). Attitudes of Owner-Managers' Children towards Family and Business Issues. *Entrepreneurship Theory and Practice*, *26*(3), 5–19. doi:10.1177/104225870202600301

Bohdanowicz, P. (2005). European Hoteliers' Environmental Attitudes. *The Cornell Hotel and Restaurant Administration Quarterly*, *46*(2), 188–204. doi:10.1177/0010880404273891

Bortz, J., & Döring, N. (2015). Forschungsmethoden und Evaluation: Für Human- und Sozialwissenschaftler (5. Aufl.). Berlin: Springer.

Brosius, H.-B., Haas, A., & Koschel, F. (2012). Methoden der empirischen Kommunikationsforschung: Eine Einführung (6., erw. und aktualisierte Aufl.). In *Studienbücher zur Kommunikations- und Medienwissenschaft.* Wiesbaden: Springer VS. Retrieved from http://gbv.eblib.com/patron/FullRecord.aspx?p=970700 doi:10.1007/978-3-531-94214-8

Brustbauer, J. K., & Peters, M. (2013). Risk perception of family and non-family firm managers. *International Journal of Entrepreneurship and Small Business*, *20*(1), 96. doi:10.1504/IJESB.2013.055695

Casillas, J. C., Moreno, A. M., & Barbero, J. L. (2011). Entrepreneurial orientation of family firms: Family and environmental dimensions. *Journal of Family Business Strategy*, *2*(2), 90–100. doi:10.1016/j.jfbs.2011.03.002

Cassia, L., de Massis, A., & Pizzurno, E. (2012). Strategic innovation and new product development in family firms. *International Journal of Entrepreneurial Behaviour & Research*, *18*(2), 198–232. doi:10.1108/13552551211204229

Chen, L. (2017). Power and ambivalence in intergenerational communication: Deciding to institutionalize in Shanghai. *Journal of Aging Studies*, *41*, 44–51. doi:10.1016/j.jaging.2017.03.004 PMID:28610754

Child, J. (1972). Organizational Structure, Environment and Performance: The Role of Strategic Choice. *Sociology*, *6*(1), 1–22. doi:10.1177/003803857200600101

Chirico, F., Sirmon, D. G., Sciascia, S., & Mazzola, P. (2011). Resource orchestration in family firms: Investigating how entrepreneurial orientation, generational involvement, and participative strategy affect performance. *Strategic Entrepreneurship Journal*, *5*(4), 307–326. doi:10.1002ej.121

Chrisman, J. J., Chua, J. H., & Litz, R. A. (2004). Comparing the Agency Costs of Family and Non-Family Firms: Conceptual Issues and Exploratory Evidence. *Entrepreneurship Theory and Practice*, *28*(4), 335–354. doi:10.1111/j.1540-6520.2004.00049.x

Chua, J. H., Chrisman, J. J., & Sharma, P. (1999). Defining the family business by behavior. *Entrepreneurship Theory and Practice*, *23*(4), 19–39. doi:10.1177/104225879902300402

Cleary, P., Quinn, M., & Moreno, A. (2019). Socioemotional wealth in family firms: A longitudinal content analysis of corporate disclosures. *Journal of Family Business Strategy*, *10*(2), 119–132. doi:10.1016/j.jfbs.2018.11.002

Cruz, C., Justo, R., & de Castro, J. O. (2012). Does family employment enhance MSEs performance? *Journal of Business Venturing*, *27*(1), 62–76. doi:10.1016/j.jbusvent.2010.07.002

Das, T. K., & Teng, B.-S. (2001). Article. *Journal of Business and Psychology*, *15*(3), 449–465. doi:10.1023/A:1007874701367

DeNoble, A., Ehrlich, S., & Singh, G. (2007). Toward the Development of a Family Business Self-Efficacy Scale: A Resource-Based Perspective. *Family Business Review*, *20*(2), 127–140. doi:10.1111/j.1741-6248.2007.00091.x

Devesa, M. J. S., & Esteban, L. P. (2011). Spanish hotel industry: Indebtedness determinants. *Applied Economics*, *43*(28), 4227–4238. doi:10.1080/00036846.2010.491446

Dickson, P. R., & Giglierano, J. J. (1986). Missing the Boat and Sinking the Boat: A Conceptual Model of Entrepreneurial Risk. *Journal of Marketing*, *50*(3), 58–70. doi:10.1177/002224298605000305

Dyer, W. G., & Whetten, D. A. (2006). Family firms and social responsibility: Preliminary evidence from the S&P 500. *Entrepreneurship Theory and Practice*, *30*(6), 785–802. doi:10.1111/j.1540-6520.2006.00151.x

Elgonemy, A. R. (2002). Debt-Financing Alternatives. *The Cornell Hotel and Restaurant Administration Quarterly*, *43*(3), 7–21. doi:10.1177/0010880402433001

Falkner, E. M., & Hiebl, M. R. W. (2015). Risk management in SMEs: A systematic review of available evidence. *The Journal of Risk Finance*, *16*(2), 122–144. doi:10.1108/JRF-06-2014-0079

Fang, H., Kotlar, J., Memili, E., Chrisman, J. J., & de Massis, A. (2018). The pursuit of international opportunities in family firms: Generational differences and the role of knowledge-based resources. *Global Strategy Journal*, *8*(1), 136–157. doi:10.1002/gsj.1197

Feltham, T. S., Feltham, G., & Barnett, J. J. (2005). The dependence of family businesses on a single decision-maker. *Journal of Small Business Management*, *43*(1), 1–15. doi:10.1111/j.1540-627X.2004.00122.x

Finucane, M. L., & Holup, J. L. (2006). Risk as Value: Combining Affect and Analysis in Risk Judgments. *Journal of Risk Research*, *9*(2), 141–164. doi:10.1080/13669870500166930

Fishbein, M., & Ajzen, I. (1975). *Belief, attitude, intention and behaviour: An introduction to theory and research*. Addison-Wesley Reading.

Fraj-Andrés, E., Martínez-Salinas, E., & Matute-Vallejo, J. (2009). Factors affecting corporate environmental strategy in Spanish industrial firms. *Business Strategy and the Environment*, *18*(8), 500–514. doi:10.1002/bse.611

French, D. P., Sutton, S., Kinmonth, A. L., & Marteau, T. M. (2006). Assessing Perceptions of Risks due to Multiple Hazards. *Journal of Risk Research*, *9*(6), 657–682. doi:10.1080/13669870600813837

Galdeano-Gómez, E., Céspedes-Lorente, J., & Martínez-Del-Río, J. (2008). Environmental performance and spillover effects on productivity: Evidence from horticultural firms. *Journal of Environmental Management*, *88*(4), 1552–1561. doi:10.1016/j.jenvman.2007.07.028 PMID:17825476

Gersick, K. E. (1997). *Generation to generation: Life cycles of the family business*. Harvard Business Press.

Getz, D., & Carlsen, J. (2000). Characteristics and goals of family and owner-operated businesses in the rural tourism and hospitality sectors. *Tourism Management*, *21*(6), 547–560. doi:10.1016/S0261-5177(00)00004-2

Glowka, G., Kallmünzer, A., & Zehrer, A. (2020). Enterprise risk management in small and medium family enterprises: The role of family involvement and CEO tenure. *The International Entrepreneurship and Management Journal*, 1–19.

Glowka, G., & Zehrer, A. (2019). Tourism Family-Business Owners' Risk Perception: Its Impact on Destination Development. *Sustainability*, *11*(24), 6992. doi:10.3390u11246992

Gomez-Mejia, L. R., Cruz, C., Berrone, P., & de Castro, J. (2011). The Bind that Ties: Socioemotional Wealth Preservation in Family Firms. *The Academy of Management Annals*, *5*(1), 653–707. doi:10.5465/19416520.2011.593320

Gómez-Mejía, L. R., Haynes, K. T., Núñez-Nickel, M., Jacobson, K. J. L., & Moyano-Fuentes, J. (2007). Socioemotional Wealth and Business Risks in Family-controlled Firms: Evidence from Spanish Olive Oil Mills. *Administrative Science Quarterly*, *52*(1), 106–137. doi:10.2189/asqu.52.1.106

Gomez-Mejia, L. R., Makri, M., & Kintana, M. L. (2010). Diversification Decisions in Family-Controlled Firms. *Journal of Management Studies*, *47*(2), 223–252. doi:10.1111/j.1467-6486.2009.00889.x

Hall, A., Melin, L., & Nordqvist, M. (2001). Entrepreneurship as Radical Change in the Family Business: Exploring the Role of Cultural Patterns. *Family Business Review*, *14*(3), 193–208. doi:10.1111/j.1741-6248.2001.00193.x

Handler, W. C. (1990). Succession in Family Firms: A Mutual Role Adjustment between Entrepreneur and Next-generation Family Members. *Entrepreneurship Theory and Practice*, *15*(1), 37–52. doi:10.1177/104225879001500105

Handler, W. C. (1994). Succession in Family Business: A Review of the Research. *Family Business Review*, *7*(2), 133–157. doi:10.1111/j.1741-6248.1994.00133.x

Henschel, K., Gruner, A., & von Freyberg, B. (2013). *Hotelmanagement (4., aktualisierte und erw. Aufl.)*. Oldenbourg. doi:10.1524/9783486753882

Hiebl, M. R. W. (2012). Risk aversion in family firms: What do we really know? *The Journal of Risk Finance*, *14*(1), 49–70. doi:10.1108/15265941311288103

Hopkin, P. (2018). *Fundamentals of risk management: understanding, evaluating and implementing effective risk management*. Kogan Page Publishers.

Japp, K. P. (2010). Risiko und Gefahr. Zum Problem authentischer Kommunikation. In C. Büscher & K. P. Japp (Eds.), *Ökologische Aufklärung: 25 Jahre "Ökologische Kommunikation* (1st ed., pp. 281–308). VS Verl. für Sozialwiss. doi:10.1007/978-3-531-92425-0_11

Kallmuenzer, A., Nikolakis, W., Peters, M., & Zanon, J. (2018). Trade-offs between dimensions of sustainability: Exploratory evidence from family firms in rural tourism regions. *Journal of Sustainable Tourism*, *26*(7), 1204–1221. doi:10.1080/09669582.2017.1374962

Kepner, E. (1991). The Family and the Firm: A Coevolutionary Perspective. *Family Business Review*, *4*(4), 445–461. doi:10.1111/j.1741-6248.1991.00445.x

La Porta, R., Lopez-De-Silanes, F., & Shleifer, A. (1999). Corporate Ownership Around the World. *The Journal of Finance*, *54*(2), 471–517. doi:10.1111/0022-1082.00115

Lee, S., & Xiao, Q. (2011). An examination of the curvilinear relationship between capital intensity and firm performance for publicly traded US hotels and restaurants. *International Journal of Contemporary Hospitality Management*, *23*(6), 862–880. doi:10.1108/09596111111153510

Lowrance, W. W. (1976). *Of Acceptable Risk*. Walliam Kaufmann.

Memili, E., Patel, P. C., Koç, B., & Yazıcıoğlu, İ. (2020). The interplay between socioemotional wealth and family firm psychological capital in influencing firm performance in hospitality and tourism. *Tourism Management Perspectives*, *34*, 100651. doi:10.1016/j.tmp.2020.100651

Molly, V., Laveren, E., & Jorissen, A. (2012). Intergenerational Differences in Family Firms: Impact on Capital Structure and Growth Behavior. *Entrepreneurship Theory and Practice*, *36*(4), 703–725. doi:10.1111/j.1540-6520.2010.00429.x

Mullins, J. W., & Forlani, D. (2005). Missing the boat or sinking the boat: A study of new venture decision making. *Journal of Business Venturing*, *20*(1), 47–69. doi:10.1016/j.jbusvent.2003.08.001

Muñoz-Bullon, F., Sanchez-Bueno, M. J., & Suárez-González, I. (2018). Diversification decisions among family firms: The role of family involvement and generational stage. *BRQ Business Research Quarterly*, *21*(1), 39–52. doi:10.1016/j.brq.2017.11.001

Nordqvist, M., Sharma, P., & Chirico, F. (2014). Family firm heterogeneity and governance: A configuration approach. *Journal of Small Business Management*, *52*(2), 192–209. doi:10.1111/jsbm.12096

Pechlaner, H., Kozak, M., Volgger, M., Zehrer, A., Raich, F., Siller, H., & Tschiderer, F. (2014). Leadership networks in destinations. *Tourism Review*.

Peters, M., & Buhalis, D. (2004). Family hotel businesses: Strategic planning and the need for education and training. *Education + Training*, *46*(8/9), 406–415. doi:10.1108/00400910410569524

Ritchie, B. W. (2004). Chaos, crises and disasters: A strategic approach to crisis management in the tourism industry. *Tourism Management*, *25*(6), 669–683. doi:10.1016/j.tourman.2003.09.004

Ritchie, B. W. (2009). *Crisis and disaster management for tourism*. Channel View Publications. doi:10.21832/9781845411077

Rohrmann, B. (2008). Risk perception, risk attitude, risk communication, risk management. *International Emergency Management Society (TIEMS) Annual Conference*.

Saldaña, J. (2015). The coding manual for qualitative researchers. *Sage (Atlanta, Ga.)*.

Schulze, W., Lubatkin, M. H., & Dino, R. N. (2003). Toward a theory of agency and altruism in family firms. *Journal of Business Venturing*, *18*(4), 473–490. doi:10.1016/S0883-9026(03)00054-5

Serrasqueiro, Z., & Nunes, P. M. (2014). Financing Behavior of Portuguese SMEs in Hotel Industry. *International Journal of Hospitality Management*, *43*, 98–107. doi:10.1016/j.ijhm.2014.09.001

Sieger, P., Zellweger, T., & Aquino, K. (2013). Turning Agents into Psychological Principals: Aligning Interests of Non-Owners through Psychological Ownership. *Journal of Management Studies*, *50*(3), 361–388. doi:10.1111/joms.12017

Sirmon, D. G., & Hitt, M. A. (2003). Managing Resources: Linking Unique Resources, Management, and Wealth Creation in Family Firms. *Entrepreneurship Theory and Practice*, *27*(4), 339–358. doi:10.1111/1540-8520.t01-1-00013

Sjoberg, L. (2000). Factors in Risk Perception. *Risk Analysis*, *20*(1), 1–12. doi:10.1111/0272-4332.00001

Slovic, P., & Peters, E. (2006). Risk Perception and Affect. *Current Directions in Psychological Science*, *15*(6), 322–325. doi:10.1111/j.1467-8721.2006.00461.x

Szydlik, M., & Künemund, H. (2009). *Generationen: Multidisziplinäre Perspektiven; Martin Kohli zum 65. Geburtstag* (1st ed.). VS Verlag für Sozialwissenschaften / GWV Fachverlage GmbH Wiesbaden. doi:10.1007/978-3-531-91499-2

Tabor, W., Chrisman, J. J., Madison, K., & Vardaman, J. M. (2018). Nonfamily Members in Family Firms: A Review and Future Research Agenda. *Family Business Review*, *31*(1), 54–79. doi:10.1177/0894486517734683

Tripsas, M., & Gavetti, G. (2000). Capabilities, cognition, and inertia: Evidence from digital imaging. *Strategic Management Journal*, *21*(10-11), 1147–1161. doi:10.1002/1097-0266(200010/11)21:10/11<1147::AID-SMJ128>3.0.CO;2-R

Vasvári, T. (2015). Risk, Risk Perception, Risk Management – a Review of the Literature. *Public Finance Quarterly*, *60*(1), 29–48.

Venkatarman, S. (2002). Ten principles of entrepreneurial creation. *Batten Briefings, 1*(1).

Weber, E. U., & Milliman, R. A. (1997). Perceived Risk Attitudes: Relating Risk Perception to Risky Choice. *Management Science, 43*(2), 123–144. doi:10.1287/mnsc.43.2.123

Wittkowski, J. (1994). *Das Interview in der Psychologie: Interviewtechnik und Codierung von Interviewmaterial.* VS Verlag für Sozialwissenschaften. doi:10.1007/978-3-322-92472-8

Zahra, S. A. (2005). Entrepreneurial Risk Taking in Family Firms. *Family Business Review, 18*(1), 23–40. doi:10.1111/j.1741-6248.2005.00028.x

Zehrer, A., & Hallmann, K. (2015). A stakeholder perspective on policy indicators of destination competitiveness. *Journal of Destination Marketing & Management, 4*(2), 120–126. doi:10.1016/j.jdmm.2015.03.003

ENDNOTES

1 "Es gibt bei Investitionen fast keinen Stillstand, weil wir sind ein gewachsenes Haus und das schon seit über sechzig Jahren."

2 "Viele nehmen da ein Risiko auf sich und sagen, ihr Haus muss so und so viel Umsatz machen, da ich so und so viel investiert habe, dafür habe ich das Schönste, das Größte und das Beste."

3 "Vielleicht kommen die Leute mal ins Denken, was dieses ‚Wettrüsten' alles anrichtet. Die Bereitschaft an diesem Wettrüsten teilzunehmen hat dazu geführt, dass ganze Tourismusregionen hohe Investitionen tätigen."

4 "Ich bin bereit, Risiko auf mich zu nehmen, aber das muss in einem verkraftbaren Bereich liegen und ein gewisser Spielraum muss bleiben."

5 "Zurzeit bin ich sehr vorsichtig, obwohl ich sagen muss, dass wir investieren. Wir investieren trotzdem eine Millionen Euro oder sowas. Aber wir hätten eigentlich Investitionen von fünf Millionen Euro geplant und haben dieses Projekt abgesagt. Also würde ich schon sagen, dass ich vorsichtiger geworden bin."

6 "Ich muss über kurz oder lang Kontrolle abgeben, wenn ich meinen Betrieb übergeben möchte."

7 "Die Entscheidungen macht jetzt schon relativ viel der *X*. Also die Gespräche mit die Architekten und Planer und andere Baufirmen macht größtenteils der *X*, aber wenn es etwas spezieller wird, dann mach ich das immer noch."

8 “Das mag wirklich situationsabhängig sein, weil die kleinen Entscheidungen überlasse ich schon meinen Kindern, aber bei den großen hätte ich noch gerne das letzte Wort.”

9 “Die jüngere Generation bringt wieder frischen Wind in die Bude. Das hat geholfen, dass wir jetzt besser dastehen.”

10 “die positive Entwicklung eines Hauses/Hotels mit Freudigkeit beim Investieren zu tun hat.”

11 “Erfahrungen bis jetzt noch wenig, weil die Investitionshoheit oder Macht haben eigentlich die Eltern. Und deswegen die Unsicherheit, weil ich selber noch keine großen Investitionen getätigt habe und weil mir einfach die Erfahrung fehlt.”

12 “Mir ist es sehr wichtig, und ich meine wirklich sehr wichtig, dass wir die richten Entscheidungen treffen, weil mir der Betrieb sehr am Herzen liegt.”

13 “Ich könnte das nicht verantworten, wenn das Unternehmen wegen dieser Investition ‚Hopps‘ geht.”

14 “Vor allem, weil ich weiß, dass meine Kinder dieses Unternehmen weiterführen werden, macht mich das auch zu einem gewissen Grad investitionsfreudiger, als ich sonst wäre.”

15 “Wir sind da ja auch eine Familie und ich bin sein Sohn. Er übergibt das Hotel keinem Fremden, wo alles passen muss. Ich führe das Unternehmen weiter und nicht er. Ich glaube, er macht sich da zu viele Sorgen, was er eigentlich nicht braucht, aber ich denke, das ist völlig natürlich.”

16 “Einen Küchenchef zu finden. Dies ist momentan ein Ding der Unmöglichkeit, da ist es auch schon egal was man bezahlt dafür, da der Markt einfach sehr trocken ist.”

17 “Nur jetzt ist es so, dass auf fünf offene Stellen ein Mitarbeiter kommt, und der sucht aus. Das heißt, mittlerweile suchen nicht mehr wir aus, sondern der Mitarbeiter sucht aus.”

18 “sie die Person kennenlernen, und vielleicht einschätzen und bei der Entscheidung mithelfen [kann], denn eine gemeinsam getroffene Entscheidung ist halt schon eine, die länger hält. Damit es nicht heißt ‚Der wurde mir auf’s Aug gedrückt‘, sondern dass derjenige Zuständige dann mitentschieden hat”

19 “Es kommt auf die Mitarbeiter darauf an. Wir haben sehr viele langjährige Mitarbeiter, und manche Mitarbeiter, obwohl sie schon längere Zeit hier arbeiten, sehen und akzeptieren mich noch nicht als Chef. In diesem Bereich gibt es dann öfters einen Konflikt.”

20 “Das ist die Hardware, bei der Software hab ich riesige Defizite.”

21 “Jetzt hat sich das Marketing und die Betriebsführung doch dahingehend geändert, dass es um eine Positionierung geht in vielen Bereichen, wir haben

diese OTA's, die man ja tagtäglich nützt. Das ist ein Zugang zu Märkten, wo man vorher nicht einmal gewusst hat, wie man die schreibt."

22 "Ja, beim Marketing. Das ist das, was meine Tochter übernimmt. Da vertraue ich ihr zu 100%. Ich habe den Betrieb in einer Zeit übernommen, da war ein Fax das Ultimative."

23 "*X* hat immer so viele Ideen, ich weiß manchmal gar nicht wohin mit denen. Aber es ist ja so, dass wir die Kosten schlussendlich tragen müssen und da habe ich das Gefühl, dass *X* das manchmal vergisst."

24 "Dieser Trend ist unaufhaltsam und wir müssen mit in diesem Strom schwimmen. Allein durch die Erderwärmung und die damit verbundenen Auflagen, die daraus entstehen werden, sind wir gezwungen zu handeln."

25 "Es ist nicht nur als Trend zu sehen, sondern ich hoffe, es ist wirklich eine nachhaltige Entwicklung, die jetzt passiert. Es wird schon leichter glaub ich, da der Markt andere Alternativen anbietet."

26 "Ich glaube auch, das sollte in unsere DNA einfließen, also wir sollten das mehr herausarbeiten und auch mehr kommunizieren."

27 "Es gibt ja viele Chancen, und das lässt sich momentan auch gut als Marketinginstrument nützen."

28 "Vor allem möchte ich mir später nicht nachsagen müssen, ich bin die Generation, die alles ‚vergeigt' hat. Das könnte ich nicht ertragen und das möchte ich nicht erleben. Keinesfalls!"

Chapter 7
Heurigen 2.0:
Succession Processes of Family-Owned Wine Taverns in Northern Burgenland

Lara Soleder
MCI Management Center Innsbruck, Austria

ABSTRACT

Traditional Heurigen and Buschenschenken have been an important part of Austrian culture for centuries and are mostly run as small family firms. In those businesses, succession is inevitable. Several factors are key for the success of transition processes within a family. Traditional concepts are confronted with digitalization, new technologies, new demand, and social changes. For the chapter at hand, incumbents and heirs of four wine making and Heurigen businesses in the Northern Burgenland were interviewed. Its aim is to investigate the perception of the succession process itself as well as the risk of implementing innovations into traditional concepts. This study shows that innovations are directed by natural circumstances rather than customer demands. The owners rely on traditional concepts with incremental changes to keep customers attracted. Thanks to trust, open communication, and succession processes that spread over long periods of time, neither generation thinks of the handover as problematic.

INTRODUCTION

The tradition of Austrian wine taverns that are run by local winemakers dates to the end of the 8th century. Carl the Great allowed the vintners to offer their products on their own premises for three months of a year, signaling opening hours by a small

DOI: 10.4018/978-1-7998-7352-5.ch007

shrub above the entry. This marks the beginning of the Buschenschank tradition. The word "Heurigen" refers to the fresh wine of the current year on the one hand, and the tavern or restaurant run by the wine producers on the other. Those institutions do not only offer their wine, but also a range of small dishes (Trenker, 2010).

Due to the long history of these taverns, most winemaking families pertain to the traditional business model and refuse to adapt to modern ways of presenting and selling their produce or running a business in general. However, when the successors take over the business, they often feel the urge to modernize the concept their parents, grandparents etc. introduced long ago when taking over the businesses (Letmathe and Hill, 2006). At different stages of the succession and on various levels of producing and selling, the members of the new generation try to implement innovative elements in order to satisfy the demand of present-day customers. Succession processes consist of two parts: the formal and legal act of handing over the business to the successor (ownership succession) and the operational part of the incumbent to transfer the management of the business (leadership succession) (Efendioglu and Muscat, 2009). The latter contains various intricacies in itself (Gbadegeshin, 2013). Adding to the uncertainty of modern business strategies, several factors can lead to disharmonies in family life, where the incumbent might act as a barrier to ideas of the successor (Bigliardi and Dormio, 2009; Letmathe and Hill, 2006).

While young companies acting in a contemporary field are aware of this, other, long-established sectors might not have enough knowledge or the willingness to adapt. It is a debatable point whether the low level of understanding is due to simple unawareness of the high relevance of this topic, or, on a psychological level, the feeling of a threat to their business (Getz et al., 2004), being inextricably connected with their personal raison d'être (Letmathe and Hill, 2006). This raises the question of how the two generations view the succession process and, if there is a clash of opinions, whether the introduction of new sales strategies constitutes risk taking or rather risk anticipation. However, research has found that modernization in some cases is the reason for succession (Letmathe and Hill, 2006) and that in the case of wineries, adding new business areas has led to greater prosperity of the family business (Getz et al., 2004).

Therefore, the research question of this chapter is the following:

Which factors influence the experience of introducing modern business strategies during succession processes in family owned Heurigen in Northern Burgenland?

The main topics of this chapter are succession processes in general, including sentiments in family enterprises and new ways of running a business, implemented by the successor generation. However, a detailed description of the strategies is not part of the goal. Another aim is the elucidation of the successors' rationale behind the decision to modernize a well-established business, respectively the thoughts of the generation handing over. If applicable, interview partners will also be surveyed about the process of them taking over the business from their parents, in order to find out what they themselves think was helpful for successfully handing over the business. As stated, some think of modernization and innovation as anticipation of external threats and risks, whereas others might perceive it as risk taking. The exploration of reasons for both points of view will also be part of the research.

Findings of this research can help in understanding current issues of family owned Heurigen businesses and offer theoretically and empirically reasoned guidance for the sector. Success factors detected in the study can help other family businesses to prepare for the new era of management. Small family businesses in rural areas are seen as accelerators of economic activity in the region, as the interaction with enterprises in the area creates strong synergetic effects (Garstenauer et al., 2012; Germanovich et al., 2020). Thus, the research at hand can have an impact on the development of the Austrian countryside, both for visitors as well as inhabitants and local companies.

BACKGROUND

Ownership and Leadership Succession in Family Businesses

As mentioned, family businesses are an important sector and represent the majority of SMEs worldwide (Bertrand and Schoar, 2006; Bigliardi and Dormio, 2009; Efendioglu and Muscat, 2009; Saiz-Álvarez et al., 2019). Despite the lack of thorough investigation about this sector (Saiz-Álvarez et al., 2019) or a common definition, similarities can be found in various approaches (Bigliardi and Dormio, 2009). They can be defined as mostly small companies that are owned and/or run by one or more members of a family (Bertrand and Schoar, 2006; Bigliardi and Dormio, 2009). According to Bigliardi and Dormio (2009), a family business has been through at least one generational change. Efendioglu and Muscat (2009) divide them into three groups according to the number of family members involved: companies, in which several members of the owner family work; companies, in which only one person of the owner family works and companies, in which none of the owner family members work. Although small family businesses have several advantages because of their size, all of them share one major risk in the course of their existence: succession.

Succession is defined as the transition of the company from one generation to the next (Bigliardi and Dormio, 2009) and is therefore also known as generational change (Rodenstock, 2015). Even though family businesses can also be handed over to external managers, this chapter is focusing on intra-succession, meaning that one or more relatives of the owner family take over the company (Gbadegeshin, 2013). Succession includes not only the legal act, but also a shift in management and control (Efendioglu and Muscat, 2009; Rodenstock, 2015) and therefore economical, psychological, socio-psychological and socio-symbolic challenges, i.e. including: hard as well as soft factors (Gbadegeshin, 2013). The term ownership succession describes the bureaucratic and legal act of inheritance of the businesses' rights and duties from the incumbent to the heir, whereas leadership or management succession means the process of transitioning responsibilities and authority to the next generation (Efendioglu and Muscat, 2009). Ownership succession is therefore accomplished by a legal act and happens in a specified moment in time. Leadership succession is achieved after a process over a longer period of time, with several critical phases that can pose a risk to the family business (Allan and Bund, 2017; Efendioglu and Muscat, 2009; Gbadegeshin, 2013; Palliam et al., 2011; Zehrer and Leiß, 2019). Some studies find succession as the biggest danger to family businesses and several studies found that roughly 30% survive the first and only 15% would successfully carry out a second generational change (Bigliardi and Dormio, 2009; Efendioglu and Muscat, 2009). However, generational change is inevitable (Bigliardi and Dormio, 2009) and therefore needs detailed planning and elaborate management decisions (Bigliardi and Dormio, 2009; Gbadegeshin, 2013).

Both types of succession can occur at the same time (Barach and Ganitsky, 1995), with the legal act happening some time during the management handover.

Handler (1991) defines three phases:

1. **Phase one:** the successor is educated and gains experience in the sector
2. **Phase two:** the incumbent introduces the heir to the responsibilities, tasks and leadership
3. **Phase three:** leadership succession is completed and the senior generation does not intervene in the business anymore.

Although Bigliardi and Dormio (2009) define five instead of three phases, the first three of them are similar. As for the second phase, they highlight the importance of the heir working side-by-side with the incumbent over a longer period, in order to get to know and fully understand each section of the business (Bigliardi and Dormio, 2009).

However, several problems can occur in the course of leadership succession and various factors, internal as well as external, influence the outcome of the process. Regarding internal factors, it is mostly psychological and social challenges a family business has to face. In many cases, the incumbent has difficulties handing over the business (Bigliardi and Dormio, 2009; Letmathe and Hill, 2006). Reasons for this can be, among others, a lack of hobbies and interests other than the business, the physical ability to still work in the business, identification with the business, fear of ageing and death or the importance of power and control for his or her personality (Letmathe and Hill, 2006). Especially the fear of handing over leads to the postponement of the succession (Bigliardi and Dormio, 2009) until relationships are strained, which does not usually provide for a good start for the succeeding generation (Gbadegeshin, 2013). A certain level of motivation to really devolve responsibilities to the heir is necessary (Gbadegeshin, 2013; González-Macías and Zizaldra-Hernández, 2016; Peters, 2005). Furthermore, the inheritor needs to show motivation and readiness, too (Gbadegeshin, 2013; Peters, 2005). On the heir's side, Gbadegeshin (2013) defines three kinds of psychological challenges:

1. **Conservative:** the heir wants to stick to the traditions and values of the business, which impedes innovative ideas and can lead to a decrease in sales
2. **Rebellious:** the heir wants to change the entire concept and eliminates the values of the former generation
3. **Wavering:** the heir wants to improve and change parts of the business in order to gain authority, but does not want to damage traditions and values, his or her incumbent has built

This can lead to a number of counterproductive behavioral patterns: indecisiveness, stagnation, inconsistency, risk aversion, weak corporate culture, focus on tradition, an entire set of new values or revolutionary changes (Gbadegeshin, 2013).

The outcome should be a stable company and harmonious family relationships (Handler, 1991). Blumentritt et al. (2012) define a harmonious transition as one, where all the family members involved agree with the choice of successor and are content with him or her leading the business. He also states that this is rather an exception than the norm (Blumentritt et al., 2013). Bertrand and Schoar (2006), Gbadegeshin (2013), González-Macías and Zizaldra-Hernández (2016) and Tobisch (2018) confirm and add that the family working relationships are essential to success or failure of the transition. Peters (2005) also mentions the relevance of approval of other family members and highlights the importance of congruent visions for the business. Allan and Bund (2017) also talk about the allocation of equity among the siblings and the personal effort that has to be remunerated, too.

The following soft factors have been abundantly discussed as success factors for leadership transition in family businesses: trust (Zehrer and Leiß, 2020), open communication (Allan and Bund, 2017; Muskat and Zehrer, 2017; Tobisch, 2018; Zehrer and Leiß, 2020) and discussions (Zehrer and Halswanter, 2010), succession planning (Bigliardi and Dormio, 2009; Efendioglu and Muscat, 2009; González-Macías and Zizaldra-Hernández, 2016), allowing a long period for the process (Allan and Bund, 2017; Gbadegeshin, 2013), thorough preparation of the heir (Efendioglu and Muscat, 2009), rational selection of the successor (Efendioglu and Muscat, 2009), gradual decrease until final termination of input from the incumbent (Efendioglu and Muscat, 2009), rules for the succession (Bigliardi and Dormio, 2009) as well as personality and the heir's expertise (Letmathe and Hill, 2006).

New Demand, Innovation and Risk Management in Wine Businesses

The importance of innovation in tourism has been thoroughly discussed in literature and research. External circumstances and the dynamics of social transition lead to the necessity of permanent, but not radical, change (Zehrer and Muskat, 2016). Especially globalization, new technologies, social change, market saturation and changes in the mentality of tourists put pressure on tourism businesses to innovate (Müller, 2006). Customers' demands are changing and are becoming more and more important for the attraction of new customers, retention of regulars and, in due course, survival of the business (Zehrer and Muskat, 2016). Even though risk management can contribute positively to the innovativeness of companies, most small family businesses cannot afford it (Glowka et al., 2020). Instead of proactively looking for ways to innovate their offers, SMEs most often react to changes after their realization (Falkner and Hiebl, 2015).

According to Schumpeter (1965), there are five types of innovation: product innovation, process innovation, innovation in the utilization of new markets, innovation regarding new resources or materials and innovation in market structures. For this chapter, mainly product and process innovations are of relevance. Attempts at innovation and their output can benefit considerably from the integration of customers into the process (Zehrer et al., 2009; Zehrer and Muskat, 2016). However, innovation has to be distinguished from invention, diffusion and imitation. Invention is simply the idea of something new and can become an innovation, once it is realized or implemented. If other companies start offering the same, diffusion or imitation starts (Peters and Pikkemaat, 2006).

Family-Owned Wine Businesses in Northern Burgenland

The tradition of serving wine in small taverns dates back to the Roman era, when the growing of wine started to spread out in the Danube area. After several intermediate steps and with various forms of wine taverns starting in the 8th century, the legal basis for Heurigen and Buschenschank was finally established in 1784 by Emperor Joseph II. Their existence played a major role in the development of regional society, as both workers as well as well-situated people got together there (Trenker, 2010). The term Buschenschank originates from the tradition of putting up a shrub ("Buschen") above the entry of the serving premises, so that customers would know they are open. The term "Heuriger" also names the first, young wine of the year (Trenker, 2010). In general understanding, there is a difference between a Heurigen and a Buschenschank, namely that a Heurigen would be open all year round and is allowed to serve cold as well as hot foods, whereas a Buschenschank only opens several times a year for short periods and offers cold dishes only (NÖ Energie- und Umweltagentur, 2019; ORF, 2020).

However, law does not mention the term Heurigen. For Buschenschank, there are separate laws for each Austrian province, with the first one issued in 1929 (Stanzer, 2012). The current law in Burgenland was established in 1979 and underwent several changes, the last one in 2019, allowing them to open for up to nine months per year, but only six months consecutively (§7 Buschenschankgesetz, 1979). It states that owners or tenants of vineyards are allowed to sell their own wine on their premises. Later on, more and more other products were added to the list of products allowed to sell (§7 Buschenschankgesetz, 1979). To run a Buschenschank, it is not necessary to get a trade licence but it will change into a restaurant once warm dishes are offered (Wirtschaftskammer Burgenland, 2019). A Heurigen on the other side, is by this definition a form of restaurant, leading to the necessity of a trade licence. Heurigen also mostly sell their own wine, but not exclusively. For this chapter, both the situations in Buschenschenken and Heurigen will be investigated.

EMPIRICAL APPROACH

Categories of Research

Based on the literature findings presented, several main issues will be investigated. One main part is about the succession processes and the current status of ownership and leadership succession in the firm. Especially the emotional and sociological aspects are of interest in this area, as the incumbent's emotional challenge in handing over (Bigliardi and Dormio, 2009; Letmathe and Hill, 2006) as well as the relationship and possible conflicts among other family members (Bertrand and Schoar, 2006; Gbadegeshin, 2013; González-Macías and Zizaldra-Hernández, 2016; Tobisch, 2018) can dramatically influence the outcome of the process. As several authors (Allan and Bund, 2017; Gbadegeshin, 2013; Muskat and Zehrer, 2017; Zehrer and Leiß, 2020) mention, time, communication and trust are important success factors, hence those topics will be addressed during the interviews. Furthermore, decision-making processes can be indicators for the degree of responsibility already handed over to the heir's generation. As both process and product innovations are relevant to the tourism sector nowadays, the interview partners will be questioned about their innovative ideas, the origins of those ideas as well as their implementation. Emphasis will be placed on the customers' new demands and how business owners react to them. As digitalization, new technologies and social changes are some of the most important innovation drivers in tourism (Müller, 2006), incumbents and heirs will be asked about their approach to those issues.

Therefore, the following categories for empirical research can be deducted from literature findings:

1. general information about the company,
2. family members involved in business,
3. the succession process itself as well as
4. its perception by both generations,
5. communication processes within the business,
6. new concepts and the origin of ideas for those,
7. decision making processes,
8. new demands on the customer side,
9. innovation and the perception of risk.

Data Selection

The empirical design consists of a qualitative, cross-sectional research. Based on the theoretical findings and categories presented, a semi-structured interview guide was established. This chapter focuses on family-owned wine production and Heurigen or Buschenschank companies of the Northern Burgenland, which consists of two official wine-growing regions: Leithaberg and Neusiedler See (Österreich Wein, 2020). Local producers in this area show a variety of very traditional businesses, run by the entire family, young producers that have recently established their wineries themselves, and family businesses that already went through the succession process. In order to get insights into various stages of the succession process and its difficulties, four family businesses have been chosen for the sample. Three of them have recently handed over the business to the next generation or are planning to do so in the upcoming years. In those families, both generations have been interviewed separately from each other. However, they were asked about the same topics, so that a distinction between the perceptions of the two generations is possible. Three of them run a wine production business as well as a Heurigen or Buschenschank. The fourth business is a special case, as the older generation passed away one year ago, so succession was finally completed already, and the third generation is yet too young, to plan further succession. This family also owns a wine production business as well as a Heurigen. However, the son and, until last year, the father run the wine business and the daughter runs the Heurigen together with the mother. Although they support each other, they are not legally connected to each other and are not participating in the management of the other area respectively. All of the family businesses investigated can be defined as small or micro businesses, as one to five family members are running the enterprises, together with up to ten employees.

Table 1 gives an overview of the interview partners and the synonyms which will be used in the course of the discussion.

Table 1. Interview partners and synonyms

Company Number	Business Areas	Winegrowing Region	Generation	Synonym	Date of Legal Succession
1	Winery and Heurigen	Leithaberg	Senior	IP1 (Sen.)	2015
1	Winery and Heurigen	Leithaberg	Junior	IP2 (Jun.)	2015
2	Winery and Heurigen	Leithaberg	Junior	IP3 (Jun.)	Planned, approx. 2025
2	Winery and Heurigen	Leithaberg	Senior	IP4 (Sen.)	Planned, approx. 2025
3	Winery and Buschenschank	Neusiedler See	Senior	IP5 (Sen.)	Planned for 2022
3	Winery and Buschenschank	Neusiedler See	Junior	IP6 (Jun.)	Planned for 2022
4	Winery	Leithaberg	Junior*	IP7 (Jun.)	2007

Source: own illustration; * Senior deceased

Data Collection

Due to the current pandemic situation, the interviews were conducted and recorded via telephone in November 2020. The interviews lasted between 15 to 30 minutes each. The recordings were transcribed and analyzed using the qualitative content analysis after Mayring (2015). Statements of the interview partners were paraphrased and analyzed. In order to detect the main topics, the nine categories mentioned above were used to analyze the statements. For each aspect of the study, the opinions of the two generations were compared, so the differences could be detected. This method allows a systemic approach and transparent analysis of collected data.

Findings

The chapter at hand will present the findings of the study, divided by the research topics at hand. It will start with an overview of the companies and how they are run. Further, it will show how family members communicate and make decisions within the business and how new demands and innovation are perceived regarding the level of risk they imply. Finally, it will discuss the various succession processes and strategies the companies applied and the (varying) perception of the two generations of this process.

General Information About the Company and Family Members Included in the Business

First, the interview partners were asked for some information about their company and how it was established. The first company was established in 1991 with a focus on wine production and in 1998 the Heurigen was opened (IP1 2020, p. 1, l. 8; IP2 2020, p. 1, ll. 4-6). The junior generation started to work in the family business in 2001 (IP1 2020, p.1, l. 16-17; IP2 2020, p. 1, ll. 8-9) and formal succession took place in 2016 (IP1 2020, p. 1, ll. 10-12). By now, the father only helps in the vineyards, whereas the mother still works in the kitchen several hours per week, although IP2 (Jun.) states that he does not want to continue like this in 2021 (IP1 2020, p. 1, ll. 20-22; IP2 2020, p. 1, ll. 16-18). In this case, the brother did not want to take over the business, although he did support the family business, especially in the field of digitalization (IP1 2020, p. 1, ll. 25-26).

The second company was started in the mid 20th century so that the third generation is already working there now and planning to take over the business in the upcoming years. This business formerly included various kinds of agriculture. However, the current senior generation changed to winemaking only and opened the Heurigen. (IP3 2020, p. 1, ll. 6-9, 12; IP4 2020, p. 1, ll. 5-8, 11-19) By now, both sons work in the business and it has not been discussed yet, what succession will look like exactly (IP3 2020, p. 1, l. 14, 16, 18, 21; IP4 2020, p. 2, ll. 40-50). Even though ownership succession from the first to the second generation took place in 1998, the transition of all responsibilities and competencies took about 25 years and the first generation is still somehow involved (IP3 2020, p. 1, ll. 22-28). IP4 (Sen.) states that this was a good way to make sure he knows how to run the business properly and is planning the same for the succession to his sons. He wants to legally hand over the business before turning 60, meaning in the next five years but he will still help them and work with them after this date (IP4 2020, p. 2, ll. 40-50, 53-54, 57- 63).

The third company also used to consist of different agricultural departments. In this case it was two separate businesses which were combined after the marriage of the current senior generation. The business as it is today has existed since 2000 (IP5 2020, p. 1, ll. 6-13). The son has always been helping out and has been responsible for most areas for a couple of years already, although ownership succession hast not yet taken place (IP6 2020, p. 1, ll. 8-10, 20-21). The senior generation still works in the family business full time. Legal succession is planned for 2022. However, IP5 (Sen.) does plan to further support his son for a couple of years after succession, as for them it is mainly a bureaucratic act (IP5 2020, p. 1, ll. 16-20, 29-30; IP6 2020, p. 1, ll. 27-31).

The last company only consists of the junior generation by now as the father and founder of the business passed away in 2019 (IP7 2020, p. 2, ll. 39-44). The company used to own various agricultural fields since the 1960s, but the current generation changed it into a winery and sold the other parts (IP7 2020, p. 1, ll. 6-18). In this case the family also owns a Buschenschank. However, the Buschenschank and the winery are legally separated and the Buschenschank is run by the sister and the mother of the owner of the winery, who started it individually and independently over 20 years ago. They support each other, and the mother helps in the vineyards occasionally, but they do not intervene in the business of one another. (IP7 2020, p. 1, ll. 21-34, p. 2, 39-44). IP7 (Jun.) started to work in the winery in 1995 and took over in 2007 (IP7 2020, p. 2, ll. 21-34). As he benefited from professional education in viticulture, it has always been evident that he would take over the business (IP7 2020, p. 2, l. 48). Today, his wife also works in the family business and their children might want to take over in the far future (IP7 2020, p. 3, ll. 78-79).

Communication and Decision-Making Processes Within the Business

Most of the interview partners have stated that communication processes are happening alongside daily business (IP2 2020, p. 2, l. 36; IP3 2020, p. 1, ll. 27-29; IP7 2020, p. 3, ll. 67-71). They confirm that due to the fact that most of them work together several hours a day (IP7 2020, p. 3, ll. 67-71) and usually have lunch together (IP7 2020, p. 3, ll. 67-71) they always find the time to discuss current problems. In the family of company number two they also talk about business at the dinner table or in urgent cases call each other (IP3 2020, p. 1, ll. 27-29). In general, none of the families and their members said that they would have certain meetings for further discussions about procedures and changes in the business.

A central topic in family businesses where succession has either happened or is planned is the question of decision making. In the first company, ownership succession has already taken place and therefore, the junior generation already takes all of the decisions (IP1 2020, p. 1, l. 31; IP2 2020, pp. 1-2, ll. 32-34). However, this only changed after the bureaucratic process was done. Regarding decisions before that, the generations seem to have different opinions about it. IP1 (Sen.) (2020, p. 2, l. 48) states that he was the one deciding upon further procedures. However, IP2 (Jun.) describes it as a joint decision-making process (IP2 2020, pp. 1-2, ll. 32-34) or that his parents were easy to convince of his new ideas most of the time (IP2 2020, p. 1, ll. 24-27). In the second company not only the current junior and senior generation, but also the very first generation takes minor part in the decision process (IP3 2020, pp. 2-3, ll. 63-65, p. 3, l. 67). IP3 (Jun.) tells the interviewer that it is not about the person, but the best arguments (IP3 2020, pp. 2-3, ll. 63-65) and that they will always work together on new ideas, as it would not be possible for a single person to implement them (IP3 2020, p. 3, l. 70). IP7 (Jun.) (2020, p. 5, ll. 143-147) recounts that although he was the one making the final decisions, he always tried to convince his father about his ideas, so that he would not be opposed to them and would help him with subsequent procedures. Furthermore, it was publicly communicated that they would lead the company together (IP7 2020, pp. 5, ll. 148-149), so it was also an issue of the image they wanted to maintain in front of the customers. However, he admits that it was never a problem to convince his father (IP7 2020, pp. 6, ll. 155-160). IP5 (Sen.) (2020, p. 2 l. 64) adds that he does not want to participate in decisions regarding the field of digitalization, marketing and the online shop, as he is not experienced and fully trusts his son in this respect. In general, he states that if somebody has an idea, they first have a closer look at it, come up with a concept and check feasibility and appropriateness, before they decide on whether to implement it or not (IP5 2020, p. 2, ll. 41-51).

New Demand, Innovation and Risk Perception

During the interviews it became evident that the interview partners do not really separate incremental changes from major innovations or the origin of innovative ideas. For them, they partly arise due to the customers' demands, their own ideas, things they see in other businesses or, most frequently, from what nature offers them. Therefore, those categories, together with the perception of the risk that comes with changing a traditional business, will be analyzed together in the chapter at hand.

The interview partners were questioned how ideas for changes and new concepts in the business emerged and whether it was the younger or the older generation that more often came up with such suggestions. In the first company, both generations said it was mainly the son who was bringing new ideas into the business (IP1 2020, p. 2, ll. 51-52; IP1 2020, p. 1, ll. 28-29). Although the older son has never really been involved in the business, it was him who first suggested and then also installed the first website and online shop (IP1 2020, p. 2, ll. 59-60; IP2 2020, p. 3, ll. 80-82, 85-87). The areas that underwent changes were mainly marketing and design, digitalization, wine production techniques, the product range and price policy in the winery as well as the food concept at the Heurigen. In order to be able to produce new assortments of wine of a higher quality and as a consequence sell it at a higher price, a new production hall was built and new machines were bought (IP2 2020, p. 2, ll. 39-44). This was a significant change and represented a major investment (IP2 2020, p. 2, ll. 39-44, p. 4, l. 105) as well as the first big step of the junior generation, only one year after legal succession, in order to ensure the success of the business in the long term (IP2 2020, p. 4, ll. 111-121). Together with other marketing campaigns, especially regarding design and merchandise (IP2 2020, p. 2, ll. 39-44), this also has a marketing impact, as they are now able to show guests around (IP2 2020, p. 4, ll. 124-126). They also started to actively sell their wine to partners, e.g. hotels and restaurants or even shops (IP1 2020, p. 2, l. 63) and even export some of it to other European countries (IP2 2020, p. 3, ll. 89-91). Although it was mostly by chance that those exporting partnerships emerged, with networks being the most important factor in this field, they have to put a lot of effort into those relationships in order to keep them (IP2 2020, p. 3, ll. 94-96, 98-102). Regarding the Heurigen, there were mainly alterations in interior design and food options (IP2 2020, p. 2, ll. 46-50). The ideas mostly arise at work during the season in summer but are implemented during the winter months, when they are closed (IP2 2020, p. 2, ll. 53-55). Both generations observed a change of the regulars since the son took over. However, they say it is due to natural circumstances, as the older generations are not going out anymore and younger people start visiting the Heurigen (IP1 2020, p. 3, ll. 66-68; IP2 2020, p. 2, ll. 56-58, p. 2-3, ll. 60-65). They still host people of all ages. According to IP1 (Sen.), demand has also changed, but they are not primarily focused on instantly responding to all of their guests' wishes (IP1 2020, p. 3, ll. 71-73, 75). Furthermore, IP2 (Jun.) states that he cannot really tell whether it is the new customers that made him offer new products, or if it is the new products that attract new customers and says he perceives it more as a bidirectional process (IP2 2020, p. 3, ll. 69-75). When it comes to risk perception, the father does not really think of those amendments as a risk, since in most cases no bigger investments are necessary and they would notice quite quickly if the new products did not sell (IP1 2020, p. 3, l. 79). The second generation, however, perceives all of those alterations as risks they are taking, as any change they make might possibly cause customers to stop visiting them (IP2 2020, p. 4, l. 108).

In the second company, the situation is similar. They also implemented an online shop, changed some of the designs, food options, product range and technique in wine production, with the two sons being the main initiators of those alterations (IP3 2020, p. 2, ll. 32-34, 37-38). They either came up with the ideas for those changes themselves or saw them in other, similar but partly more innovative, businesses and then adapted those ideas fully or in parts into their own business (IP3 2020, p. 2, ll. 42-45). Especially in the field of digitalization, the father does not intervene because he does not know enough about this topic and trusts his sons as, in his opinion, they know how to handle investments (IP4 2020, p. 3, ll. 67-69, 73-77). They only recently received a certificate of sustainability as a result of their efforts to develop a more future-oriented business model and are looking for more ways to improve their use of resources (IP3 2020, p. 2, ll. 56-59). Regarding the Heurigen and the new demand in this business field, IP4 (Sen.) was glad that his son started working with them, as he considers him to know best how to satisfy the customers at the Heurigen (IP4 2020, pp. 3-4, ll. 83-96). Also, he states that it is not really possible or economically feasible to put high investments or make drastic changes in a business that only operates about 50 days per year (IP4 2020, pp. 3-4, ll. 83-96). Furthermore, IP4 (Sen.) does not want to invest in a business he will not be responsible for anymore in about five years, since his sons might not like his changes and start undoing them later (IP4 2020, p. 4, ll. 100-111). He nonetheless appreciates his sons' input and supports them in their efforts to improve or invest in the business once they have taken over, as the Heurigen will then be about 50 years old and will need some new ideas in order to be successful in the long run (IP4 2020, p. 4, ll. 114-122). As most of the changes so far have been minor improvements that did not require big investments for implementation, they are not perceived as a high risk (IP3 2020, p. 2, ll. 48, 51). Although they do not specifically identify it as such, they do in fact have a system of risk management. They continuously monitor sales and customer behavior and analyze it periodically (IP3 2020, p. 3, ll. 82-84). In doing so, they always notice if new products sell or customers prefer other products (IP3 2020, p. 3, ll. 74-79). Furthermore, they also notice that every couple of years the number of customers decreases and sales slow down unless they offer something new (IP3 2020, p. 3, ll. 74-79). For this reason, they are constantly changing minor things and offer new products, while simultaneously keeping the traditional offers (IP3 2020, p. 3, ll. 90-93, p. 4, ll. 96-98). According to IP3 (Jun.), it is essential to always keep the business in motion, just as their competitors do, but says that he does not consider it necessary to constantly grow in size (IP3 2020, p. 4, ll. 106-112). IP4 (Sen.) also noticed that even regular customers are appreciative of new offers, too, although there is only up to 15% new demand and it is still mainly the traditional products that are asked for (IP4 2020, p. 4, ll. 100-111).

The third company looks for new ideas both in other businesses (IP6 2020, p. 2, ll. 36-43) as well as actively looking for them in guides or similar informational materials, but mostly decides based on their natural resources (IP5 2020, p. 2, ll. 54-60). IP5 (Sen.) used to be part of an initiative that was established years ago, named "Eating healthy at Heurigen", for the Burgenland. However, this club only existed for about 10 years, as the number of Buschenschenken decreased and the remaining ones were not able to agree on standards anymore. Especially the differences in opinions between people in the northern versus the southern Burgenland were too big (IP5 2020, p. 3, ll. 79-87, pp. 3-4, ll. 90-100). Nevertheless, this was part of the movement aiming at the satisfaction of customers' different food preferences (IP5 2020, p. 3, ll. 79-87). IP5 (Sen.) has also observed a change of the regulars but does not relate it to current events or his son joining the business, but rather describes it as general fluctuations about every five to six years (IP5 2020, p. 3, ll. 67-75). In his opinion, this is due to natural circumstances, as most people tend to change their taste or try new products from time to time – be it new varieties of wine or a new winery in general (IP5 2020, p. 3, ll. 67-75). He considers this to be the main reason that they did not observe any radical differences regarding their customer segment (IP6 2020, p. 3, ll. 70-71). Similar to the second company, they are trying to keep an eye on sales and transformations of customer demands, so that they can better evaluate the success of new products or the traditional products that have to be eliminated (IP5 2020, p. 4, ll. 106-116). For him it is essential to stick to the traditional concept, but to always make minor changes so that customers stay interested in the products they offer (IP5 2020, p. 4, ll. 106-116). Nonetheless, major investments were made with the goal of bringing their winemaking technology up to current standards, which he considered necessary as well as opening up a function room (IP5 2020, p. 4, ll. 121-126). In addition to this, his son believes that they do not have to make bigger changes in the business concept itself but is adjusting the business strategy by trying to spread out from only catering private customers to also contacting hotels, restaurants and wine stores in order to sell their wine there (IP6 2020, p. 2, ll. 36-43). In his view, the main factor when deciding on a new product is their conviction of its quality (IP6 2020, p. 2, ll. 52-59). He first surveys interesting new products and then evaluates whether it would be feasible and suitable for their own business, economically as well as personally and regarding their values (IP6 2020, p. 2, ll. 52-59). Other than the assortment of wines, most of the changes he has realized were in the area of design and marketing, where it is essential to show the personality of the family and its business (IP6 2020, p. 2, ll. 61-63, 65). In general, they did not implement any major changes or invest higher amounts of money in the recent years (IP6 2020, p. 3, ll. 75-82) and currently have no plans to do so in the upcoming years (IP6 2020, p. 3, ll. 84-86). IP6 (Jun.) states that he does not want to grow in numbers and size anymore, as he wants to be able to always keep an eye on all of the processes himself (IP6 2020, p. 3, ll. 75-82). As they do not make any major changes to the concept, he also does not perceive the changes they have realized as risks, but rather as a way to attract a broader range of customers (IP6 2020, pp. 3-4, ll. 92-96).

In the last company investigated, the junior started with significant alterations in the wine assortment, right after coming back from his education (IP7 2020, pp. 4-5, ll. 102-141). When taking over business in 2000, they invested in new winemaking technologies, which he considered necessary, but says that his father would not have done that without him (IP7 2020, p. 4, ll. 102-121). Another big step was the start of the online shop, which his father did not take seriously (IP7 2020, p. 5, ll. 120-125). Only after he saw the amount of wine being sold without anyone having visited the winery, he realized the importance of the online shop (IP7 2020, p. 5, ll. 130-141). IP7 (Jun.) claims that nature has the highest importance when deciding about the assortment of wines (IP7 2020, p. 6, ll. 181-190). Although he sometimes responds to the demands of customers, he usually does not follow the hype of certain varieties of produce (IP7 2020, p. 6, ll. 164-180). Furthermore, he is not willing to produce any variety he does not enjoy himself (IP7 2020, pp. 7-8, ll. 203-224). He says it took him some years to understand that you do not have to respond to all requests and that he became more relaxed over time (IP7 2020, pp. 7-8, ll. 203-224). He always tries to keep the customers interested by offering new varieties or producing limited editions (IP7 2020, pp. 7-8, ll. 203-224). Other than the investment back in 2000, he does not consider his new varieties or changed assortment to be a high risk, as the worst that can happen is that the wine does not sell as fast as others would (IP7 2020, p. 8, ll. 230-246). He mentions that although he sells his wine to various countries via his online shop, he is not interested in starting to export his wine in a larger scale (IP7 2020, p. 8, ll. 249-255). He also says that he is sometimes asked for higher amounts of wine but can neither currently cater to those wishes nor has any plans to do so in the future and would rather stick to the amounts he is currently producing, keeping the company the same size (IP7 2020, p. 9, ll. 257-261). Similar to other interview partners, he did not observe an extreme shift in customer segments, but only the natural changes due to the average age of the customers (IP7 2020, p. 9, ll. 265-271). However, he does note that young people more commonly order online whereas the elder ones that still come to visit his winery in Burgenland tend to buy their wine at the yard sale (IP7 2020, p. 9, ll. 271-281).

Succession Processes and Their (Varying) Perception of the Two Generations

In the first company, succession already took place in 2015, after the implementation of the cash register (IP1 2020, p. 2, ll. 39-43). Currently the senior supports and offers advice, but does not intervene unless asked to do so (IP1 2020, p. 2, ll. 33-36). For IP1 (Sen.) it was not a problem to hand over, as it was clear that it would happen eventually and he actually believes that it was already due (IP1 2020, p. 3, ll. 85-86, 88). Only in 2012 it was obvious for the junior that he and his wife would take over the business (IP2 2020, p. 1, ll. 20-21). He also believes that it was the right time for his parents to stop working in the business, as for example, they would not have been able to handle the cash register or any other new technology (IP2 2020, p. 5, ll. 139-144). For him, the main challenge of the succession was not to gain his father's respect and trust regarding the management of the business, but the challenge of running the daily business, especially when it came to finance (IP2 2020, p. 5, ll. 130-135).

The second company has planned the succession. However, there is no specific date and details are not yet set. Both generations cannot think of any psychological or personal problems when it comes to the succession (IP3 2020, p. 4, l. 116; IP4 2020, p. 5, ll. 127-144). IP4 (Sen.) believes that every generation can contribute in their own way to this traditional business, and he is thankful that his sons want to take over and continue running the business his father once established (IP4 2020, p. 5, ll. 127-144). He therefore is not planning to interfere in their decisions, once the legal succession has taken place, but does want to support them for several years after that (IP4 2020, p. 5, ll. 127-144).

In the third family, the daughter visited a school for winemaking. However, she does not want to take over, but the son will (IP5 2020, p. 2, ll. 33-36). Succession is planned for 2022, although the father plans on working in the business until about 2025 (IP5 2020, p. 1, ll. 29-30). He himself had problems with his father and father-in-law, when he and his wife took over the businesses, since running the company the way they wanted to lead to friction (IP5 2020, p. 5, ll. 131-141). He therefore does not want to bother his son with things like this and rather wants to show him that he is supposed to enjoy his work and make a comfortable work-life-balance possible (IP5 2020, p. 5, ll. 131-141). Since they initially thought that their son did not want to continue the business, they are very happy that he has eventually decided to do so (IP5 2020, p. 5, ll. 144-149). The son (IP6 2020, p. 1, ll. 13-17) states that there was no specific point in time when he decided to take over, but he rather grew into the business and enjoys it now. He is also confident that his parents will not give him a hard time handing over the business and expects them to let him run it the way he wants to, as he is the decision leader in most cases already now (IP6 2020, p. 4, ll. 103-105, 109-112).

IP7 (Jun.) talks about his late father and how he wanted to influence the daily business. For him, his father's influence was not a problem as he understood that the winery business had been his fathers' job his entire life and therefore, he accepted letting his father decide when it came to working in the vineyards (IP7 2020, p. 2, ll. 53-63). When it came to production, bottling, assortment and other operational aspects however, he did try to convince his father of the decisions he had already made, so it was an implicit division of responsibilities (IP7 2020, p. 2, ll. 53-63). His own kids are still too young to be able to decide whether they want to continue the business, but he is already making an effort to make succession as attractive as possible for them (IP7 2020, pp. 3-4, ll. 78-98). He understands that running this kind of business has become riskier in recent years, not only because of the climate change and its effects on wine growing, but also due to administrative obligations (IP7 2020, pp. 3-4, ll. 78-98).

DISCUSSION

The interviews show a consensus with literature findings in several areas. Ownership succession has already taken part in two out of four businesses. In the fourth business, leadership succession is also completed, primarily due to the decease of the senior generation. In the first company however, ownership succession is finalized, but the incumbents have not yet fully stopped contributing. As in the other two businesses ownership succession is still in the planning phase, leadership succession is in the process, too. In all four enterprises, the three stages of leadership succession according to Handler (1991) can be found, with all characteristics mentioned. Therefore, the second company is still preparing the heirs for the handover of the business, by educating them and letting them gain experience. The third business is already in the second stage, with the successor already obtaining more and more responsibilities and the father stepping back. In all four cases, the senior generations state that the succession process needs to take place over several, up to 25, years in order to guarantee that the successors fully understand all areas of the business.

When it comes to decisions about daily business as well as new ideas, it is noticeable that decision-making power slowly shifts from the incumbent to the heir's generation, with a long period of mutual decisions. During the second stage of succession, decision-making is based on the arguments made by all contributors, whereas in the first phase it is mostly the incumbent and in the last phase mostly the heir that takes the final decision.

Regarding the incumbents' fears or emotional challenges of the succession, all interview partners state that this was not or will not be an issue in the family. One of the incumbents says that he was given a hard time shaping the business the way he wanted by his parents and parents-in-law when he took over and does not want to do the same to his children. All of them say that the legal act of ownership succession will not make a major difference in the daily business at first and that they will continue to support their successors. However, all agree that they will try not to intervene if not asked to. They justify the reason for this intention in the trust they have in their offspring and the good communication within the family and business. They mostly see the challenges in new administrative tasks and environmental change affecting the wine growing and producing sector.

All of them have established an online shop by now and notice differences in the customers that order online and those that come to visit the winery. However, the customers and regulars of the Heurigen or Buschenschank have not changed drastically. Although the average customer segment is becoming younger, all of them believe it is a natural process. Societal change brings with it new demands and they are, at least partly, reacting to this. Nevertheless, it is not the customers demand that decides the innovation of products, but rather natural circumstances. They are aware of customers asking for certain products on a regular basis but are not willing to bend to all of their wishes. Also, some of them look for new ideas in other, similar businesses, but most of the ideas just arise during daily business.

In all of the businesses, innovations were rather incremental than radical, with the biggest investments being the implementation or building of new cellar technologies in the wine production. Other than that, no major investments have been made. Therefore, most of them do not perceive the small changes in assortment or offer as risks they have taken. However, it is rather the junior generations that are concerned about the possible outcomes of their innovations. Two interview partners of separate businesses specifically mentioned the periodical observation of product sales in order to gain in-depth understanding of which products sell and which do not. This helps to decide about which traditional products to eliminate and which new ones to keep. This can help evaluating the risk of future innovations. One business already made an effort to receive a certificate of sustainability, whereas the others mainly see improvements in their marketing, design, processes, products and price policies as their innovations.

SOLUTIONS AND RECOMMENDATIONS

Although it seems that all of the businesses investigated have had or are having a successful succession process, they do not show aspects of succession planning. Also, two out of four businesses do not use any kind of controlling regarding their products and sales. Those factors can have either a positive or negative influence on the overall success of the business and should therefore not be neglected. Efendioglu and Muscat (2009) conducted research in this area with the focus on unexpected life-changing events that resulted in sudden, major changes of successor processes, which most of the family businesses would not be able to survive. Therefore, planning and observation should be implemented in both areas, succession and innovation. The findings showed, that neither the heir's nor the incumbent's generation are strategically planning on how to further improve or develop the companies or parts of it, but rather spontaneously work on problematic areas of the business. Focus lies on the daily business as well as minor improvements of processes or partly broadening the product range. However, major innovations do not seem to be of interest for owners of traditional wine businesses and Heurigen or Buschenschenken. Reason for this mindset is found in the long tradition of the family business and the success in running the company so far. However, possible threats or possibilities this sector might be confronted with in the future are not seen by the owners.

FUTURE RESEARCH DIRECTIONS

As a lot of families struggle with the rational decision of which child will inherit the business, further research could be done on the main factors that guides a senior generation in this decision. Successful examples of co-leadership by siblings in this sector should be investigated, in order to be able to offer guidance to other family businesses. Furthermore, follow-up research of how the businesses investigated continue their succession process and how successful the heirs lead the company might bring additional knowledge about success factors of leadership transition in this business sector.

CONCLUSION

To sum it up, the family businesses investigated show minor incremental but nonetheless steady innovation processes. Due to the low costs of those innovations, most of them do not perceive it as risk taking. As most changes come as a response to the circumstances of the natural environment rather than new demands of existing and potential customers, their product and process innovations are mainly based on their own ideas. None of them is concerned about the emotional and sociological aspects of the succession process. For them, the long years of gradually transferring responsibilities and power from the senior to the junior generation marks the key success factor of the handover. The aspect of keeping the traditions and values of the businesses the first generations established together with the permanent incremental changes of the following generations provide the basis for resilient family enterprises.

The research question was the following:

Which factors influence the experience of introducing modern business strategies during succession processes in family-owned Heurigen in Northern Burgenland?

According to the arguments and research findings at hand, it has to be answered as follows: trust in the incumbent's generation, several years of working side-by-side with a gradual shift of all responsibilities from the senior to the junior generation, steady and incremental rather than radical innovations and the coordination with natural resources rather than satisfaction of new customer demands leads to successful transitions in family-owned Heurigen and winery businesses.

REFERENCES

§7 Buschenschankgesetz. (1979). *Buschenschankgesetz*. https://www.ris.bka.gv.at/GeltendeFassung.wxe?Abfrage=LrBgld&Gesetzesnummer=10000132

Allan, S., & Bund, R. (2017). Passing on a wine business that your kids want to own. *Wine & Viticulture Journal*, *1*, 56.

Barach, J. A., & Ganitsky, J. B. (1995). Successful succession in family business. *Family Business Review*, *8*, 131–155.

Bertrand, M., & Schoar, A. (2006). The role of family in family firms. *The Journal of Economic Perspectives*, *20*(2), 73–96. https://doi.org/10.1257/jep.20.2.73

Bigliardi, B., & Dormio, A. I. (2009). Successful generational change in family business. *Measuring Business Excellence*, *13*(2), 44–50.

Blumentritt, T., Mathews, T., & Marchisio, G. (2013). Game Theory and Family Business Succession: An Introduction. *Family Business Review*, *26*(1), 51–67. https://doi.org/10.1177/0894486512447811

Efendioglu, A. M., & Muscat, E. (2009). Internal disruptions in family business succession: Death, Divorce, and Disability. *The Journal of Industrial Relations and Human Resources*, *11*(1), 17–31.

Falkner, E. M., & Hiebl, M. R. W. (2015). Risk management in SMEs: A systematic review of available evidence. *The Journal of Risk Finance*, *16*(2), 122–144.

Garstenauer, R., Schwarz, U., & Tod, S. (2012). Alles unter einen Hut bringen: Bäuerliche Wirtschaftsstile in zwei Regionen Niederösterreichs 1945-1985. *Historische Anthropologie: Kultur, Gesellschaft. Alltag*, *20*(3), 383–426.

Gbadegeshin, S. A. (2013). Awareness of "Ownership Succession" and Family Business Continuity. *International Journal of Business and Management Invention*, *2*(9), 72–87.

Germanovich, A. G., Vasilieva, O. N., Ordynskaya, M. E., Allanina, L. M., & Gorokhova, A. E. (2020). Impact of Tourism on Sustainable Development of Rural Areas: International Experience. *Journal of Environmental Management and Tourism*, *11*(4), 965–972.

Getz, D., Carlsen, J., & Morrison, A. (2004). Family-owned tourist attractions and wineries. In The family business in tourism and hospitality (pp. 153–160). CABI Publishing.

Glowka, G., Kallmünzer, A., & Zehrer, A. (2020). Enterprise risk management in small and medium family enterprises: the role of family involvement and CEO tenure. *International Entrepreneurship and Management Journal*. doi:10.100711365-020-00682-x

González-Macías, C. J., & Zizaldra-Hernández, I. (2016). La unión en empresas familiares restauranteras : Una aproximación a la sucesión generacional en empresas con actividad turística de San Cristóbal de las Casas. *Revista Iberoamericana de Turismo*, *6*, 18–31.

Handler, W. C. (1991). Key interpersonal relationships of the next-generation family members in family firms. *Journal of Small Business Management*, *29*(3), 21–32.

Letmathe, P., & Hill, M. (2006). Strukturumbrüche der Unternehmensnachfolge. *Zeitschrift Für Betriebswirtschaft*, *76*(11), 1113–1138.

Mayring, P. (2015). *Qualitative Inhaltsanalyse- Grundlagen und Techniken* (12th ed.). Beltz.

Müller, H. (2006). Qualitätsmanagement als Triebfeder von Innovationsprozessen und ihre Grenzen. In B. Pikkemaat, M. Peters, & K. Weiermair (Eds.), *Innovationen im Tourismus* (pp. 109–117). Erich Schmidt Verlag.

Muskat, B., & Zehrer, A. (2017). A power perspective on knowledge transfer in internal succession of small family businesses. *Journal of Small Business and Entrepreneurship*, *29*(5), 333–350. https://doi.org/10.1080/08276331.2017.1345208

NÖ Energie- und Umweltagentur. (2019). *Niederösterreichische Heurigenkultur*.

ORF. (2020, May 11). *Buschenschanken bis Jahresende geöffnet*. https://wien.orf.at/stories/3048113/

Österreich Wein. (2020). *Weinbaugebiete Burgenland*. https://www.oesterreichwein.at/unser-wein/weinbaugebiete/burgenland

Palliam, R., Cader, H. A., & Chiemeke, C. (2011). Succession Issues among Family Entrepreneurs in Countries of the Gulf. *International Journal of Business Administration*, *2*(2), 25–34.

Peters, M. (2005). Succession in Tourism Family Business: The motivation of succeeding Family Members. *Tourism Review*, *60*(4), 12–18.

Peters, M., & Pikkemaat, B. (2006). Innovationsforschung im Tourismus: ein Ausblick. In B. Pikkemaat, M. Peters, & K. Weiermair (Eds.), *Innovationen im Tourismus. Wettbewerbsvorteil durch neue Ideen und Angebote* (pp. 303–306). Erich Schmidt Verlag.

Rodenstock, B. (2015). Wie sichere ich Familie, Unternehmen und Eigentum für die Zukunft? In A. Nagl (Ed.), *Wie regele ich meine Nachfolge? Leitfaden für Familienunternehmen* (2nd ed., pp. 37–58). Springer Gabler.

Saiz-Álvarez, J. M., Palma-Ruiz, J. M., & Leitão, J. (2019). Introduction to Entrepreneurship and Family Business Vitality. In J. M. Saiz-Álvarez, J. M. Palma-Ruiz, & J. Leitão (Eds.), *Entrepreneurship and Family Business Vitality. Surviving and Flourishing in the long term* (pp. 1–5). Springer.

Schumpeter, J. A. (1965). *The theory of economic development: An inquiry into profit, credit, interest, and the business cycle*. Oxford University Press.

Stanzer, M. (2012). *Rechtsfragen des Buschenschankes in der Steiermark* (Diploma thesis). Karl-Franzens-Universität Graz.

Tobisch, K. K. (2018). *Generationenwechsel im Familienunternehmen unter dem Gesichtspunkt der wandelnden Organisation (Master thesis)*. Karl-Franzens-Universität Graz.

TrenkerD. (2010). *Heurigenkultur*. https://www.bmlrt.gv.at/land/lebensmittel/trad-lebensmittel/spezialkulturen/heurigenkultur.html

Wirtschaftskammer Burgenland. (2019). *Das freie Gastgewerbe*. Academic Press.

Zehrer, A., & Halswanter, J. (2010). *Management of change in tourism - The problem of family internal succession in family-run tourism SMEs*. Academic Press.

Zehrer, A., & Leiß, G. (2019). Family entrepreneurial resilience – an intergenerational learning approach. *Journal of Family Business Management*. doi:10.1108/JFBM-09-2018-0037

Zehrer, A., & Leiß, G. (2020). Intergenerational communication barriers and pitfalls of business families in transition—A qualitative action research approach. *Corporate Communications, 25*(3), 515–532. https://doi.org/10.1108/CCIJ-03-2020-0056

Zehrer, A., & Muskat, B. (2016). Der Treiber Innovation in touristischen Unternehmen. In H. Siller & A. Zehrer (Eds.), *Entrepreneurship und Tourismus: Unternehmerisches Denken und Erfolgskonzepte aus der Praxis* (2nd ed., pp. 29–46). Linde.

Zehrer, A., Pechlander, H., & Reuter, C. (2009). Innovativeness in tourism: The perception of innovation awards participants. *Journal of Tourism*, *14*(1), 11–29.

KEY TERMS AND DEFINITIONS

Austrian Winegrowing Region: Austria is divided into several winegrowing regions with their own specialties and certificates.

Buschenschank: A Buschenschank is legally restrained and only allowed to sell the wine of own production and cold foods. Buschenschenken may only be opened during nine months of the year, of which only six may be consecutive.

Heurigen: The term "Heurigen" in this context has two different meanings. It can either be the fresh, new wine produced in the current year, available soon after the grape harvest, or the tavern of the company, where they sell their wine and warm dishes.

Imitation: An invention that is copied by competitors.

Innovation: An invention that is realized.

Invention: A new idea that represents the solution to a certain problem.

Leadership Succession: The process of transferring all responsibilities, power, and authorities from one person to another, in this chapter it always regards two generations of a family.

Ownership Succession: The legal act of transferring a business from one person to another, in this chapter it always regards two generations of a family.

Chapter 8
Crisis Favors the Prepared Firm:
How Organizational Ambidexterity Relates to Perceptions of Organizational Resilience

Laura Bechthold
Zeppelin University, Germany

Maximilian Lude
Technical University Munich, Germany

Reinhard Prügl
Zeppelin University, Germany

ABSTRACT

This chapter examines the interplay of the fundamental constructs of perceived organizational resilience and organizational ambidexterity by providing insights from a case study of an integrated service company in the tourism and leisure industry located in the Swiss Alps. The authors theorize that organizational ambidexterity (i.e., the balance between adaptability and alignment) leads to higher levels of resilience by boosting the capability of the firm to mitigate and adapt to major disruptions in its environment. Data from a survey with non-family employees show that establishing structures of alignment and adaptability and imprinting an "and/or"-mindset into the company culture has enabled all employees to constantly navigate new challenges while staying aligned on priorities throughout the hardships of the COVID-19 pandemic. However, the positive effects of alignment and adaptability can only unfold if they are combined with clear communication and transparency by the company's leadership.

DOI: 10.4018/978-1-7998-7352-5.ch008

INTRODUCTION

The nature, frequency and impact of global risks such as exponentially spreading pandemics have tremendously changed during the past decades. Next to an unsettled geopolitical landscape and cyber-threats, global risks associated with climate change and pandemics pose the greatest danger to people and the planet today, both in terms of likelihood as well as impact (World Economic Forum, 2019). With rising levels of environmental complexity and uncertainty, however, the question of how to achieve organizational resilience as well as understanding the role of ambidexterity (i.e., the balance between adaptability and alignment) becomes increasingly important. Especially the tourism industry as a climate-sensitive and innovation-pressurized industry faces a conundrum of strategic complexity (Martínez-Ros and Orfila-Sintes, 2012).

The current situation illustrates this complexity very well: On the one hand, the recent COVID-19 pandemic and related travel bans have put entire business models at stake and require quick reactions to ensure short-term business survival. On the other hand, long-term sustainability risks, such as climate change, become increasingly tangible and urge the industry to protect, support, and enhance the natural resources that will be needed for sustaining the firm in the long-term future. Hence, developing organizational resilience, i.e., adaptive capabilities to cope with adversity and external shocks, has become a prerequisite for businesses to ensure short-term survival and future viability. A literature review by Annarelli and Nonino (2016) shows, however, that research dealing with the question of how to reach organizational resilience remains scant. In particular, there is a lack of research that explores how small and medium-sized companies (Doern, Williams, and Vorley, 2019; Morgan, Anokhin, Ofstein, and Friske, 2020; Williams, Gruber, Sutcliffe, Shepherd, and Zhao, 2017; Youssef and Luthans, 2007) as well as family-controlled businesses (for a recent exception see Kammerlander, Patzelt, Behrens, and Röhm, 2020) can build organizational resilience and ambidexterity. The term ambidexterity thereby refers to the capacity to exploit and align current business operations while simultaneously enabling the organization as well as the employees to adapt to changes in the environment (Gibson and Birkinshaw, 2004). Investigating organizational ambidexterity in the family firm context is of particular interest because of the unique ownership and management constellations as well as the family involvement in the firms' top management, both affecting ambidextrous behavior (Hiebl, 2015; Kammerlander et al., 2020; Stubner et al., 2012). For example, family firms are unique due to the relevance of non-economic goals like transgenerational intentions (Kammerlander and Ganter, 2015) and idiosyncratic structures (Duran et al., 2016), which underlines the relevance of furthering our understanding of ambidextrous behavior in that particular context.

Thus far, we lack knowledge about the perception of organizational resilience and ambidexterity by employees in general (e.g., Kim, 2020; Shin, Taylor, and Seo, 2012) and among non-family employees in family firms in particular (Tabor et al., 2018). This is surprising, as family firms are not only particularly prevalent in the tourism industry (Hauck and Prügl, 2015; Zehrer and Leiß, 2019), but employee perceptions may also be particularly relevant in this sector, considering that the service sector is highly dependent on the mindset, motivation, and ability of employees, especially in times of change (e.g., Spiess and Zehrer, 2020).

Prior research on ambidexterity has mainly explored the antecedents of creating ambidextrous behaviors and structures (e.g., Kammerlander et al., 2020) as well as the linkage between ambidexterity and firm performance outcomes (Gibson and Birkinshaw, 2004; Lubatkin et al., 2006). In this chapter, we extend prior literature and investigate a novel outcome of ambidexterity: the perception of organizational resilience by employees. The interplay of organizational ambidexterity and perceived organizational resilience from the perspective of the non-family employees in the tourism industry represents a contemporary linkage which offers several contributions for theory and practice alike. We theorize that establishing ambidextrous organizational structures helps to decrease volatility in family firm resilience stemming from environmental complexity, which is particularly important for coping with sudden external shocks. Furthermore, we argue that adopting ambidextrous thinking as a strategic priority elevates the general level of family firm resilience in the eyes of non-family employees, thus fostering long-term survival of those organizations.

We present a case study of an integrated service company in the tourism and leisure industry located in the Swiss Alps. The presented case is unique in that the family-owned company operates around 90 business units along the tourism value chain, including hotels and restaurants, rental shops, sport schools, as well as alpine infrastructure facilities (e.g., mountain ropeways and slopes). While being economically stable, the company faces several strategic challenges. On the one hand, noticeable fluctuations in natural conditions, such as an increasing lack of snow in the winter season, have created a sense of urgency to address the long-term risks of climate change and make corporate sustainability, i.e., balancing between financial, environmental, and social goals, one of the company's strategic priorities. On the other hand, the recent COVID-19 pandemic has caused a sudden, profound crisis that required short-term operational changes to adapt to the new situation and secure economic survival. In this in-depth case study, we set out to explore how the company deals with different levels of uncertainty both on a strategic as well as an operational level. By drawing on empirical results from an employee survey in 2020, we derive theoretical insights and practical recommendations on how ambidextrous structures, and related perceptions of organizational resilience might help the family firm to ensure long-term survival.

BACKGROUND

(Perceived) Organizational Resilience

As range, likelihood and intensity of exogenous shocks increase (Williams et al., 2017), the question how organizations can prepare for, deal with and recover from sudden or gradual disruptions becomes paramount. Whereas crisis management and research during the 1980s and 1990s primarily focused on intra-organizational disruptions, i.e., in case of large-scale industries accidents like the Chernobyl nuclear power station disaster in 1986, events like the 9/11 terrorist attacks in the United States or the 2008 global financial crisis led to increased interest in researching and dealing with exogenous shocks (Doern et al., 2019; Linnenluecke, 2017) and their impact on organizational survival. While crisis management research mainly investigates how organizations reach normal functioning as fast as possible after a disruption (Doern et al., 2019), the concept of resilience does not only concern the capacity to respond to adversity, but also to mitigate harm before, during, and after a crisis (Williams et al., 2017).

In business research, resilience is treated as a fractal concept that has been explored on several levels: (1) on the micro-level, as an individual psychological trait of entrepreneurs or managers (e.g., Kossek and Perrigino, 2016; Zehrer and Leiß, 2019), (2) on the meso-level, as a property of single organizations (e.g., Doern et al., 2019; Herbane, 2015, 2019; Morgan et al., 2020), or (3) on the macro-level, as factors that enhance the capacity of larger socio-economic systems to mitigate and adapt to major disruptions, such as ecological or economic shocks (Korber and McNaughton, 2018). In an attempt to provide an inclusive definition, Williams et al. (2017) describe resilience as "the process by which an actor (i.e., individual, organization, or community) builds and uses its capability endowments to interact with the environment in a way that positively adjusts and maintains functioning prior to, during, and following adversity" (p. 742). The authors thereby emphasize the idea of resilience as a dynamic and emergent property of interactions between individuals, organizations and their broader socio-economic environment. This is also in line with other management scholars, such as Doern et al. (2016) or Herbane (2015) who conceptualize resilience as an organization's behavior before, during, and after a crisis. Thereby, special attention is paid to the idea that resilience means not only rapid restoration of pre-crisis functioning, but the ability to maintain 'reliable' functioning throughout the disruption (Doern et al., 2019, p. 4), or even to transform to a new state of balance when needed (Korber and McNaughton, 2018). While the definitional question about resilience looms large in recent literature, a review by Annarelli and Nonino (2016) reveals that research regarding the question of how to reach long-term organizational resilience is still scarce. One general assumption is, however, that high levels of organizational resilience are not achieved through a passive-reactive attitude, but rather require proactive strategic measures to enhance adaptive capabilities of the organization at stake (Conz, Lamb, and De Massis, 2020). Hence, organizational resilience should be 'deliberately nurtured' through a combination of intentional strategic planning and the development of an organization's ability to learn, reinvent and transform itself when needed (Chrisman, Chua, and Steier, 2011; Herbane, 2019; Johnson, Elliott, and Drake, 2013).

Thus, we understand organizational resilience as a "firm's ability to absorb complexity and emerge from a challenging situation stronger and with a greater repertoire of actions to draw from than were available before the disruptive event" (Lengnick-Hall et al., 2011, p. 244). When investigating organizational resilience in terms of 'reliable' functioning throughout a disruption or even the stronger emergence from a disruptive situation, prior research has mainly focused on a retrospective assessment of organizations' resilience from a post-crisis perspective. We introduce a new perspective by investigating organizational resilience from the perspective of the non-family employees (perceived organizational resilience, POR) during a disruptive event (i.e., COVID-19) as they represent a key stakeholder group for the long-term survivability of an organization. This perspective is of particular interest for organizations operating in service-oriented industries such as the tourism industry.

Building (Perceived) Organizational Resilience in Family Firms

In the realm of family business research, there is little research directly addressing questions of organizational resilience (Conz et al., 2020). However, based on an analysis of 98 pairs of family and non-family firms, Amann and Jaussaud (2012) find that family firms are more resilient than non-family firms, both during and after an economic crisis, as they mitigate crises better, recover quicker and maintain higher performance over time. Also, a survey among 215 SMEs in the UK reveals that firm age – rather than size – is a predictor for companies' threat orientation, meaning that the older the company, the higher the likelihood to establish strategic plans for crisis prevention instead of relying on ad-hoc response structures (Herbane, 2015). This finding connects well to research about family businesses as one of their main characteristics is that they are typically designed for long-term survival. In fact, the 500 largest family-owned companies in Germany are on average 101 years old – a number that speaks for itself when it comes to long-term orientation. The longevity of family firms is driven by their idea of keeping the company in the family's possession for future generations, which shapes their organizational values, strategic planning, and operational management (Le Breton-Miller and Miller, 2006; Miller and Le Breton-Miller, 2003). Succession plans, however, can disrupt the longevity and thus the resilience of family firms as a substantial number of family firms face the challenge to find and successfully integrate family-internal successors which leads to the fact that less than 10% of family firms are still in the hands of the founding family beyond the third generation (Le Breton-Miller, Miller, and Steier, 2004). Furthermore, family firms usually attach great importance to maintaining long, stable, and trusting relationships with all their stakeholders, such as strategic partners or employees (Le Breton-Miller and Miller, 2006; Wimmer, Domayer, Oswald, and Vater, 2005). Hence, due to their lasting social capital and their intergenerational orientation, it is reasonable to assume that family businesses are particularly crisis-proof and inherently more resilient than non-family firms. At the same time, given their focus on tradition and long-term commitments to employees and suppliers, family firms might exclude common means of crisis management, such as strong cost savings through layoffs or contract terminations, which might limit their ability to react to demands of environmental disruptions (Bauweraerts, 2016). Also, due to their often strongly hierarchical organizational structure, dealing with crises in family firms highly depends on the behavior of the owners and top management (Conz and Magnani, 2020). Hence, especially for family firms, organizational resilience requires finding a balance between robustness and flexibility, including the capacity to innovate, transform and renew themselves during a crisis. This idea is underlined by a study in the wine industry, which finds that experimentation and innovation are particularly suitable fields of strategic action to counteract increasing entropy and build organizational resilience (Golicic, Flint, and Signori, 2017).

Based on the above-mentioned considerations, we are particularly interested in investigating the impact and interplay of perceived organizational resilience and organizational ambidexterity. We consider organizational ambidexterity to be a highly relevant factor to achieve family firm resilience in the tourism industry. We theorize that establishing ambidextrous organizational structures helps to decrease volatility in family firm resilience stemming from environmental complexity, which is particularly important for coping with sudden external shocks.

Organizational Ambidexterity and the Perception of Non-Family Employees

Creating and developing an organization which embodies the flexibility as well as the innovation mindset to identify and occupy new business opportunities is one thing. Doing this while simultaneously maintaining and further optimizing the performance of the daily business activities is another. This idea can be conceptualized as organizational ambidexterity (e.g., Gibson and Birkinshaw, 2004), which describes firms that are efficient in their daily business operations and exploit what they already know while also being adaptive to respond to changes in the environment and constantly explore new business opportunities (Duncan, 1976; Tushman and

O'Reilly, 1996; Gibson and Birkinshaw, 2004). Thus, ambidextrous organizations can be defined by balancing alignment (exploitation) and adaptability (exploration) within one organization.

To shed light on this balance, prior literature differentiates between structural (Duncan, 1996), contextual (Gibson and Birkinshaw, 2004), and hybrid ambidexterity (Ossenbrink, Hoppmann, and Hoffmann, 2019). Contextual ambidexterity refers to the "behavioral capacity to simultaneously demonstrate alignment and adaptability across an entire business unit" (Gibson and Birkinshaw, 2004, p. 209). Structural ambidexterity, on the other hand, refers to organizations who build up dual structures for both alignment and adaptability, thus following more of a trade-off ("either/or") logic. In contrast, contextual ambidexterity leans more towards paradoxical thinking ("and/or" logic), which leads to the capacity to pursue both alignment and adaptability simultaneously across an entire business unit (Gibson and Birkinshaw, 2004).

As an organization, developing such behavioral capacities is complex and time-consuming (Amit and Schoemaker, 1993). Thereby, organizations try to create a culture through structures and processes, which enables individuals, regardless of job function, to self-select how to subdivide their time into alignment and adaptability efforts. This is why the contextual ambidexterity is closely linked to the concept of individual ambidexterity, which are the cognitive abilities necessary to balance efforts between the two modes of alignment and adaptability (Good and Michel, 2013). Scholars state that "the more dynamic and unpredictable the context, the more individual ambidexterity is thought to be necessary for success" (Good and Michel, 2013, p. 436). In other words, the individual capacity of balancing between the two modes is an important determinant for contextual ambidexterity. Although prior research on organizational ambidexterity in family firms mainly focuses on the role of the family (Kammerlander et al., 2020) and top management teams, a few scholars have emphasized the role of operational managers and applied a bottom-up perspective (Mom et al., 2018; Zimmermann et al., 2018). Overall, organizational ambidexterity represents a multilevel approach, which involves top-down as well as bottom-up perspectives (Mom et al., 2018). To assess ambidexterity within an organization, prior literature mainly "relied on single informants to answer questions on behalf of an entire organization" (Gibson and Birkinshaw, 2004, p. 215), which is contradictory as the main actors of ambidextrous behavior of an organization are the respective non-family employees who interact within an organizational context.

The development of an ambidexterity-supportive organizational context requires more than leadership and building up formal organizational structures. It is the creation and development of a set of systems, processes, and shared beliefs (Gibson and Birkinshaw, 2004) that enable the simultaneous activation of the two factors of ambidexterity, both on an individual as well as on a business unit level. Based on the study of Gibson and Birkinshaw published in the Academy of Management

Journal (2004), antecedents of contextual ambidexterity are manifold, showing different paths to successful contextual ambidexterity (Gibson and Birkinshaw, 2004, p. 223). This means, depending on the heritage and values of the organization, contextual ambidexterity can be achieved in different ways. Some organizations build their focus around adaptability skills, some around alignment skills. The common ground is threefold. First, applied systems to create the co-existence of both modes of ambidexterity are rather simple and involve little formality (Gibson and Birkinshaw, 2004). Second, the role and mindset of senior executives, the Top Management Team (TMT), as well as the family involvement are crucial as they are responsible for implementing systems and processes that allow and support the individual judgement of balancing between alignment and adaptability (Gibson and Birkinshaw, 2004; Kammerlander et al., 2020; Venugopal et al., 2019; Cao, Simsek, and Zhang, 2009). Third, and complementary to the latter, the importance of decentralized structures, shared vision, and supportive leaders is a key success factor of ambidexterity (Tushman and O'Reilly, 1996; Cao et al., 2009). This is in line with recent publications stating that the behavior of the TMT significantly influences organizational ambidexterity in general (Venugopal et al., 2019) as well as in the context of family firms (Kammerlander et al., 2020).

Most organizations, however, do not simply apply either structural or contextual ambidexterity. These two modes rather represent two extrema of an ambidextrous continuum. Organizations often strive for a hybrid mode depending on the firms' environment (Ossenbrink et al., 2019). Facing an environment which is coined by numerous opportunities, a high degree of uncertainty (e.g., high likelihood of being exposed to external shocks), as well as the requirement of a novel culture, organizations are more likely to pursue initiatives that lead to hybrid ambidexterity (Ossenbrink et al., 2019). Thereby, some environmental characteristics (e.g. new culture and high degree of uncertainty) call for structural ambidexterity while others (e.g. numerous opportunities) motivate for contextual initiatives. Facing a global pandemic, which reshuffles the entire basic assumptions of supply and demand, may lead organizations to chase both structural and contextual ambidexterity to fully unfold the potential of adapting and aligning their organization in this situation.

Previous literature regarding ambidexterity mainly explored the different modes as well as the antecedents of ambidexterity (e.g., Venugopal, 2020; Ossenbrink et al., 2019; Good and Michel, 2013). Referring to the beginning of this book chapter, the core challenge of organizations affected by the current global pandemic is to find paths and strategies to become more resilient. Hence, we turn the perspective and raise the question whether the ability of ambidextrous behavior of an organization (from the perspective of non-family employees) can be considered as a predictive variable for perceived resilience of an organization in the eyes of non-family employees. The perspective of non-family employees in family firms, especially the benefits and challenges associated with non-family members, has become an increasingly intriguing topic in family firm research in the past years (Tabor et al., 2018). Pioneering research has found non-family employees to be important for family firm success (Carney, 1998; Chua, Chrisman, and Sharma, 2003; Sciascia and Mazzola, 2008) as non-family employees are found to be instrumental in strategic decision making (Mitchell, Morse, and Sharma, 2003) and for the expansion into new markets (Chung and Luo, 2008; Graves and Thomas, 2006). However, there is still a lot of room for inquiry (Tabor et al., 2018), in particular when it comes to the roles and perceptions of non-family employees in the context of ambidexterity in family firms and its relation to perceived organizational resilience. To uncover the proposed relationship, we analyze data from an employee survey in a world leading and family-owned ski resort located in the Swiss Alps, which activated a significant number of initiatives with the aim to increase ambidexterity right before the pandemic hit Europe.

CASE STUDY: ORGANIZATIONAL AMBIDEXTERITY IN A WORLD-LEADING FREESTYLE ALPINE RESORT

Case Introduction and Context

Our case represents a family-owned organization which is an integrated service company operating in the tourism and leisure industry. It is located in Switzerland and provides four season Alpine leisure experience, comprising the operation of hotels, gastronomy, ski, snowboard, and freestyle schools, as well as wellness and cultural offers. The company employs between approx. 230 (summer) and 1,100 (winter, including seasonal workers) employees. The alpine resort covers more than 100 square km and is ranked among the largest ski areas in Switzerland. The family-controlled resort has repeatedly demonstrated pioneering initiatives and has well recognized and integrated the demands of the target market: alpine enthusiasts of all ages and at any season of the year. They have won several innovation and freestyle awards and thus belong to the leading alpine resorts in Europe with a high awareness from international guests (e.g., Japan, USA). Compared to other organizations in the alpine industry, the uniqueness of this organization stems from the integration and bundling of many often separately owned and operated and thus stand-alone offerings, e.g., mountain infrastructure, hotels and hostels, restaurants, ski and snowboard schools, digital platforms, as well as event locations. This leads to an integrated customer experience with multiple customer interaction touch points, which are designed and operated by one single organization. This all-from-one-source concept has been made possible by two entrepreneurial families who represent the majority of the publicly traded company and who developed, acquired, and integrated all the single business units (e.g., ski schools) over the past decades. One family member serves as long-time president of the organization, constantly steering and mapping the future visions of the alpine resort.

The Transformation Process

Starting in summer 2019, the president and co-owner of the organization started to question the until then dominant organizational structure as well as the management style of his organization due to imminent challenges of the tourism industry such as climate change, evolving consumer behaviors and needs, digitization as well as potential external shocks. At that time, he is aware of the strengths (creating new ideas, creativity, events) as well as the weaknesses (lack of prioritization, interdisciplinary cooperation, internal communication) of the organization. Transforming the organization with the objective to increase the focus and performance in daily business, i.e., alignment, while at the same time fostering adaptability and creating a nutrient soil for ideas and opportunity recognition among his non-family employees represented a major challenge and a watershed in the company's history. Based on a joint endeavor with a consultancy specializing in family firms and innovation, a transformation process was initiated. The main objective was to create an organizational structure which enables contextual as well as structural ambidexterity to systematically build and balance adaptability and alignment capacities. In addition to transforming the organization, a newly hired non-family CEO (who was installed in early 2020 to manage the daily business) together with a non-family TMT implemented objectives and key results (OKR) as an established management tool to increase alignment and cross-functional collaboration in daily business operations. The new organization structure as well as the OKR process was co-created with a task force (with existing key employees) and activated in March 2020 right before the global pandemic COVID-19 hit Europe and led to temporary closures of mountain infrastructure services as well as restaurants and hotels.

Implementing an Ambidextrous Organizational Structure

In order to enable both, the alignment and performance of the daily business as well as the ability to constantly recognize business opportunities and the adaptability to react to changes in the environment, the newly implemented organizational structure comprises a dual structure (see Fig. 1). The dual structure reflects the ambidexterity strategy and is visualized with two hands symbolizing the interplay between exploitation (alignment) and exploration (adaptability). The organization, and thus the daily business operations, has now transformed into an organizational structure comprising seven performance circles, one performance board, and one CEO (see left hand in Figure 1). The seven performance circles represent operational areas: (1) mountain infrastructure, (2) accommodation (hotels and hostels), (3) food and beverage, (4) stores and rentals, (5) internal services (accounting and finance), (6) marketing and communications, and (7) education (ski and snowboard schools). The performance circles fulfill the purpose of gathering the business units in their efforts towards a common goal. The business units are managed as teams in different areas of the business and design their work path and objectives by themselves. The head of each performance circle is part of the performance board, which is the strategic steering committee led by the CEO. Following this structure, the organization's goal is to better align their daily operations, improve the focus as well as the information flow across and throughout performance circles. To increase adaptability, the new structure and management architecture allows a cross functional and interdisciplinary temporary team building to solve emerging challenges. In addition to that, the organization implemented a systematic process to enable all employees to present new ideas for business models, products or services (see right hand in Figure 1) in front of an innovation board, which represents the structural manifestation of adaptability in the context of the aspired ambidexterity.

Figure 1. Conceptualization of ambidextrous organizational structure
Source: Own illustration.

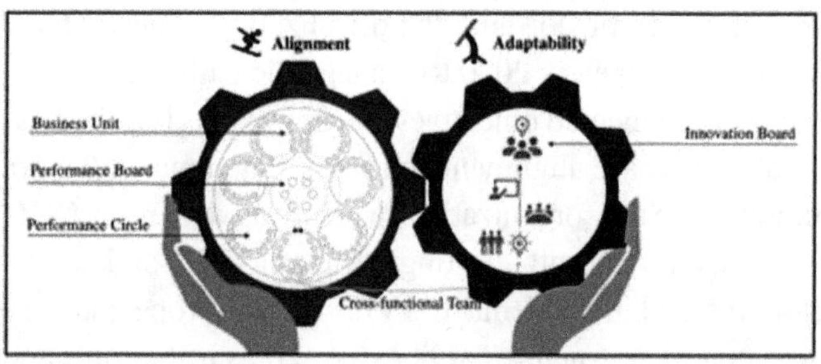

Proposed Relationship Between Ambidextrous Structure and Resilience

We argue that in an increasingly uncertain world, organizational resilience manifests itself as a dynamic interplay along three faultlines (see Fig. 2): (1) balancing and improving efficiency of established operations (alignment), (2) keeping short-term flexibility to respond to unforeseen events while fostering opportunity recognition (adaptability), and (3) the perceived organizational resilience among existing non-family employees as result of (1) and (2).

Figure 2. Proposed relationship between organizational ambidexterity and resilience
Source: Own illustration

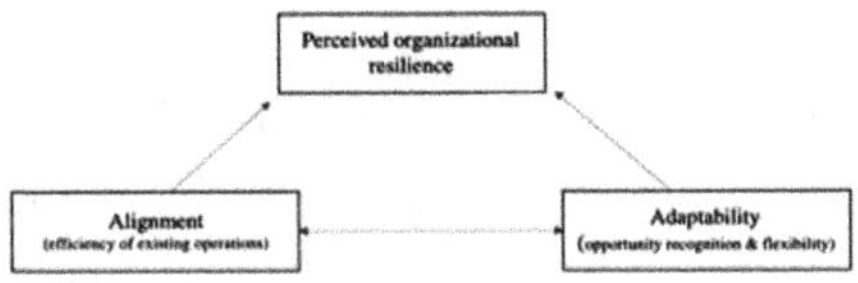

In this model, perceived resilience builds on the concept of adaptive capacity (Korber and McNaughton, 2018) as well as innovation and incorporates the organizational ability of behaving flexibly and adjusting the business model to new circumstances. This is in line with Morgan et al. (2020), who investigate strategies for small businesses to bounce back from exogenous shocks with a focus on the role of pivoting, i.e., structured processes of adapting business models as well as product and service offerings when facing an exogenous shock. While "young" organizations are trained to frequently pivot and to deal with changes in the business environment leading to a high flexibility, established structures and routines in more mature organizations often hinder their adaptability. Given the previously mentioned increasing likelihood of external shocks, it becomes a fundamental task for established firms to set up structures and systems that allow for experimentation while not putting the existing performance and efficiencies at stake. In fact, the organizational structure of a company has proven to be one of the key success factors in turning the organizations into an entrepreneurially thinking and resilient ecosystem.

Our case study offers a unique example to explore the effect of ambidextrous organizational modes on the perceived resilience among employees following an external shock. As a corollary, we are deeply interested in the perception of existing non-family employees regarding the effects of a transformation towards ambidexterity on the perceived resilience of the organization in unpredictable and challenging times.

Data Collection

Data were collected via an online survey among all non-family employees in fall 2020 which comprises a workforce of approximately 230 people in total. The survey was sent out by the CEO encouraging all employees to inform the company's leadership about their perception of the transformation process and the management of the current COVID-19 crisis. After deleting all observations that included missing values in our variables of interest, we yielded 170 valid answers from employees across all seven performance circles, with the most respondents working in the realm of mountain infrastructure (34%). The survey comprised several closed and open-ended questions. The open-ended questions provided participants with the opportunity to elaborate on their evaluations and give direct feedback to the leadership team.

All variables were based on and derived from relevant literature in the fields of crisis management as an indicator for resilience, and organizational ambidexterity. We first translated the items with the support of bilingual experts and using the translation-back-translation method (Brislin, 1970). Second and where necessary, we adapted the items to the company's jargon to ensure that all employees understood them. Unless indicated otherwise, all items were measured by using a 7-point Likert scale.

Dependent Variables: Crisis Preparedness and Perceived Effectiveness

To operationalize the perception of organizational resilience, we defined two distinct outcome variables: perceived *crisis preparedness* and perceived *effectiveness* during the crisis. Each of the constructs was measured by three items. Perceived *crisis preparedness* (Cronbach's α= 0.76) included (1) "The new organizational structure (structure in circles and implementation of OKRs as steering tool) has made us more resilient"; (2) As a company, we were well prepared for a crisis like the COVID-19 pandemic."; (3) "We were able to respond quickly to changing environmental conditions during the COVID-19 pandemic". Perceived *effectiveness* (Cronbach's α= 0.87) included (1) "So far, our business unit has been able to realize its full potential in 2020"; (2) "In my Business Unit, we are satisfied with our performance."; (3) "Our Business Unit empowers me to do my best."

Independent Variables: Alignment and Adaptability

To assess the effect of organizational ambidexterity on *effectiveness* and *crisis preparedness*, we based our approach on Gibson and Birkinshaw (2004) and included two distinct measures. *Alignment* (Cronbach's α= 0.89) was measured by four items indicating in how far the employees perceive the new organizational structure to be beneficial for focused collaboration and working towards the same goals (i.e., "Due to the new organizational structure, we are working better together"). *Adaptability* was measured by three items (Cronbach's α= 0.88, i.e., "Our organizational structure is flexible enough to respond quickly to changes") indicating employees` perception about the firm's general capability to adapt to new challenges.

Control Variables

To account for heterogeneity within the group of employees, we include four control variables. First, we control for age group and gender. Second, we include a dummy for tenure differentiating between those employees who have been employed for five years or less, and those who have been employed for more than five years. Third and in line with Ossenbrink et al. (2019), we include a binary variable that differentiates between frontline employees (frontline = 1), i.e., those employees working in service-oriented performance circles that directly interact with the customers such as hotels and restaurants, and those who are less likely to interact with customers on a frequent basis, such as internal services (e.g., accounting and finance).

Especially during times of change and crises, clear communication by the leadership is one of the key success factors for firms as employees may experience high levels of strain and uncertainty. To make sense of their situation and handle changes best, employees are in need of clearly communicated messages and knowing what's expected of them (Adamu and Mohamad, 2019). Hence, we also assume *communication* to be a driving force in employees' perceptions in the presented case. Accordingly, we draw from Adamu and Mohamad (2019) and include five items concerning internal crisis *communication* (Cronbach's α= 0.91; i.e. "I have been clearly informed by management internally about external responses to be used during crisis".)

Results

Descriptive Statistics

After deleting all observations that included missing values in our variables of interest, we yielded 170 valid answers. In the next step, we dropped five questionnaires submitted by members of the company's performance board, leaving us with a sample of n=165 observations as a basis for the analysis. Means, standard deviations, and correlations for the main variables are displayed in table 1. 61% participants were male, and 39% were female. Furthermore, 45% indicated having worked for the company for more than 5 years. As age was measured as a group variable, the mean (4.56) indicates that on average participants are around 35 years old and 64% were 40 years or younger by the time of the data collection.

Empirical Analysis

To estimate the effects of adaptability and alignment – as measures accounting for organizational ambidexterity – on our two dependent variables, namely perceived

Table 1. Descriptive statistics and correlations

	Mean	S.D.	Effectiv./	CrisisPrep	Adaptability	Alignment	Comm./	Frontline	Tenure	Gender	Age
Effectiveness	4.95	1.33	1.00								
Crisis Prep./	4.93	1.29	0.58***	1.00							
Adaptability	5.06	1.37	0.59***	0.68***	1.00						
Alignment	4.38	1.43	0.58***	0.76***	0.72***	1.00					
Communication	4.74	1.45	0.65***	0.78***	0.68***	0.75***	1.00				
Frontline	0.43	0.50	0.12	-0.02	0.12	0.04	0.07	1.00			
Tenure	0.45	0.50	0.01	-0.09	-0.07	-0.08	-0.09	0.20***	1.00		
Gender	0.39	0.49	0.00	-0.06	0.03	-0.04	0.02	0.12	-0.13	1.00	
Age	4.56	1.86	0.03	-0.01	0.05	0.08	-0.01	-0.01	0.35***	-0.25***	1.00

Note: n = 165. Significance levels at *** $p<0.01$. ** $p<0.05$. * $p<0.1$.

crisis preparedness and perceived *effectiveness*, as two measures jointly representing the construct of perceived organizational resilience, we apply OLS regressions with heteroskedasticity-robust standard errors clustered on the performance circle level. In the first step, we include the key variables and the four controls accounting for employee heterogeneity. In the second step, we use a hierarchical linear regression to add the control variable for communication. The results are reported in table 2.

As regards the first outcome variable, we observe that both *adaptability* and *alignment* indicate a positive and highly significant effect on perceived *crisis preparedness* (column 1a). This means that those non-family employees who experience high levels of *adaptability* and *alignment* in their work, also assume the company to be better prepared to handle disruptions. Adding *communication* (column 1b), we observe a positive and significant effect, too. Thereby, communication explains an additional 8% of the outcome variable and leads to a decrease of the effect size of both alignment and adaptability compared to the previous model specification. Following Baron and Kenny (1986), this indicates that the effects of *adaptability* and *alignment* on *crisis preparedness* might be partially mediated via *communication*.

Moving to the second outcome variable, perceived *effectiveness*, we observe a similar pattern (column 2a). In the first model specification, both *adaptability* and *alignment* display a positive and highly significant effect on the outcome variable. When adding communication, both effect sizes diminish in size with *alignment* tending towards 0 (column 2b). This calls again for a mediating effect of communication with a full mediation for *alignment* and a partial mediation for *adaptability*.

Looking at the control variables, we observe that belonging to the group of frontline workers negatively impacts the perceived crisis preparation. This is could be explained by the fact that constant interaction with customers in the midst of a crisis and dealing with their inquiries and questions could reinforce a sense of

Table 2. Regressions

	(1a)	(1b)	(2a)	(2b)
	Crisis Preparedness	**Crisis Preparedness**	**Effectiveness**	**Effectiveness**
Adaptability	0.270***	0.150**	0.336**	0.222*
	(0.023)	(0.054)	(0.123)	(0.096)
Alignment	0.475***	0.266***	0.285**	0.087
	(0.026)	(0.026)	(0.106)	(0.135)
Communication		0.390***		0.370*
		(0.095)		(0.161)
Controls				
Frontline	-0.168*	-0.180*	0.138	0.126
	(0.082)	(0.083)	(0.113)	(0.116)
Tenure	0.003	-0.011	0.170	0.157
	(0.092)	(0.114)	(0.227)	(0.206)
Gender	-0.112	-0.160	0.036	-0.010
	(0.101)	(0.111)	(0.157)	(0.130)
Age	-0.040	-0.022	-0.016	0.001
	(0.021)	(0.044)	(0.050)	(0.029)
Constant	1.788***	1.409**	1.936***	1.576***
	(0.357)	(0.529)	(0.184)	(0.386)
Observations	159	159	159	159
R-squared	0.595	0.673	0.380	0.445
ΔR-squared		0.08***		0.07***
VIF	1.46	1.75	1.46	1.75

Robust standard errors (in parentheses) clustered on the performance circle level. *** $p<0.01$, ** $p<0.05$, * $p<0.1$

stress and uncertainty, and thus attenuate the levels of perceived crisis preparedness among the respective employees.

Combining these empirical results with exemplary insights from the open questions, we can draw compelling insights on the positive relationship between organizational ambidexterity and perceived organizational resilience as a firm's capacity to mitigate harm before and during a crisis. First, as shown in the empirical analysis, experiencing high levels of alignment leads to a higher perception of *crisis preparedness* and perceived *effectiveness* during the crisis. This is also underlined by two quotes from employees:

"The restructuring has worked out well. We were able to use the opportunity and made the most of the summer. Main success factors were that we had more contact persons than before and worked in a more targeted way thanks to the OKRs. In addition, the OKRs helped us work together as a team to achieve goals."

"The collaboration in our team is better than ever. It's more like a family than a team."

Furthermore, the capacity to react quickly to changes in the external environment, such as constantly changing regulations due to the COVID-19 pandemic, has kept the perceived effectiveness high despite all obstacles. Again, as two non-family employee quotes exemplify, this can be also partially attributed to the organizational structure that allows for flexibility and adaptability:

"Agility, fast response time, short distance reporting, transparency, good team communication."

"Collaboration and synergy with team colleagues. Our collaborative and open-minded way of working contributed to fun, better and creative solutions. "

At the same time, and this is also indicated in the empirical analysis, structures only become effective if they are activated and accompanied by effective communication. In other words, the positive effects of alignment and adaptability can only unfold if they are combined with clear communication and transparency by the company's leadership. Especially during times of uncertainty, employees need regular reassurance that what they are doing is still the right thing given the current situation. This need becomes particularly clear when looking at the survey entries to the open questions of the questionnaire. There, one question asked employees to reflect upon the previous months and to name their three most critical success factors for managing crises as a team. Figure 3 shows the Top 10 results clustered by topic and how often they have been mentioned in absolute numbers. It is evident that communication is perceived as the single most important factor, whereby the respondents elaborated that communication should be fast, clear, and honest – or as one employee mentions:

"If the leadership does not communicate, we cannot solve problems."

Discussion and Recommendations

This chapter examines the interplay of the fundamental constructs of organizational resilience and ambidexterity within family-owned organizations in the tourism

Figure 3. Success factors displayed as absolute number of mentions
Source: Qualitative data collection (open answers)

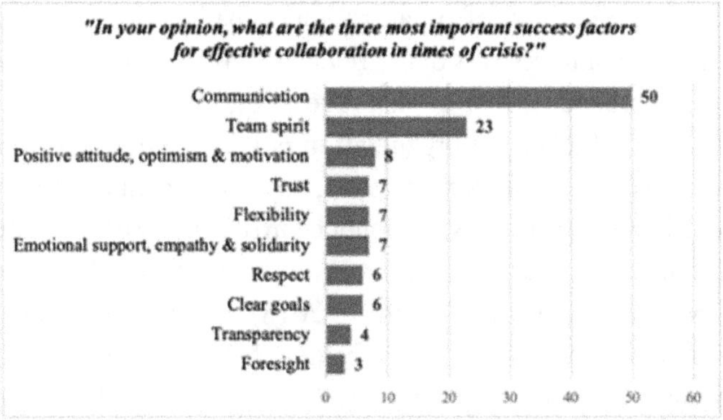

industry, an industry which currently faces the most impactful disruptions in the post-war era. The global COVID-19 pandemic coupled with an economic crisis has led to fundamental upheavals in the way we live, work and travel. Many organizations are forced to temporarily close their services and to rethink their entire business model. Especially in the travel industry, it remains unpredictable to which extent the recent disruptions will induce long-term changes in consumer behavior and needs, potentially threatening the basic assumptions of touristic activities. Whether individuals will plan, spend and evaluate their leisure experiences as they did before the pandemic remains an open question. In times like these, organizations in industries that highly depend on environmental conditions and economic purchasing power experience a shift in paradigms with regards to their overall company objectives. Abruptly and without warning, the unconditional pursuit of higher margins, greater profits and optimized customer lifetime as the dominating business paradigms in the past prosperous years has been replaced by the existential question of short-term business survival and mitigating unpredictable risks. In this chapter, we present the case study of a company that was hit hard by the pandemic by having to shut down all business activities during the high season in March 2020 but was able to adapt quicky to all adversities and managed to keep performance levels as well as employee engagement high. This was possible due to the introduction of an ambidextrous organizational structure prior to the crisis. Establishing structures of alignment and adaptability and imprinting an "and/or"-mindset into the company culture has enabled all employees to constantly navigate new challenges while staying aligned on priorities and main objectives. In other words, the case exemplifies how organizational ambidexterity leads to higher levels of resilience by boosting the capability of the firm to mitigate and adapt to major disruptions in its environment.

Theoretical Implications

Prior literature in the domain of organizational resilience has so far highlighted and explored different antecedents of resilience like the influence of the company's leadership (e.g., Herbane, 2019) or firm age (e.g., Herbane, 2015). Turning the perspective towards whether the organization is perceived as resilient by its own employees and what drives this perception is, however, an untouched yet important area of research. This is particularly pronounced in the family business context where especially the roles and perceptions of non-family employees lack attention so far (Tabor, Chrisman, Madison, and Vardaman, 2018). In this chapter we contribute to furthering our understanding of how firms balance between contextual and structural ambidexterity and how this balance is perceived and translated into resilience perceptions by non-family employees. Adding to recent literature in family business research focusing on the antecedents of and ability for organizational ambidexterity (e.g., Kammerlander et al., 2020; Veider and Matzler, 2016), we are among the first who explore the influence of ambidexterity on perceived resilience of an organization. It has been shown that organizational ambidexterity leads to increased performance of firms (Gibson and Birkinshaw, 2004) and family firms in particular (Stubner, Blarr, Brands, and Wulf, 2012). Turning the perspective towards resilience instead of performance covers the economic need of these times. Examining ambidexterity as antecedent of resilience adds to the current research on ambidexterity in family firms as well as on organizational resilience. Our findings demonstrate that adaptability as well as alignment are important drivers for perceived organizational resilience among non-family employees. Furthermore, prior research on organizational ambidexterity so far rather focused on investigating the structural and leadership components of ambidexterity (e.g. Kammerlander et al., 2020). Our research extends this view by exploring the main actors of ambidextrous behavior, the existing non-family employees. Thus, we contribute to furthering our understanding of a perceptual perspective of non-family employees in family firms (Tabor et al., 2018). While non-family employees have been recognized as important for family firm success (Carney, 1998; Chua, Chrisman, and Sharma, 2003; Sciascia and Mazzola, 2008), much less is known when it comes to a) the roles and perceptions of non-family employees in general and b) in the context of ambidexterity and resilience in family firms in particular.

The presented case demonstrates that by transforming the organization towards a more decentralized, focused, and vertical-interacting organization results in achieving ambidexterity. On the continuum between structural and contextual ambidexterity (Ossenbrink et al., 2019), our case represents a hybrid mode. The company's transformation created a structural duality of alignment and adaptability but did not formally separate units for exploitation and exploration. The transformation rather created a structure as nutrient soil for a context that allows non-family employees to exploit existing business within their own field of excellence as well as to explore new opportunities with cross-functional team members.

Beyond that, our findings strongly indicate communication as an important concept in successfully turning ambidexterity into perceived resilience. Facing two distinct outcome dimensions, namely perceived *crisis preparedness* as well as the perceived effectiveness, communication plays a crucial role to activate ambidexterity in terms of daily business operations and mindset among employees, which confirms the important role of the entire TMT's behavior (Venugopal et al., 2019; Cao et al., 2009). Furthermore, Cao and colleagues (2009) underline the importance of CEO-TMT interaction including communication richness and power decentralization to achieve ambidexterity.

Managerial Implications

Our findings suggest two major practical implications: First, crises seem to favor the prepared family firm, i.e., by building organizational structures and processes that enable organizational ambidexterity in terms of balancing alignment and adaptability. Second, the mindset, motivation and ability of non-family employees in family firms is of utmost importance to navigate organizational disruptions and high levels of environmental uncertainty.

In sum, our findings clearly demonstrate a strong link between an ambidextrous structure and perceived resilience, which underlines the importance to integrate employees in the process of organizational development towards ambidexterity. Non-family employees are found to be highly committed towards their organizations as for example reflected in their long tenure. Hence, they present a key resource when it comes to ensuring the long-term survivability of the family firm. At the same time, organizational identities reflect the "who we are" as an organization, which influences employees' perspectives, motivations, and actions. In other words, being perceived as resilient by employees will increase their trust in the company's survivability during a crisis, resulting in higher levels of confidence in their business practice, and ultimately increase effectiveness. In turn, a lack of perceived resilience could potentially result in high levels of insecurity and in the worst case in chaotic, uncoordinated actions driven by confusion and fear. Consequently, considering the employees' perspective is an important starting point to build organizational resilience in general, which is why it should be key priority for organizations in the tourism industry.

Our study provides initial insights in shaping the perceived resilience of an organization through fostering ambidextrous behavior. To achieve organizational ambidexterity, this case study presents several recommendations how to simultaneously increase the focus and performance in daily business (alignment) as well as the capability to quickly react to changes and disruptions (adaptability). First, while the general organizational transformation was initiated by the company's leadership, an employee task force was installed to shape and co-create the new organizational structure. This integration of a top-down and a bottom-up approach was essential to promote acceptance across the entire organization. Second, in terms of alignment, the introduction of objectives and key results (OKR) as a management system enabled effective collaboration across all organizational levels, both hierarchically as well as vertically. However, new management systems are only as good as their adoption rate in daily business practice. Hence, the company partnered with an external consultancy to train employees on how to work with OKRs and accompany them in the first year after their introduction. External expertise, however, should also always be backed up by employees, such as the previously mentioned employee task force, who act as ambassadors to drive change from within. Third, while the new company structure and the OKR system allowed for a better alignment of daily operations, this study also illustrates the importance of leadership communication in times of crises. In particular, we see that the perception of employees seems to be an important and at the same time manageable factor. Especially in times of rapid changes and high levels of uncertainty, effective communication can help to activate ambidexterity and elevate the company's capacity to mitigate a crisis. Therefore, it is important for the leadership to establish mutual communication channels that allow frequent and transparent exchange with the staff. In this way, the leadership stays informed about the challenges of employees while the employees receive information about new developments, which in turn increases their level of confidence for their daily operations. This leads also to higher levels of adaptability as changes in the company's direction can be communicated and adopted quickly and efficiently. In the presented case, the company leadership frequently communicated via video messages, distributed via the firm's intranet, while the employee perspective was taken in by regular townhall meetings or an employee survey.

Limitations and Future Research Directions

As any empirical work, this study has some caveats that have to be considered. With regard to this study's sample and empirical approach, the generalizability of the model is limited. First, although our sample size is consistent with many studies in the family business literature (e.g., Arzubiaga, Kotlar, De Massis, Maseda, and Iturralde, 2018; Hauck and Prügl, 2015; Kammerlander et al., 2020), we acknowledge that it is relatively small. While our model specifications are consistent with the 1-in-10 rule of thumb (Harrell, Lee, Califf, Pryor, and Rosati, 1984), we contend that in an optimal setting, one should use a sample size that allows for a higher statistical power and the inclusion of further control variables. Thus, we hope that future research replicates our results with larger data sets. Second, our sample is restricted to one organizational, industrial (tourism) and cultural context (Switzerland), and the results might be different across cultures. Moreover, we focus on a specific type of family firm in the tourism industry – namely, comparably large ones and those with family ownership and non-family CEOs and TMT members. Future research should replicate our study design in other family firm contexts and with different family and firm contexts. Third, our research design is cross-sectional, thus inhibiting conclusions about the causality of the relationships between the predictor and outcome variables. Thus, future research could follow up with longitudinal as well as experimental research designs (e.g., Lude and Prügl, 2020) in order to uncover direction and causality of the relationships uncovered in this study.

REFERENCES

Adamu, A. A., & Mohamad, B. (2019). A reliable and valid measurement scale for assessing internal crisis communication. *Journal of Communication Management (London)*, *23*(2), 90–108. doi:10.1108/JCOM-07-2018-0068

Amann, B., & Jaussaud, J. (2012). Family and non-family business resilience in an economic downturn. *Asia Pacific Business Review*, *18*(2), 203–223. doi:10.1080/13602381.2010.537057

Amit, R., & Schoemaker, P. J. (1993). Strategic assets and organizational rent. *Strategic Management Journal*, *14*(1), 33–46. doi:10.1002mj.4250140105

Annarelli, A., & Nonino, F. (2016). Strategic and operational management of organizational resilience: Current state of research and future directions. *Omega (United Kingdom)*, *62*, 1–18. doi:10.1016/j.omega.2015.08.004

Arzubiaga, U., Kotlar, J., De Massis, A., Maseda, A., & Iturralde, T. (2018). Entrepreneurial orientation and innovation in family SMEs: Unveiling the (actual) impact of the Board of Directors. *Journal of Business Venturing*, *33*(4), 455–469. doi:10.1016/j.jbusvent.2018.03.002

Baron, R. M., & Kenny, D. A. (1986). The Moderator-Mediator Variable Distinction in Social Psychological Research: Conceptual, Strategic, and Statistical Considerations. *Journal of Personality and Social Psychology*, *51*(6), 1173–1182. doi:10.1037/0022-3514.51.6.1173 PMID:3806354

Bauweraerts, J. (2016). How Do Family Firms Manage Risky Situations? An Organizational Resilience Perspective. *International Business Research*, *9*(5), 1. doi:10.5539/ibr.v9n5p1

Brislin, R. W. (1970). Back-Translation for Cross-Cultural Research. *Journal of Cross-Cultural Psychology*, *1*(3), 185–216. doi:10.1177/135910457000100301

Cao, Q., Simsek, Z., & Zhang, H. (2010). Modelling the joint impact of the CEO and the TMT on organizational ambidexterity. *Journal of Management Studies*, *47*(7), 1272–1296.

Carney, M. (1998). A management capacity constraint? Obstacles to the development of the overseas Chinese family business. *Asia Pacific Journal of Management*, *15*, 137–162.

Chrisman, J. J., Chua, J. H., & Steier, L. P. (2011). Resilience of family firms: An introduction. *Entrepreneurship Theory and Practice*, *35*(6), 1107–1119. doi:10.1111/j.1540-6520.2011.00493.x

Chua, J. H., Chrisman, J. J., & Sharma, P. (2003). Succession and nonsuccession concerns of family firms and agency relationship with nonfamily managers. *Family Business Review*, *16*(2), 89–107. doi:10.1111/j.1741-6248.2003.00089.x

Chung, C. N., & Luo, X. R. (2013). Leadership succession and firm performance in an emerging economy: Successor origin, relational embeddedness, and legitimacy. *Strategic Management Journal*, *34*, 338–357.

Conz, E., Lamb, P. W., & De Massis, A. (2020). Practicing resilience in family firms: An investigation through phenomenography. *Journal of Family Business Strategy*, *11*(2), 100355. doi:10.1016/j.jfbs.2020.100355

Conz, E., & Magnani, G. (2020). A dynamic perspective on the resilience of firms: A systematic literature review and a framework for future research. In European Management Journal (Vol. 38). doi:10.1016/j.emj.2019.12.004

Doern, R., Williams, N., & Vorley, T. (2019). Special issue on entrepreneurship and crises: Business as usual? An introduction and review of the literature. *Entrepreneurship and Regional Development*, *31*(5–6), 400–412. doi:10.1080/08985626.2018.1541590

Duncan, R. B. (1976). The ambidextrous organization: Designing dual structures for innovation. *The Management of Organization, 1*(1), 167-188.

Duran, P., Kammerlander, N., van Essen, M., & Zellweger, T. (2016). Doing more with less: Innovation input and output in family firms. *Academy of Management Journal, 59*(4), 1224–1264. doi:10.5465/amj.2014.0424

Gibson, C. B., & Birkinshaw, J. (2004). The antecedents, consequences, and mediating role of organizational ambidexterity. *Academy of Management Journal, 47*(2), 209–226. doi:10.2307/20159573

Golicic, S. L., Flint, D. J., & Signori, P. (2017). Building business sustainability through resilience in the wine industry. *International Journal of Wine Business Research, 29*(1), 74–97. doi:10.1108/IJWBR-02-2016-0005

Good, D., & Michel, E. J. (2013). Individual ambidexterity: Exploring and exploiting in dynamic contexts. *The Journal of Psychology, 147*(5), 435–453. doi:10.1080/00223980.2012.710663 PMID:24003589

Graves, C., & Thomas, J. (2006). Internationalization of Australian family businesses: A managerial capabilities perspective. *Family Business Review, 19*, 207–224.

Harrell, F. E., Lee, K. L., Califf, R. M., Pryor, D. B., & Rosati, R. A. (1984). Regression modelling strategies for improved prognostic prediction. *Statistics in Medicine, 3*(2), 143–152. doi:10.1002im.4780030207

Hauck, J., & Prügl, R. (2015). Innovation activities during intra-family leadership succession in family firms: An empirical study from a socioemotional wealth perspective. *Journal of Family Business Strategy, 6*(2), 104–118. doi:10.1016/j.jfbs.2014.11.002

Herbane, B. (2015). Threat orientation in small and medium-sized enterprises. *Disaster Prevention and Management: An International Journal, 24*(5), 583–595. doi:10.1108/DPM-12-2014-0272

Herbane, B. (2019). Rethinking organizational resilience and strategic renewal in SMEs. *Entrepreneurship and Regional Development, 31*(5–6), 476–495. doi:10.1080/08985626.2018.1541594

Hiebl, M. R. W. (2015). Family involvement and organiza- tional ambidexterity in later-generation family businesses: A framework for further investigation. *Management Decision, 53*(5), 1061–1082. doi:10.1108/MD-04-2014-0191

Johnson, N., Elliott, D., & Drake, P. (2013). Exploring the role of social capital in facilitating supply chain resilience. *Supply Chain Management, 18*(3), 324–336. doi:10.1108/SCM-06-2012-0203

Kammerlander, N., & Ganter, M. (2015). An attention-based view of family firm adaptation to discontinuous technolog- ical change: Exploring the role of family CEOs' noneco- nomic goals. *Journal of Product Innovation Management, 32*(3), 361–383. doi:10.1111/jpim.12205

Kammerlander, N., Patzelt, H., Behrens, J., & Röhm, C. (2020). Organizational Ambidexterity in Family-Managed Firms: The Role of Family Involvement in Top Management. *Family Business Review, 33*(4), 393–423. doi:10.1177/0894486520961645

Kim, Y. (2020). Organizational resilience and employee work-role performance after a crisis situation: Exploring the effects of organizational resilience on internal crisis communication. *Journal of Public Relations Research, 32*(1–2), 47–75. doi: 10.1080/1062726X.2020.1765368

Korber, S., & McNaughton, R. B. (2018). Resilience and entrepreneurship: A systematic literature review. *International Journal of Entrepreneurial Behaviour & Research, 24*(7), 1129–1154. doi:10.1108/IJEBR-10-2016-0356

Kossek, E. E., & Perrigino, M. B. (2016). Resilience: A Review Using a Grounded Integrated Occupational Approach. *The Academy of Management Annals, 10*(1), 729–797. doi:10.5465/19416520.2016.1159878

Le Breton-Miller, I., & Miller, D. (2006). Why Do Some Family Businesses Out-Compete? Governance, Long-Term Orientations, and Sustainable Capability. *Entrepreneurship Theory and Practice, 30*(6), 731–746. doi:10.1111/j.1540-6520.2006.00147.x

Le Breton–Miller, I., Miller, D., & Steier, L. P. (2004). Toward an integrative model of effective FOB succession. *Entrepreneurship Theory and Practice, 28*(4), 305–328. doi:10.1111/j.1540-6520.2004.00047.x

Lengnick-Hall, C. A., Beck, T. E., & Lengnick-Hall, M. L. (2011). Developing a capacity for organizational resilience through strategic human resource management. *Human Resource Management Review, 21*(3), 243–255. doi:10.1016/j.hrmr.2010.07.001

Linnenluecke, M. K. (2017). Resilience in Business and Management Research: A Review of Influential Publications and a Research Agenda. *International Journal of Management Reviews, 19*(1), 4–30. doi:10.1111/ijmr.12076

Lubatkin, M. H., Simsek, Z., Ling, Y., & Veiga, J. F. (2006). Ambidexterity and performance in small-to medium-sized firms: The pivotal role of top management team behav- ioral integration. *Journal of Management, 32*(5), 646–672. doi:10.1177/0149206306290712

Lude, M., & Prügl, R. (2020). Experimental studies in family business research. *Journal of Family Business Strategy*, 100361.

Martínez-Ros, E., & Orfila-Sintes, F. (2012). Training plans, manager's characteristics and innovation in the accommodation industry. *International Journal of Hospitality Management, 31*(3), 686–694. doi:10.1016/j.ijhm.2011.09.004

Mehr Grün im weissen Skiparadies - PET-Recycling Schweiz. (2019, February 24). *Petrecycling.Ch.* https://www.petrecycling.ch/de/entdecken/detail/mehr-gruen-im-weissen-skiparadies

Mitchell, R., Morse, E., & Sharma, P. (2003). The transacting cognitions of nonfamily employees in the family businesses setting. *Journal of Business Venturing, 18*(4), 533–551. doi:10.1016/S0883-9026(03)00059-4

Mom, T. J., Chang, Y. Y., Cholakova, M., & Jansen, J. J. (2019). A multilevel integrated framework of firm HR practices, individual ambidexterity, and organizational ambidexterity. *Journal of Management, 45*(7), 3009–3034. doi:10.1177/0149206318776775

Morgan, T., Anokhin, S., Ofstein, L., & Friske, W. (2020). SME response to major exogenous shocks: The bright and dark sides of business model pivoting. *International Small Business Journal: Researching Entrepreneurship, 38*(5), 369–379. doi:10.1177/0266242620936590

Ossenbrink, J., Hoppmann, J., & Hoffmann, V. H. (2019). Hybrid ambidexterity: How the environment shapes incumbents' use of structural and contextual approaches. In Organization Science (Vol. 30). doi:10.1287/orsc.2019.1286

Schulze, G., & Mori, S. (1993). Increases in environmental entropy demand evolution. *Acta Biotheoretica, 41*, 149–164. doi:10.1007/BF00712163

Schweiz: größte Skigebiete - Größtes Skigebiet in der Schweiz. (n.d.). *Skiresort. De.* Retrieved December 3, 2020, from https://www.skiresort.de/skigebiete/schweiz/sortiert-nach/laenge-pisten/#:%7E:text=Gr%C3%B6%C3%9Ftes%20Skigebiet%20in%20der%20Schweiz%20(580%20km),Die%20Pistenl%C3%A4nge%20betr%C3%A4gt%20580%20km

Sciascia, S., & Mazzola, P. (2008). Family involvement in ownership and management: Exploring nonlinear effects on performance. *Family Business Review, 21*, 331–345.

Shin, J., Taylor, M. S., & Seo, M.-G. (2012). Resources for Change: The Relationships of Organizational Inducements and Psychological Resilience to Employees' Attitudes and Behaviors toward Organizational Change. *Academy of Management Journal*, *55*(3), 727–748. doi:10.5465/amj.2010.0325

Shin, J., Taylor, M. S., & Seo, M.-G. (2012). Resources for Change: The Relationships of Organizational Inducements and Psychological Resilience to Employees' Attitudes and Behaviors toward Organizational Change. *Academy of Management Journal*, *55*(3), 727–748. doi:10.5465/amj.2010.0325

Spiess, T., & Zehrer, A. (2020). Employees' change-oriented and proactive behaviors in small and medium-sized family businesses. In J. M. Saiz-Àlvarez, J. Leitao, & J. M. Palma-Ruiz (Eds.), *Entrepreneurship and Family Business Vitality - Surviving and Flourishing in the Long Term* (pp. 49–64). Springer. doi:10.1007/978-3-030-15526-1_4

Stubner, S., Blarr, W. H., Brands, C., & Wulf, T. (2012). Organizational Ambidexterity and Family Firm Performance. *Journal of Small Business and Entrepreneurship*, *25*(2), 217–229. doi:10.1080/08276331.2012.10593570

Tabor, W., Chrisman, J. J., Madison, K., & Vardaman, J. M. (2018). Nonfamily Members in Family Firms: A Review and Future Research Agenda. *Family Business Review*, *31*(1), 54–79. doi:10.1177/0894486517734683

Tushman, M. L., & O'Reilly, C. A. III. (1996). Ambidextrous organizations: Managing evolutionary and revolutionary change. *California Management Review*, *38*(4), 8–29. doi:10.2307/41165852

Veider, V., & Matzler, K. (2016). The ability and willingness of family-controlled firms to arrive at organizational ambidexterity. *Journal of Family Business Strategy*, *7*(2), 105–116. doi:10.1016/j.jfbs.2015.10.001

Venugopal, A., Krishnan, T. N., Kumar, M., & Upadhyayula, R. S. (2019). Strengthening organizational ambidexterity with top management team mechanisms and processes. *International Journal of Human Resource Management*, *30*(4), 586–617. doi:10.1080/09585192.2016.1277369

Weissearena. (2020, October 5). *Portrait*. WAG.

Williams, T. A., Gruber, D. A., Sutcliffe, K. M., Shepherd, D. A., & Zhao, E. Y. (2017). Organizational response to adversity: Fusing crisis management and resilience research streams. *The Academy of Management Annals*, *11*(2), 733–769. doi:10.5465/annals.2015.0134

Williams, T. A., Gruber, D. A., Sutcliffe, K. M., Shepherd, D. A., & Zhao, E. Y. (2017). Organizational response to adversity: Fusing crisis management and resilience research streams. *The Academy of Management Annals*, *11*(2), 733–769. doi:10.5465/annals.2015.0134

Wimmer, R., Domayer, E., Oswald, M., & Vater, G. (2005). Familienunternehmen — Auslaufmodell oder Erfolgstyp? Familienunternehmen — Auslaufmodell oder Erfolgstyp? doi:10.1007/978-3-322-82895-8

World Economic Forum. (2019). *The Global Risks Report*. Retrieved from http://wef.ch/risks2019

Youssef, C. M., & Luthans, F. (2007). Positive Organizational Behavior in the Workplace. *Journal of Management*, *33*(5), 774–800. doi:10.1177/0149206307305562

Zehrer, A., & Leiß, G. (2019). Family entrepreneurial resilience – an intergenerational learning approach. *Journal of Family Business Management,* (ahead-of-print). doi:10.1108/JFBM-09-2018-0037

Zimmermann, A., Raisch, S., & Cardinal, L. B. (2018). Managing persistent tensions on the frontline: A configurational perspective on ambidexterity. *Journal of Management Studies*, *55*(5), 739–769. doi:10.1111/joms.12311

Section 4

Digitalizing the Tourism Product: Risk or Opportunity?

The fourth and final section of this book, 'Digitalizing the Tourism Product: Risk or Opportunity?' consists of three chapters exploring the communication of family firms to potential guests in a digitalized word. Therefore, Chapter 9 analyzes how family-run hotels can use social media as a communication tool in times of crisis. Chapter 10 addresses the resilience of the traveler and strategies for family firms to communicate with them. Finally, Chapter 11 examines the risk of information overload in a digitalized world and how holiday farms could address this risk in communication with their guests.

Chapter 9
Physically Apart, Emotionally Close:
How Family-Run Hotels Used Instagram During the COVID-19 Pandemic - Social Media Crisis Communication During the COVID-19 Pandemic

Denise Fecker
MCI Management Center Innsbruck, Austria

Monica Nadegger
https://orcid.org/0000-0002-2489-7065
MCI Management Center Innsbruck, Austria

Stefanie Haselwanter
MCI Management Center Innsbruck, Austria

ABSTRACT

This chapter aims to explore the use of social media channels in the crisis communication of Austrian family-led hotels during the COVID-19 pandemic. Building on current research on family businesses and crisis communication on social media, the authors investigated seven Instagram profiles of family-owned hotels in Tyrol and its borders. The data were collected in spring and summer 2020 and then analyzed using a mixed-method approach. First, the researchers did a quantitative analysis of the hashtags and then conducted a qualitative content analysis of the pictures and respective captions. The results show that information

DOI: 10.4018/978-1-7998-7352-5.ch009

relating to the crisis is rarely communicated and the focus of communication rather lies on positive attitudes, emotions, and classic alpine marketing topics. During the crisis, the hotels present themselves as resilient and anchored in their family values. This chapter adds new insights to current research on family firms' crisis communication and provides valuable findings for the development of a successful communication strategy for family businesses.

INTRODUCTION

In light of the COVID-19 pandemic, the tourism industry suffered from an unprecedented economic crisis, resulting in a decrease of -70% in international tourist arrivals (World Tourism Organization, 2020). The global crisis disrupted global travel, but also most of the social interactions. Tourism firms and entrepreneurs have to adapt and mitigate the negative effects of the crisis (Zenker and Kock, 2020). One way to stay close to potential travelers is to communicate via digital channels (Wong, Ou et al., 2021). On the one hand, the continuous exchange between tourism firms and (potential) visitors facilitates the transmission of vital information and safety measures (Sano and Sano, 2019). On the other hand, it enables tourism organizations like hotels, restaurants, or destinations to stay emotionally close to their visitors despite the physical distance and thereby increase the intention to visit the region after the crisis (Hang, et al., 2020).

This book chapter investigates the crisis communication of family businesses during the COVID-19 pandemic in spring and summer 2020. The researchers focus on the peculiarities of family businesses, the decisive influence of the family as manager-owners, and their values. It situates the crisis communication in the context of family firms and investigates how family-led businesses communicate during the first COVID-19 lockdown and the reopening in the summer season. Theoretically, the research is anchored in the field of crisis management and crisis communication (Faulkner, 2001; Pillmayer and Scherle, 2018; Ritchie and Jiang, 2019), with a focus on family values (and the resulting challenges) of family firms (Rau et al., 2019). As Nadegger et al. (2020) have argued in their analysis of tourism destinations in the COVID-19 crisis, strategic and authentic communication can foster engagement with (potential) guests, deliver crucial information in time, and create emotional intimacy despite physical distance. With the steady rise of ICTs (Information and Communication Technologies), there is a need to investigate the basic principles of the new ways to communicate (e.g. spontaneous, emotional, interactive) (Oliveira and

Huertas, 2019; Roshan et al., 2016) also within the family firm context. Therefore, this study investigates the major themes used by family-owned hotels in their crisis communication on Instagram during the lockdown and the reopening of the Austrian tourism sector in spring and summer 2020.

With a multi-modal framework (Leeuwen, 2011; Ravelli and van Leeuwen, 2018), this study analyzes 598 Instagram posts of seven family-led hotels in Austria. The complex nature of Instagram posts is analyzed by adopting a mixed-method approach, regarding the hashtags, the visual elements, and the captions. The underlying themes are inductively discovered and discussed according to two different periods: the lockdown from March to May and the reopening from June to September 2020. This book chapter discusses how family-run hotels cope with various stages during a global pandemic by assessing their crisis communication on Instagram. The researchers identify underlying strategies and themes that are used to keep their community close despite being unable to travel or to visit the region. The paper concludes by discussing the relevance of social media platforms for small family businesses to achieve this kind of emotional closeness through crisis communication on social media platforms. The insights should help family business firms to embrace their family values (e.g. solidarity, persistence among others, for details see Rau et al., 2019) and also include them for a successful communication strategy - in times of crisis and afterwards.

BACKGROUND

Corporate Culture and Values in Family Businesses

Holland and Oliver (1992, p. 262) define family-led firms as businesses where "[...] ownership & management are concentrated within a family unit, and to the extent its members strive to achieve, maintain and/or increase intra-organizational based relatedness." Due to the concentration of ownership & management by the family members, the family has a decisive influence and control over the decision-making processes (Sharma et al., 1997). It becomes apparent that family-led businesses are characterized by a coexistence of two systems, the system "family" on one side and the system "business" on the other side. It is argued that these two systems in a family-led business have a reciprocal influence on each other (Burns, 2001). Family structures and the entrepreneurial mechanisms are thus inevitably intertwined. Therefore, family businesses are characterized by the intensive family structures and the intra-organizational relationships among the family members, which have been established possibly also across several generations (Pikkemaat and Zehrer, 2016; Ramadani and Hoy, 2015, Sharma et al., 1997).

It becomes apparent that especially the cross-generational relatedness, the intertwined family structures and the inter-organizational ties distinguish family businesses from non-family-led businesses (Sharma et al, 1997). This distinguishing character reflects best in the following definition: "A business is a family business when it is an enterprise growing out of the family's needs, built on the family's abilities, worked by its hand and minds, and guided by its moral and spiritual values" (Lea 1998, p. 1 in Zehrer and Haslwanter, 2011, S. 149). Reviewing these definitional components of family businesses, it becomes evident that especially 'softer' factors form a special corporate culture which in turn is based on certain family beliefs and values.

Due to the coexistence of these two systems, the corporate culture is formed by the core family values, which stem from the family system and manifest within the business context (Hall et al., 2006). Cross-generational family stories and the personalities of the involved family members have a lasting impact on the cultural patterns in family businesses (Baumgartner, 2009; Bridge et al., 1998). Based on the particular families' traditions and histories, each family develops a certain set of core values and beliefs (Baumgartner, 2009, Ward, 2004). It is argued that family members internalize these lived values through socialization already in their childhood and thus adopt these core principles (Hall, 2003). Value systems in a family are dynamic and evolve over time and develop over generations (Ward, 2004).

Values in an institutional context can be defined as "[…] social principles, goals and standards […] (that) define what members of an organization care about." (Hatch, 1997, p. 214). Due to the interrelatedness of the systems family and business, the personal values of the family itself are embedded in the business context (Hall et al., 2006; Sorenson, 2011). Carlock and Ward argue that (2001, p.35): "[…] the family's commitment and vision of itself are shaped and by what the family holds as important […]. For these reasons, core family values are the basis for developing a commitment to the business". Thus, family businesses are guided by a vision which is explicitly or implicitly acknowledged by the involved family members (Sharma et al., 1997). The vision for the future of the family business is a manifestation of personal family values (Ward, 2004). According to Ward (2004), family values form the foundation for the vision, ownership structure and the formulation of the strategy in family businesses. Values play a crucial role in the corporate culture of a family business and guide the management practices and leadership principles within it (Baumgartner, 2009; Hall et al., 2006).

Rau et al (2019) argue that these values and beliefs represent the core of family businesses and their distinctiveness. Hence, the cultural patterns in family-led businesses are an inimitable bundle of knowledge, stories, traditions, and resources (Habbershon and Williams, 1999; Poza, 2007). In literature, it is acknowledged that the core values resulting in a certain corporate culture can form a source of competitive advantage in family businesses (Dana and Smyrnios, 2010; Habbershon et al., 2003; Habbershon and Williams, 1999; Gibb Dyer, 1994). As the values and beliefs are rooted in the unique traditions and histories of family firms, they are quite impossible to imitate by their competitors (Dana and Symrnios, 2011). In addition to that, it is assumed that a value-based management approach can support the long-term survival of family firms (Rau et al., 2019). Values thus, can provide an essential guideline for family businesses to direct their strategic actions and underline their competitiveness (Sorenson, 2011).

Value Framework According to Rau et al. (2019)

As each family is unique (Zehrer and Leiß, 2019), it is assumed that also the embedded cultural patterns and value systems in family-led businesses are unique and heterogeneous (Habbershon and Williams, 1999; Rau et al., 2019). Due to this fact, Rau et al. (2019, p. 203) developed a six factor framework for values in family firms (see Table 1), which is generic in nature and applicable to family businesses in a wide range of sectors. Their framework is theoretically anchored in values theory according to Kabanoff, Waldersee, and Cohen (1995) and Schwartz (1992). Their research is based empirically on a sample of 170 family-owned firms in Germany operating in the manufacturing sector. The authors analyzed the respective mission statements, company philosophies, and firms' histories through an exploratory quantitative content analysis approach followed by an exploratory factor analysis. The identified values are constituted by six general value categories, namely, universal responsibility, benevolence, family (firm) persistence, cohesion & solidarity, power & innovation, and embeddedness. These value categories are differentiated in further detail by more specific value dimensions (Rau et al., 2019, p. 199-203). The six value categories are defined and depicted with the value dimension in the following table.

Resilience in Family Business and Connection to Values in Family Firms

In literature, it is argued that companies – regardless of their ownership structure and size – must continually adapt to the uncertain environments to sustain their competitiveness and longevity in a sustainable way (Burnard and Bhamra, 2011). Changing factors, whether internal or external, have an impact on the business and thus can make any organizational system vulnerable (Beech et al.; 2020). Especially unforeseeable external changes in the economic, social, and ecological context can result in disruption and discontinuity. A volatile and uncertain environment therefore can threaten the existence of companies (Astrachan, 2010; Bhamra et al, 2011; Burnard and Bhamra, 2011). Reacting, adapting, coping, and recovering with

Table 1. Six factor framework for values in family firms by Rau et al., 2019

Value Category	Definition	Value Dimensions
Universal Responsibility	Understanding one's responsibility for and protecting the welfare of all people and nature	• Reliability • Sustainability • Community • Competence • Responsibility
Benevolence	Understanding and appreciating the dignity of human beings	• Equality • Tolerance • Respect
Family (Firm) Persistence	Respecting and being proud of and committed to the history, customs, and achievements of family	• Family • Succession • Tradition
Cohesion & Solidarity	Accepting challenges and standing together, even in times of struggle	• Being challenged • Loyalty
Power & Innovation	Valuing and aspiring independence and control over people and resources	• Dominance • Creativity & Exploration • Ambition
Embeddedness	Being proud of, appreciating, and wishing to preserve one's origins	• Home • Solidity • Reputation

Source: (Rau et al., 2019, p. 203)

and from these changes in a resilient way is hence crucial in a volatile and turbulent environment (Beech et al., 2020; Chrisman et al., 2011).

Resilience in general can be defined as "[...] the capability and ability of an element to return to a stable state after a disruption" (Burnard and Bhamra, 2011, 2011, p. 5376). Within the organizational context, resilience can be described as a "[...] system's robustness to change and is an element of both response and vulnerability, which shapes an adaptive capacity to respond and recover thereby continuing to achieve its purpose" (Beech et al., 2020, p. 163). As family businesses tend to focus on a sustainable continuation and succession of the business over several generations (Astrachan, 2010; Nordqvist and Melin, 2000), it is argued that a certain degree of resilience in family businesses is essential for cross-generational longevity (Astrachan, 2010; Chrisman et al., 2011; Conz et al., 2020). Burnard and Bhamra (2011) argue that organizational resilience resides in both, the individual and organizational level. Applying this argument to the context of family-led businesses, it can be claimed that resilience also resides within the family itself.

Hanson et al. (2019, p. 2) argue that family resilience in a family-led firm represents the "core of functionally strong families". A certain degree of resilience in a family business enhances the family and business system to cope and adapt to risky and stressful events (Hanson et al., 2019; Zehrer and Leiß, 2019). Familial resilience is embedded in the intertwined relationships of the involved family members and is developed over time on the cultural and interpersonal patterns in a family business (Hanson et al., 2019). Walsh (2002) claims that family resilience is fostered by shared beliefs, norms, and values. As values provide fundamental guidelines for family businesses and support (Sorenson, 2011), it is especially relevant for family businesses to reconsider their core value system also during turbulent times of crisis and change and thus to align their decisions and processes.

The current COVID-19 pandemic is an example of how volatile external factors can threaten the existence of family-led companies. The pandemic poses major challenges especially for family-led businesses in the tourism sector (Schwaiger and Zehrer, 2020, Shapoval et al., 2002). Due to the COVID-19 outbreak, travel bans have been issued and locked borders were the consequence, thus, traveling has been limited severely (Zenker and Kock, 2020). This not only impacts the jobs generated by family businesses in tourism but also the liquidity and, thus, the long-term survival of tourism businesses (Bajrami et al., 2020). In light of the turbulent circumstances, it is especially relevant for family businesses in tourism to cope with the resulting challenges in a resilient way (Schwaiger and Zehrer, 2020). Moreover, it can be observed that the pandemic also brings a shift in consumer behavior. Due to isolation and lockdown, consumers are increasingly using social media for communicating and informing themselves (Donthu and Gustafsson, 2020). Tourism in general is a service-oriented industry, which involves a direct host-guest interaction; this becomes especially apparent in family-led tourism businesses. To maintain and enhance this host-guest relationship also during the pandemic, social media marketing is particularly suitable (Wong, Ou et al., 2021).

Crisis Communication on Social Media

Crisis communication reflects a central part of every crisis emergency plan which belongs, next to identifying potential danger sources, to the first of three phases of crisis management (Racherla and Hu, 2009; Santana, 2004). Facing the networked/cross-linked world of today, tourism organizations cannot allow themselves not to deal with the topic of crisis management (Pillmayer and Scherle, 2018). An overview on tourism risk, crisis, and disaster management is given by Ritchie and Jiang (2019) who conducted a literature review of articles discussing these themes published between 1960 and 2018. Of all crisis management efforts, crisis communication is seen as one of the most important factors for an organization because the organization gets judged for its behavior towards the current situation (Civelek et al., 2016; Edworthy et al., 2015). While Ritchie (2004, p. 670) argues that "crisis and disaster management should be a core competency for tourism destination managers" and Coombs (2015) notes crisis communication needs to be strategic, both seem to be equally important for hotel managers as well (Campiranon and Scott, 2014).

In the process of crisis management, the first phase "crisis precaution" is followed by the second phase "crisis" during which the crisis emergency plan is activated to minimize any damage. Finally, the third phase "resolution of crisis" contains reconstruction and rehabilitation as well as evaluating and adjusting the crisis management plan (Santana, 2004). A more extensive crisis process for tourism destinations has been described by Faulkner (2001) who defined six phases (Figure 1). Following a pre-event phase (1) he added a prodromal phase (2) in which a crisis becomes unavoidable apparent, the emergency (phase 3), intermediation (phase 4), long-term recovery (phase 5), and resolution (phase 6). Even though this framework has been approached and modified in a few studies, Ritchie and Jian (2019) claim for more current testing of the model.

Figure 1. Tourism disaster management framework
(Source: based on Faulkner, 2001)

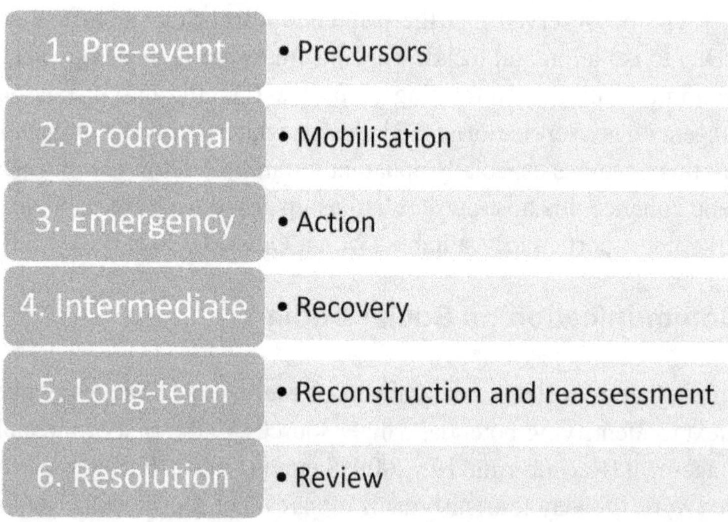

Previous research into classical crisis communication follows an approach which defines recipients of crisis communication as mostly passive with organizations using mass media. In contrast to this approach, recipients in social media communication are more engaged and enabled to answer organizations' messages directly and thus, be part of interactive and real-time communication (Chawinga, 2017; Schultz et al. 2011). Since social media marketing has become an essential part of the marketing mix, family-led businesses might use the advantages of social media channels for crisis communication. Posting images and text to social media channels not only transports information but also emotion to the followers and creates a relationship with followers. In addition, social media facilitates early and flexible communication as electronic word-of-mouth enables new forms of rapidly spreading current information (Lis and Korchmar, 2013). In terms of emergencies, for example, an earthquake or terrorist attacks, providing information about problems and needs or calling for help can cause immediate supportive action worldwide (Alexander, 2014; Lazreg et al., 2018). It has been shown that social media has been more effective than traditional forms of communication during crisis and emergency events (Curry et al., 2015).

In the tourism sector, advantages, and values of sharing visual content have been recognized long ago (Chalfen, 1979). Today, by the use of social media, travelers are able to stay in touch with friends, family, and other travelers and share their experiences in real-time and may engage other people to travel (Germann Molz and Paris, 2015). By creating, expanding, or re-shaping the perception of a destination, photographs and snapshots have a huge influence on the destinations' image (Volo, 2020). In addition, destination management organizations acknowledge the increasing influence of social media on the customer journey in tourism. Regarding social media activities, Tyrolean destination management organizations claim more cooperation of all stakeholders within a destination to ensure communication and coordination (Bosio et al., 2018).

In order to make full use of social media advantages in crisis communication, it is necessary to identify not only a crisis communication plan but also a social media strategy, which defines content management on various channels (Gilsbach, 2018; World Tourism Organization, 2011). In general, travelers use digital media to inform themselves about the latest news concerning the crisis in a destination, safety issues, and current precautionary measures in potential accommodations. Respecting such behavior, professionally working with social media by hotels can be helpful to manage information seeking, improve customer relationships and attract new guests. Moreover, by using social media, communication hotels may embed image-promoting measures to build up trust and credibility (Ritchie, 2004; Jeong and Lee, 2017; Triantafillidou and Yannas, 2020).

As social media differ from conventional media in terms of its functional logic, it is crucial to keep in mind its special features when using it for crisis communication. Social media were created to bring people together and connect them by their interests. That is why the logic of communication does not follow the traditional top-down principle of mass media but a peer-to-peer logic and enables users to actively take part in the information process instead of being passive recipients (Fournier & Avery, 2011). Consequently, regarding their crisis communication on social media, hotels must recognize their limited ability to control and influence information rates. Following the recent developments, social media became an object of interest for recent studies in Tyrolean tourism. For example, Bosio et al. (2018) investigated the usage of different social media channels of tourism organizations in Tyrol. Regarding the use of social media in crisis situations, several researches have focused on different types of crises. In general, DMOs, as well as economic organizations tend to not fully take advantage of the potential of social media for crisis communication, no matter if it is in times of terroristic attacks, natural disasters, or human-error accidents (Oliveira and Huertas, 2019; Roshan et al., 2016).

With the rapid rise of social media in the last two decades, Instagram, following Facebook and Youtube, has become one of the most popular social networks where people share images and videos (Caliandro and Graham, 2020; Livingstone and Brake, 2010; We Are Social, 2019). In Germany, which is a source market for Tyrolean tourism, Instagram ranks fourth of the top social networks and messengers. In January 2020, 29% of the German population over 13 years were using Instagram (DataReportal, 2020). Considering these recent developments, Instagram has become an important communication tool for tourism stakeholders in Tyrol. Beyond that, and particularly interesting for family-led businesses, Instagram as a free of charge online application offers low barriers to entry and uncomplicated operating (Volo, 2020). In addition, the transport of emotions by visual content on Instagram can support family businesses in communicating their important family values or any other emotional thoughts and current information during crisis events.

Based on the presented current state of literature, this book chapter aims to investigate what modes of communication are used by Tyrolean family-owned hotels for what kind of messages on Instagram. Moreover, it uncovers the major themes used to make sense of the crisis in front of a digital audience, considering any changes in the communication patterns between lockdown and reopening of the hospitality sector in Tyrol in 2020.

METHOD

Case Selection

As mentioned in the beginning of this article, family-led businesses play a particularly important role in the Austrian economy and this especially applies to the tourism industry in Austria. From the 315,900 registered businesses across all industries, roughly 50% are family-owned. With 67%, this number is even higher in the tourism industry (Gavac et al., 2020). The close and personal host-guest relationship is a central aspect of family businesses and builds the basis for an authentic tourism experience (Getz and Carlson, 2005; Pikkemaat and Zehrer, 2016). Based on the importance of family businesses for the tourism sector in Austria, this study chooses family-led hotels in Austria to assess the crisis communication on social media during the COVID-19 pandemic. The researchers argue that for family businesses, which rely on emotional closeness and personal host-guest relationships, social media channels are an important channel to mitigate the negative impacts of the crisis. As the European Union is also dominated by family-run businesses (European Union, 2020), the selected case study is not only of great relevance in the context of Austria but provides valuable insights for crisis communication and resilience for a broader public.

Data Collection and Data Set

As this study focuses on the crisis communication of family-owned hotels, the selection of our cases is oriented towards the biggest hotels in Tyrol that are run by families, according to top.tirol (2019). The researchers carefully examined the posting activity of the respective hotels on Instagram and included one more established family-owned hotel (Naturhotel Forsthofgut) just at the border of Salzburg and Tyrol with an extraordinarily insightful social media presence to complete our data set. As can be seen in Table 2, the Naturhotel Forsthofgut is characterized by an active posting behavior (136 Posts) by an interactive engagement with its audience with more than average 1.400 likes and more than 20 comments on average per post. In addition, they heavily emphasize the family-entrepreneurial background of their hotel on their Instagram page.

The selected hotels and respective Instagram accounts are the following:

- **Stanglwirt** (https://www.instagram.com/stanglwirt/)
- **Alpenresort Schwarz** (https://www.instagram.com/alpenresortschwarz/)
- **Stock ***** resort** (https://www.instagram.com/stockresort/)
- **Trofana Royal ***** Superior** (https://www.instagram.com/trofanaroyal/)
- **Wellnessresidenz Alpenrose** (https://www.instagram.com/wellnessresidenz.alpenrose/)
- **Schlosshotel Fiss** (https://www.instagram.com/schlosshotel_fiss/)
- **Naturhotel Forsthofgut** (https://www.instagram.com/naturhotelforsthofgut/)

Data was collected from Mid-March 2020 to Mid-September 2020 to cover the communication during the first COVID-19 lockdown, the reopening, and the main holiday season in summer 2020 in Tyrol. In the end, the data collection encompassed the Instagram channels of seven family-run hotels in Tyrol from 13th March to 30th September 2020. In total, the data set consists of 598 individual Instagram posts. Table 2 provides an overview of the selected hotels, the number of Instagram posts per hotel, and the average likes and comments per post.

Table 2. Overview and description of the collected data.

	Nr. of Posts (absolute)	Nr. of Posts (%)	Nr. of Likes (mean)	Nr. of Comments (mean)
STOCK***** resort	160	26.8	581.76	14.74
Naturhotel Forsthofgut	136	22.7	1444.46	21.16
Schlosshotel Fiss	82	13.7	130.12	1.56
Alpenresort Schwarz	68	11.4	383.91	9.43
Trofana Royal ***** Superior	53	8.9	99.83	0.91
Bio-Hotel Stanglwirt	52	8.7	2226.81	40.21
Wellnessresidenz Alpenrose	47	7.9	251.82	8.13
Total Amount of Posts	**598**	**100**	**767.9**	**14.26**

Methodological Approach

To fully grasp the communication during and shortly after the lockdown, this study investigates seven Instagram profiles of large family-run hotels in Tyrol and Salzburg from a multi-modal perspective on social media content (Leeuwen, 2011; Ravelli and van Leeuwen, 2018; Small, 2011). The researchers follow a similar approach as Nadegger et al. (2020) to analyze complex data on multiple levels. The guiding research questions for the analysis are:

- **RQ1:** What modes of communication (pictures, hashtags, captions) are used for what kind of messages on Instagram?
- **RQ2:** What are the major themes and values used by family-owned hotels to make sense of the crisis in front of a digital audience?
- **RQ3:** How do these themes change during the lockdown and shortly after the reopening of the hospitality sector in Tyrol?

As the respective data set consists of rather rich and complex data (i.e. Instagram posts), the researchers decided to analyze the different modes of communication (hashtags, captions, pictures) with different approaches. The researchers adopted a mixed-method approach with, first, a quantitative analysis of the hashtags (to crudely uncover general themes in the crisis communication) following Mayring (2001) and Merten (1995) and second, a combination of qualitative text (Mayring, 2015) and visual analysis (Rose, 2016). All three steps of analysis were assisted by the use of the software MAXQDA.

For the analysis of the captions and visual content posted, the researchers conducted a qualitative content analysis of the posted pictures and the respective captions with the aim to explore basic topics, values, and communication strategies. In a first step, the data was approached independently with an open, inductive coding strategy and, in a second step, the codes were discussed and grouped into overarching categories for the pictures and the captions. In a third step, the interrelations between the evolving themes in the captions and pictures was discussed.

With the independent open coding by all three researchers and the mixed-method approach combining qualitative and quantitative analyses, the validity and strength of analysis is ensured following the principles of researcher and method triangulation (Decrop, 1999; Flick, 2011). Following the open coding, the individual codes were grouped into major themes and discussed among the researchers.

FINDINGS

The following paragraphs discuss the findings from the quantitative hashtag analysis and the qualitative analysis of posted pictures and the Instagram captions. Regarding the qualitative coding, the findings were clustered into two major periods: the first

Figure 2. Top 15 hashtags in the data set with the percentage of occurrence in the total number of posts (n=598).

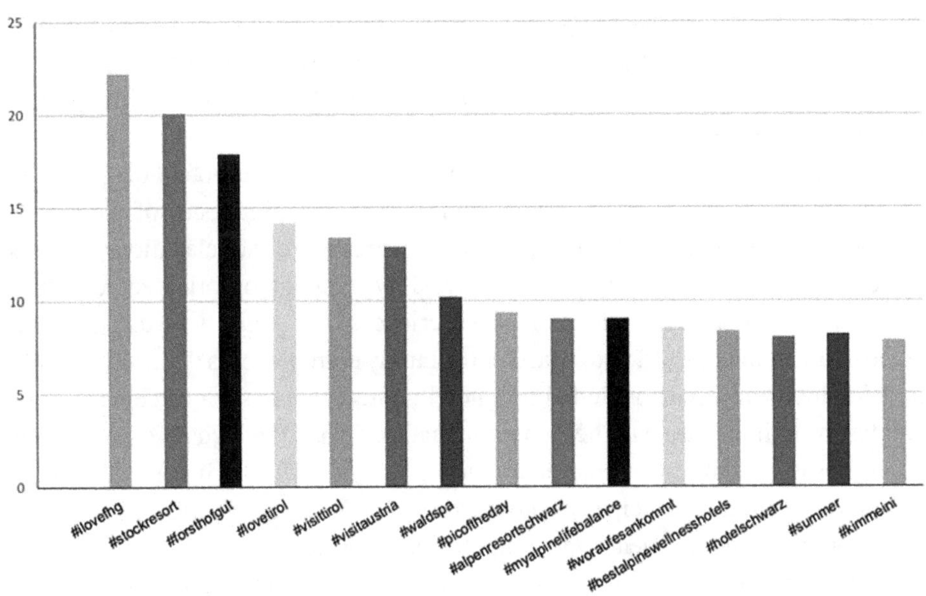

COVID-19 lockdown (March to end of May 2020) and the reopening phase in the summer season (end of May to end of September 2020).

Hashtag Analysis

Assisted by the software MAXQDA, a total list of hashtags in the data set was generated. In a second step, the most popular hashtags were uncovered and their occurrence in % of total posts is illustrated in Figure 2.

As can be seen, most of the hashtags are directly linked to the branding (#waldspa or #bestalpinewellnesshotel) and the name of the respective hotels (e.g. #forsthofgut or #stockresort). Other popular hashtags are the standardized DMO hashtags (#visittirol, #lovetirol) and more general Instagram hashtags (#picoftheday or #summer). Surprisingly, hashtags with a clear relation to COVID-19 (e.g. #stayhealthy, #corona, #stayhome) are rarely used in the posts of the hotels. This stands in contrast to the posting activities of Tyrolean DMOs, who used the predefined COVID-19 hashtags (such as #mitabstandnah or #dahoambleibn) of the government or the national destination agency Tirol Werbung (for more detail see Nadegger et al. 2020). Overall, the findings suggest that the hashtags do not indicate a strong relation to COVID-19 or respective safety or health topics when observing the social media communication of the chosen hotels.

Qualitative Text Analysis

After the individual coding by all three researchers, the emerging codes were discussed and summarized into overarching topics: general themes of the text, people referred to or speaking and ventriloquized emotions as well as distinct family values. Each of the general topics will be described in general and with the respective importance during each of the two phases.

First, the general themes that are used by the family-run hotels describe the central holiday activities and the management and marketing focus of the hotels. The most prominent topics displayed in the posts are nature and relaxation (in terms of wellness, spa, and re-energizing), followed by more action-oriented activities, fitness, and outdoor sports, and culinary experiences. The people who tell stories in the posts are mainly locals, hotel staff, or the family entrepreneurs themselves. Only the Bio-Hotel Stanglwirt included non-hotel related people such as celebrities or athletes from time to time in their communication. When the focus is on particular employees or the family entrepreneurs, they mainly provide a glimpse behind the scenes by showing new recipes or their favorite spots, as seen in Figure 3.

Important announcements or symbolic gestures (e.g. when communicating readiness or responsibility) in the reopening was often expressed in relation to the

Figure 3. A glimpse behind the scenes into everyday life of family entrepreneurs.
Retrieved November 17, 2020 from https://www.instagram.com/p/CCs_XDJHuk/.

Figure 4. Percentage of use of code categories in the textual analysis according to the two phases. Blue representing the reopening phase (from 29th May to 30th Sep) and orange representing the lockdown phase (13th March to 18th May).

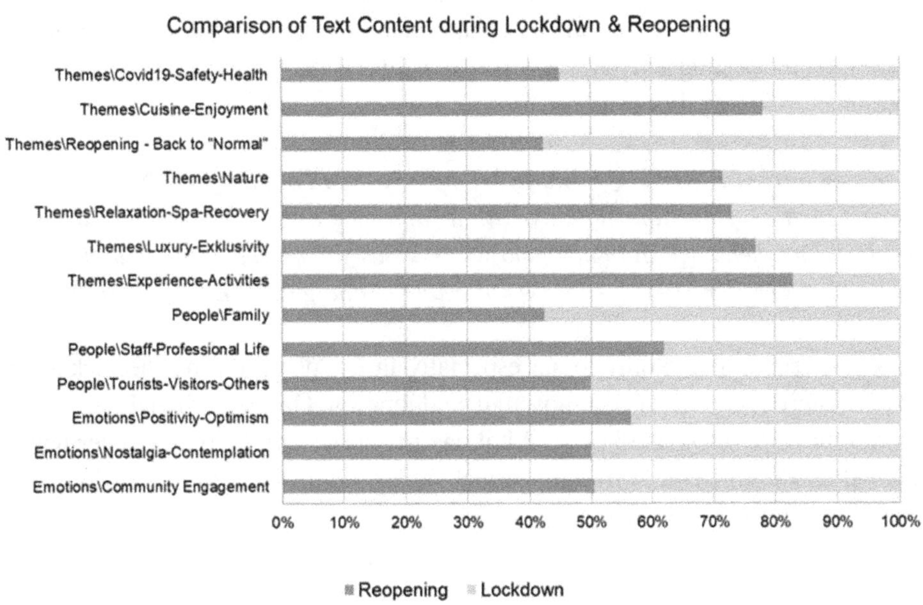

family entrepreneurs of the hotel. Hospitality, affection, and longing for a reunion were central emotions used in these announcements. Taking a look at the occurrence of coding categories in both phases (see Figure 4), there is only little difference when it comes to the use of themes "experience-activities", "cuisine-enjoyment" (which dominates the reopening period). However, examining the captions more closely, various enactments and nuances of the major themes were uncovered over time.

A majority of the posts do not reflect the COVID-19 crisis directly, but many of the themes implicitly express the increased need for space (and how the hotels ensure this new requirement), recreation after "hard times" and a "new" normality. During the lockdown, most of the hotels clearly articulated or named the COVID-19 pandemic only at the very beginning in Mid-March 2020. Afterwards, most of the captions just included implicit references to "challenging times" or worldly wisdoms such as "There's light at the end of the tunnel" or "Everything comes in time to him who knows how to wait". These references to challenging times were also used to express values of solidarity and loyalty to the guests as shown in the following example:

"Thank you for this wonderful summer! You have filled our house with love, happiness and warmth especially during these challenging times. These warm and unforgettable encounters with you make us feel so thankful and proud. We looked after each other so well and we will keep on doing this in the future as well. We promise! […]" (@ alpenresortschwarz, 15th September 2020)[1]

Although the hotels could not welcome guests on-site, many engaged with their (potential) visitors via social media channels. Using raffles, questions, or photo challenges, the posting activity was geared towards fostering connections or creating emotional proximity. For example:

"ForsthofgutHAPPINESS 🏵 Happy moments at the Forsthofgut 😍 Share your favorite pictures from your holiday with us and let's think back together to those wonderful moments 🙌 Join now and win - a stay for 2 people and 2 nights incl. ForsthofgutKITCHEN 🤗 [---] " (@naturhoteforsthofgut, 27th March 2020) 2

A[n] interesting dichotomy found especially in the phase during the lockdown was the alternating reference to nostalgia and renewal. On the one hand, the hotels referred to better days and beautiful holiday memories, while, on the other hand, channeling the difficult period as perfect timing for renovation, expansion, and renewal, as this example shows:

"Good morning from the STOCK resort! Today we have another #magicmemory for you: A photo from 1983 - a lot has happened since then and we are currently busy with rebuilding and expanding." (@stockresort, 29th April 2020) 3

In the transition from lockdown to the summer season, most hotels referred to the season start with positivity and information about the travel bans, safety measurements in the hotels themselves and the newly established hotel facilities. It is in this transition, where most hotels explicitly refer to COVID-19 related information, but as a means to celebrate the starting season.

"+++ BREAKING GOOD NEWS!! +++ 😃 The borders to Germany, Liechtenstein, Switzerland, Slovakia, Slovenia, the Czech Republic and Hungary will be open again for travel from tomorrow, the 4th of June! 🥳 This means a reunion is possible much earlier than expected for our many dear guests from these neighboring countries 🥰 We are really looking forward to seeing you and wanted to share this joyful news with you immediately. See you soon ... at home at the Stanglwirt! [...]" (@ stanglwirt, 3rd June 2020) 4

During this time, the reference to COVID-19 related information was used additionally to express universal responsibility for each other. The following quotation highlights how the family hotels showed their sense of responsibility for their guests' health.

"Place of longing. 🏞♀ To be safe and with each other: Our wellness area offers entire 50 square meters per guest, that's how to relax. 🌿." (@deralpinekraftplatz/ Wellnessresidenz Alpenrose, 15th May 2020)5

By implicitly referring to their huge space offer as an option to keep distance as well as explicitly referring to "responsibility", "safety" and "health" as in the next quotation, the hotels communicated the value of responsibility for their guests as well for their family members and employees.

"Dear fans & friends! For us, healthy employees, guests and locals are at the top of the agenda. Even though we are thankful for the high response and pleasant booking situation in the current circumstances, we decided to cancel or postpone the 30th anniversary of our legendary "Weißwurstparty", the traditional meeting of singers, as well as the Stanglwirt Christmas market. We decided this independently from possible upcoming loosening or restrictions by the government. Facing the urgent need to interrupt infection chains, we are sure that you understand this decision guided by our high sense of responsibility [...]" (@stanglwirt, 30th September 2020)6

Furthermore, in the reopening phase, the family-led businesses spent time to express both power and innovation values as well as family firm persistence. By telling that the hotel owners had not had any rest day for a long period, ambition and dominance as a dimension of power becomes obvious.

„[…] no rest day for more than 250 years! We stay true to our motto! After the "compulsory time of closure", we are happy from the bottom of our heart to open the doors of our traditional "Gasthof Stangl" to usual opening hours, daily from 11-22 o'clock." (@stanglwirt, 15th May 2020)[7]

Another value category that emerged with reopening was family firm persistence. In the following two quotations, the explicit reference to "tradition" underlines the importance of sticking to the attitude of being a welcoming host.

"That is how nice it was on reopening day at our traditional guesthouse. Our friends Tobias and Julia Moretti also came over to have fresh asparagus and of course, we made some music – as it is tradition at the Stanglwirt. We are happy and thankful to be able to be there for our guests again finally. After all, "happiness in life is happiness in giving"… to be host! 😊" (@stanglwirt, 15th May 2020)8

"Is it possible to reflect on traditional values and thus take a pioneering role? Yes, it is possible! It's a great honor to us that @voguebusiness highlights our "principle of nest warmth" as a special feature. We are looking forward to pass you this nest warmth in the only luxury resort with integrated organic farm soon. […]" (@ stanglwirt, 6th May 2020)[9]

In addition to this, the hotels expressed on Instagram the value of embeddedness by focusing on the relation to local farm supply. The support of their home region is explained in this example:

"Local. Fresh. Rich. On a regular basis, we receive seasonal products from the Gasslbauer in Ladis and many more close farms in this area. For us, it is an affair of the heart to cooperate with farmers in order to serve a piece of home on the plate to our guests." (@schlosshotel_fiss, 9th June 2020)[10]

Apart from these themes and values, positive emotions were often in the spotlight: hospitality, joy, motivation, and anticipation are key themes in the transition from the lockdown to the summer season. The summer season was dominated by posts about experiences and activities, embracing a holiday experience free of worries and paid little to no attention to the current crisis or measures taken to prevent future outbreaks.

Qualitative Visual Analysis

Similar to the textual analysis, the researchers also coded all Instagram posts (n=598) individually and subsequently grouped the codes into overarching categories: scenery shown, themes used and people depicted. However, in contrast to the captions, the visuals of the Instagram posts only slightly vary during the different phases. Overall, the use of themes remained rather constant and universal (see Figure 5).

Figure 5. Percentage of use of code categories in the visual analysis according to the two phases. Blue representing the reopening (from 29th May to 30th Sep) and orange representing the lockdown (13th March to 18th May).

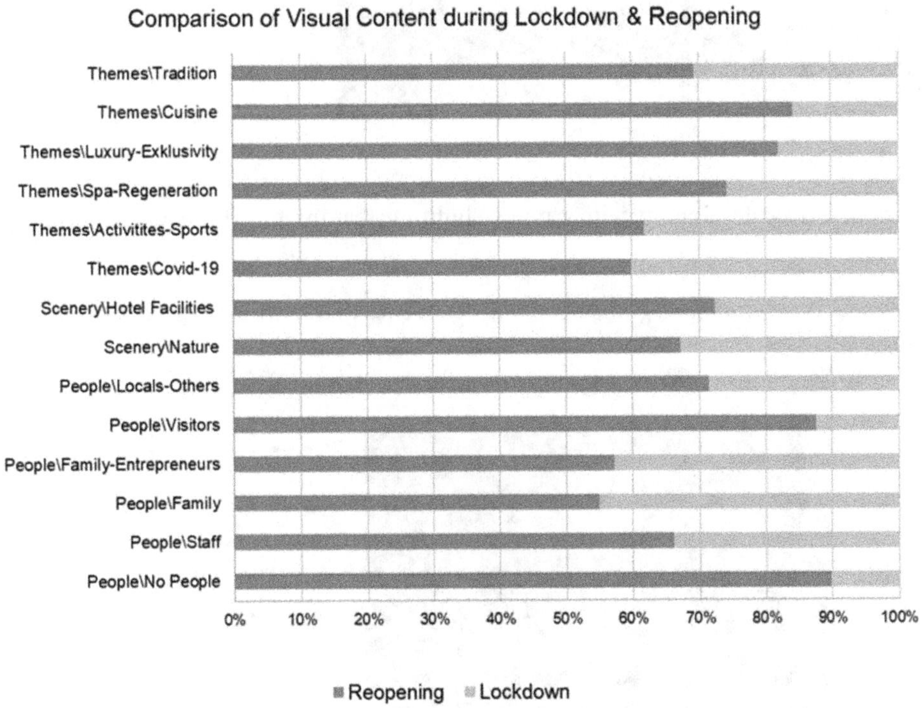

The majority of visuals display either natural surroundings (mountains, meadows, alpine scenery, forest, or close-ups of flowers) or hotel facilities (hotel architecture, rooms, suites, restaurant, spa area, or interior). The majority of pictures are carefully

Figure 6. Stylized shot of the exterior facilities.
Retrieved November 17, 2020 from https://www.instagram.com/p/CC1I5NhpEuJ/.

crafted, shot professionally with arranged lighting, feature staged situations including

Figure 7. A spontaneous glimpse behind the scenes of a hotel during the COVID-19 pandemic.
Retrieved November 17, 2020 from https://www.instagram.com/p/CBdKrGFnIXC/.

models or hotel staff, and apply extensive post-production with filters and design elements. They feature core facilities or USPs of the respective hotel (such as the spa area or the outdoor pool), but also focus on carefully arranged decorative elements (for examples see Figure 6).

Only a few images are clearly taken spontaneously (e.g. with the smartphone or household cameras) and posted without the use of filters or post-production. Most of the more "amateurish" pictures feature renovation activities, a glimpse behind the scenes, or lockdown pastimes like garden work (see Figure 7).

Similar to the topics in the captions, the major themes or activities promoted revolve around relaxation, (gentle) sportive activities, and culinary indulgence, with a tendency towards luxurious arrangements and exclusivity. Nevertheless, many pictures incorporate traditional and authentic Tyrolean elements such as wood or metal ornaments, cozy interior in the style of the "Tiroler Stubn" (a certain type of snuggery parlor, see Figure 8), or traditional folk costumes and dresses.

Figure 8. Wooden interior and a traditional fireplace - a typical "Tiroler Stubn".
Retrieved November 17, 2020 from https://www.instagram.com/p/CD_ib8XpaZ3/.

However, what is striking in terms of the visuals used, are the elements that are missing. There was practically no reference to COVID-19 in the pictures (such as face masks or floor markings for keeping distance). Moreover, there were only a couple of pictures which featured people, especially in relation to tourists, at all. Most visuals focused on interior, architecture, or nature. Family entrepreneurs and hotel staff are a few exceptions. They are either portrayed to provide a glimpse behind the scenes during the lockdown or, together with staff, show the team spirit and the readiness to start into the summer season in rare group photos, such as Figure 9. Other pictures featuring picture-perfect couples or families are arranged professionally for marketing purposes. Authentic pictures or insights of the hotel stay in the summer season from the guests' perspective are completely absent.

Figure 9. Team photos (in front of the hotel or outdoors) posted during the summer season in 2020.
Retrieved November 17, 2020 from https://www.instagram.com/p/CFKQhuNpTLx/.

Summing up the findings and the examples above, we discovered surprisingly few examples of explicit crisis topics that relate directly to the COVID-19 pandemic. Instead, the open, inductive approach revealed several themes that explicitly or implicitly linked to family values. Visual coding categories like nature or traditional settings symbolized embeddedness and an appreciation of the surrounding environment and the persistent traditions in the hotel, whereas renovation activities provided insights into the innovative power of the family hotels in crisis times. The textual analysis revealed a deep sense of community, responsibility, and connectedness to visitors and hotel staff by referring to emotions like challenging times, solidarity but also hope for the future. As a whole, Instagram communication utilized the complex dynamic of family values to connect to the audiences through emotions like solidarity, hope, and empathy, but also to display resilience and innovative power through insights into the renovation activities and a glimpse behind the scenes.

DISCUSSION

The qualitative text and visual analysis found that positive attitudes and emotions, teamwork, and motivational factors were central themes in the hotels' communication on Instagram. In terms of product marketing, typical topics of alpine summer vacation such as nature, being active, having new experiences, and wellness were communicated. Overall, the hotels constantly seemed to consider the relationship to their guests. This includes expressing lots of positive emotion and confidence in their statements. Moreover, by engaging, sending wishes and greetings to the community over all investigated phases, the hotels aimed to create closeness and somehow include the guests in their virtual family.

Crisis Communication in General

Regarding the practical features and advantages offered by Instagram as a social media platform (Ritchie 2004, Volo 2020), the hotels drew on a few, but not all, possible resources in terms of their crisis communication management. Especially in their statements, emotions and the longing for relationships were transported in order to stay in touch with the community. The hotels could give their guests the opportunity to mentally escape into the desired destination which was currently not accessible in real life. Thus, on the one hand, the hotels did take advantage of social media features by using them to transport emotions and create a positive atmosphere. To a certain extent, staying in touch and sharing real-time experiences as one main aim of social media (Germann Molz and Paris, 2015) is fulfilled. On the other hand, this study found very few incidences of where hotels took the chance to provide transparent, early, and real-time information about what was going on during the crisis in the summer season. Furthermore, the family-led hotels did not show a lot of insights and behind the scenes experiences except for occasionally giving a few glimpses of the family life.

Contrary to expectations, the results of the hashtag analysis did not indicate any strong relation to the COVID-19 crisis or health themes. As DMOs were found to make high use of COVID-related organized hashtags (#stayhome, #mitabstandnah, #farbutnear, #staypositive; see Nadegger et al. 2020), family-led hotels concentrated on their branding and product offers and used mostly their own personalized hashtags or standardized DMO hashtags (#lovetirol).

To sum up, the key findings from the analysis are that the investigated hotels did not actively communicate their crisis management or any health-related concepts via the social media channel Instagram. Apart from initial explicit references to the crisis (concerning closure and lockdown) and occasional hints during reopening (invitations to recover after hard times and pictures showing lots of space and less people to suggest safety), opportunities to inform guests about current issues relating to the crisis situation remained unused. Although the family firm might have reacted to and learned from the crisis, the visibility of their resilience seems to be located in the background as the external communication is mostly marketing orientated and following the concept of "business as usual". Following the insight that social media is one of the most effective forms of communication during crisis events (Curry et al., 2015), family-owned hotels are called to engage more actively in their communication during such situations.

Crisis Communication Regarding Family Values and Resilience

Apart from the general themes in crisis communication, a special focus of the analysis focused on the evidence of distinct family values in the hotels' communication. It becomes evident that the majority of postings are directly or indirectly anchored in the distinctive corporate culture of the respective family-led hotels. By referencing, for example, the cross-generational existence of the hotel, the familial personalities behind the scenes, and insights into the special hospitableness of the hotel, it becomes evident that the values embedded in the system "family" also reflect within the business context. This value-based social media communication is a sign that team and family members of family-led hotels internalize the core principles and beliefs and thus reflect them in their daily business and practices (Hall, 2003; Sorenson, 2011). The values reflected were the following: responsibility for the safety of guests and staff, cohesion and solidarity despite physical distance, traditions and family history, and innovation and improvement in terms of hotel infrastructure and strategic positioning. Thus, the hotels underpin their postings by communicating this special uniqueness of the respective hotel, which emerges from the intertwined family ties, their stories, and traditions (Zehrer and Leiß, 2019; Habbershon and Williams, 1999). This in turn suggests that the hotels are quite aware that a distinctive corporate culture and values can form a competitive advantage, which is of high interest for potential and existing visitors.

Overall, the results indicate that family-led hotels invoke family values especially in times of crisis. Instead of withdrawing from the public, the investigated family firms continue to rely on their traditional and supporting values and thus represent high resilience and robustness. Following the approach of the six factor framework of Rau et al. (2019), a few of the described values were identified in the Instagram posts of the family-led hotels. In the following, it is shown how the distinct value categories were communicated by the hotels.

First, the topic of universal responsibility came up during the first lockdown phase in spring 2020 due to the closure of all Tyrolean destinations including all stakeholders. Appealing to mutual responsibility and safety, the hotels represented themselves as highly responsible for their guests, employees, and family members. By explicitly referencing "responsibility" in their statements, a straight and strong appearance of the family businesses was transmitted around the 13th of March 2020. Towards the summer season, reliability was shown when the hotels referred to safety measures and hygiene concepts to show their community that they were prepared for the upcoming summer season.

In addition to being responsible in crisis times, "cohesion and solidarity" was another value category revealed by the analysis. As a response to the pandemic-related uncertainties, all investigated hotels portrayed themselves as staying positive in challenging times and solidarizing with their staff and their guests. According to this value, the hotels do not show any weakness outwardly and thus try to appear resilient to any kind of threat.

Another important value carried out in the Instagram statements was the family firms' persistence, including the dimensions "family" and "tradition". In particular, the Stock Resort intended to turn the spotlight on its family history by showing old photographs in combination with the hashtag #magicmemory. Displaying information about the family's history in nostalgic ways expresses the pride of the family members as well as the commitment to keep the family's achievements active after the crisis.

Moreover, to the family-led hotels, local traditions are just as important as family traditions. These Tyrolean traditions form local embeddedness as a further value and generate feelings of pride and community which they want to preserve. The expression of this embeddedness in the regional environment is achieved by depicting traditional Tyrolean furnishing, typical culinary highlights, and local clothing.

Finally, power and innovation were identified as one more topic in relation to family values. By showing pictures of big hotel facilities and luxury settings the hotels demonstrate a certain form of power and independence. They also implicate their control over resources by providing insights on renovating the facilities for example. In a time where lots of control is missing due to the health crisis, actively improving things or creating something new brings back the feeling of power and dominance as well as it shows ambition.

Summarizing the discussion of communicated family values, it is claimed that family-led hotels in Tyrol show high resilience based on their family values like responsibility and family tradition. With respect to Hanson et al. (2019), it can be summed up that the families' traditional values play a supportive role in crises and that the values are a safety net in difficult times.

LIMITATIONS AND FUTURE RESEARCH

The results of this study have to be considered with limited meaning. In fact, only a small sample of family-led hotels in Tyrol was investigated, which showed similar communication behavior regarding their Instagram posts. Though the qualitative analysis did not include any analysis of engagement in the form of comparing numbers of likes or followers, it still provided valuable new information to the research gap. Furthermore, the analysis included only the hotels' external communication on one social media channel and didn't cover all options on this channel. For example, the content of Instagram stories has not been included in the analysis. This could provide more insights into the channel owners' spontaneous communication. Therefore, the analysis of the Instagram posts does not reflect the whole communication extent on Instagram. Nevertheless, as Instagram is one of the most important communication channels of the respective target group, the results of this qualitative analysis still add value to the understanding of family-led hotels' communication. Thus, a further investigation on how the target group on Instagram can be characterized and subdivided would be very valuable for family businesses to improve their presence on Instagram and create more specific content.

In the future, research should increase this knowledge by examining the whole channel and other forms of external communication (hotel websites etc.) during a crisis situation. Due to the fact that internal communication was not included in the analysis, no strategy-building conclusions can be drawn. Following this, future research could focus on the existence of underlying communication strategies by analyzing internal communication processes in family-led hotels. Finally, applying the approach of Rau et al. (2019), which is originally based on family firms in the manufacturing sector, in the context of family firms in tourism, their framework is found to be very useful and with high practice-relevance for the understanding of family firms' values.

CONCLUSION

Overall, by using a mixed-method approach to analyze the external communication on Instagram during the COVID-19 pandemic some valuable insights on family firms' behavior in crisis times were provided. After taking note of the crisis situation and its consequences, all observed family-led hotels concentrated consistently on promoting their offers as usual and posted very little information about the crisis itself and the measurement taken. However, family-led hotels evoked their family values to mitigate the negative consequences of the pandemic and to stay emotionally close to their customers despite being physically apart.

REFERENCES

Alexander, D. E. (2014). Social media in disaster risk reduction and crisis management. *Science and Engineering Ethics*, *20*(3), 717–733. doi:10.100711948-013-9502-z PMID:24306994

Astrachan, J. H. (2010). Strategy in family business: Toward a multidimensional research agenda. *Journal of Family Business Strategy*, *1*(1), 6–14. doi:10.1016/j.jfbs.2010.02.001

Baumgartner, B. (2009). *Familienunternehmen und Zukunftsgestaltung: Schlüsselfaktoren zur erfolgreichen Unternehmensnachfolge*. Gabler Verlag. doi:10.1007/978-3-8349-8541-5

Bajrami, D., Terzić, A., Petrović, M. D., Radovanović, M., Tretiakova, T. N., & Hadoud, A. (2020). Will we have the same employees in hospitality after all? The impact of COVID-19 on employees' work attitudes and turnover intentions. *International Journal of Hospitality Management*, (October). Advance online publication. doi:10.1016/j.ijhm.2020.102754

Beech, N., Devins, D., Gold, J., & Beech, S. (2020). In the family way: An exploration of family business resilience. *The International Journal of Organizational Analysis*, *28*(1), 160–182. doi:10.1108/IJOA-02-2019-1674

Bhamra, R., Dani, S., & Burnard, K. (2011). Resilience: The concept, a literature review and future directions. *International Journal of Production Research*, *49*(18), 5375–5393. doi:10.1080/00207543.2011.563826

Bosio, B., Haselwanter, S., & Ceipek, M. (2018). Social Media Nutzung in Tiroler Destinationen. In Wandel im Tourismus. Internationalität, Demografie und Digitalisierung. Berlin: ESV.

Burnard, K., & Bhamra, R. (2011). Organisational resilience: Development of a conceptual framework for organisational responses. *International Journal of Production Research*, *49*(18), 5581–5599. doi:10.1080/00207543.2011.563827

Caliandro, A., & Graham, J. (2020). Studying Instagram Beyond Selfies. *Social Media + Society*, *6*(2), 205630512092477. doi:10.1177/2056305120924779

Campiranon, K., & Scott, N. (2014). Critical Success Factors for Crisis Recovery Management: A Case Study of Phuket Hotels. *Journal of Travel & Tourism Marketing*, *31*(3), 313–326. doi:10.1080/10548408.2013.877414

Carlock, R. S., & Ward, J. L. (2001). *Strategic planning for the Family Business*. Palgrave. doi:10.1057/9780230508750

Carlock, R. S., & Ward, J. L. (2001). *Strategic Planning for the Family Business – Parallel Planning to Unifythe Family and Busines*. New York Press. doi:10.1057/9780230508750

Chalfen, R. M. (1979). Photography's role in tourism: Some unexplored relationships. *Annals of Tourism Research, 6*(4), 435–447. doi:10.1016/0160-7383(79)90006-9

Chawinga, W. D. (2017). Taking social media to a university classroom: Teaching and learning using Twitter and blogs. *International Journal of Educational Technology in Higher Education, 14*(1), 3. doi:10.118641239-017-0041-6

Chrisman, J. J., Chua, J. H., & Steier, L. P. (2011). Resilience of family firms: An introduction. *Entrepreneurship Theory and Practice, 35*(6), 1107–1119. doi:10.1111/j.1540-6520.2011.00493.x

Civelek, M. E., Cemberci, M., & Eralp, N. E. (2016). The Role of Social Media in Crisis Communication and Crisis Management. *International Journal of Research in Business and Social Science, 5*(3), 111–120. doi:10.20525/ijrbs.v5i3.279

Conz, E., Lamb, P. W., & De Massis, A. (2020). Practicing resilience in family firms: An investigation through phenomenography. *Journal of Family Business Strategy, 11*(2), 100355. doi:10.1016/j.jfbs.2020.100355

Coombs, W. T. (2015). The value of communication during a crisis: Insights from strategic communication research. *Business Horizons, 58*(2), 141–148. doi:10.1016/j.bushor.2014.10.003

Curry, G. D., Leflar, J. J., Glasser, M., Loyear, R., Grey, B., Tim, J., Ong, L., Preining, W., & Sobron, J. M. (2015). How Social Media is Transforming Crisis Management and Business Continuity. *Journal of Physical Security, 8*(2), 15–36.

Dana, L. E., & Smyrnios, K. X. (2010). Family business best practices: Where from and where to? *Journal of Family Business Strategy, 1*(1), 40–53. doi:10.1016/j.jfbs.2009.12.002

DataReportal. (2020). *Digital 2020: Germany 2020*. https://datareportal.com/reports/digital-2020-germany

Decrop, A. (1999). Triangulation in qualitative tourism research. *Tourism Management, 20*(1), 157–161. doi:10.1016/S0261-5177(98)00102-2

Donthu, N., & Gustafsson, A. (2020). Effects of COVID-19 on business and research. *Journal of Business Research, 117*, 284–289. doi:10.1016/j.jbusres.2020.06.008 PMID:32536736

Edworthy, J., Hellier, E., Newbold, L., & Titchener, K. (2015). Passing crisis and emergency risk communications: The effects of communication channel, information type, and repetition. *Applied Ergonomics*, *48*, 252–262. doi:10.1016/j.apergo.2014.12.009 PMID:25683552

European Union. (2020). *Family business.* https://ec.europa.eu/growth/smes/supporting-entrepreneurship/family-business_en

Faulkner, B. (2001). Towards a framework for tourism disaster management. *Tourism Management*, *22*(2), 135–147. doi:10.1016/S0261-5177(00)00048-0

Flick, U. (2011). *Triangulation*. Beltz Verlag. doi:10.1007/978-3-531-92864-7

Fournier, S., & Avery, J. (2011). The uninvited brand. *Business Horizons*, *54*(3), 193–207. doi:10.1016/j.bushor.2011.01.001

Gavac, K., Heckl, E., & Petzlberger, K. (2020). *Familienunternehmen in Österreich 2019: Endbericht*. Vienna: KMU Forschung Austria. https://news.wko.at/news/oesterreich/kmu-forschung-bericht-familienunternehmen-2019.pdf

Germann Molz, J., & Paris, C. M. (2015). The Social Affordances of Flashpacking: Exploring the Mobility Nexus of Travel and Communication. *Mobilities*, *10*(2), 173–192. doi:10.1080/17450101.2013.848605

Getz, D., & Carlsen, J. (2005). Family business in tourism: State of the art. *Annals of Tourism Research*, *32*(1), 237–258. doi:10.1016/j.annals.2004.07.006

Gibb Dyer, W. Jr. (1994). Potential contributions of organizational behavior to the study of family-owned businesses. *Family Business Review*, *7*(2), 109–131. doi:10.1111/j.1741-6248.1994.00109.x

Gilsbach, C. (2018). Krisenkommunikation im Tourismus: Kommunikationskanäle, -instrumente und Medienmix. In S. Hahn & Z. Neuss (Eds.), Krisenkommunikation in Tourismusorganisationen (pp. 61–82). Springer Fachmedien.

Habbershon, T. G., Williams, M., & MacMillan, I. C. (2003). A unified systems perspective of family firm performance. *Journal of Business Venturing*, *18*(4), 451–465. doi:10.1016/S0883-9026(03)00053-3

Habbershon, T. G., & Williams, M. L. (1999). A resource-based framework for assessing the strategic advantages of family firms. *Family Business Review*, *12*(1), 1–25. doi:10.1111/j.1741-6248.1999.00001.x

Hall, A. (2003). *Strategizing in the context of genuine relations*. JIBS Dissertation Series No. 08, Jönköping International Business School, Jönköping, Sweden.

Hall, A., Melin, L., & Nordqvist, M. (2006). Understanding Strategizing in Family Business Contexts. In P. Poutziouris, K. Smyrnios, & S. Klein (Eds.), *Family Business Research Handbook* (pp. 253–268). Edward Elgar. doi:10.4337/9781847204394.00025

Hang, H., Aroean, L., & Chen, Z. (2020). Building emotional attaching during COVID-19. *Annals of Tourism Research*, 83. PMID:32834221

Hanson, S. K., Hessel, H. M., & Danes, S. M. (2019). Relational processes in family entrepreneurial culture and resilience across generations. *Journal of Family Business Strategy*, *10*(3), 100263. doi:10.1016/j.jfbs.2018.11.001

Hatch, M. J. (1997). *Organisation Theory: Modern, Symbolic and Post-modern Perspectives*. Oxford University Press.

Holland, P. G., & Oliver, J. E. (1992). An empirical examination of stages of development of family business. *Journal of Business & Entrepreneurship*, *4*(3), 27–38.

Jeong, M., & Lee, S. A. (2017). Do customers care about types of hotel service recovery efforts? *Journal of Hospitality and Tourism Technology*, *8*(1), 5–18. doi:10.1108/JHTT-09-2016-0049

Kabanoff, B., Waldersee, R., & Cohen, M. (1995). Espoused values and organizational change themes. *Academy of Management Journal*, *38*, 1075–1104.

Lazreg, B. M., Chkraborthy, N. R., Stieglitz, S., Potthoff, T., Ross, B., & Majchrazk, T. (2018). Social Media Analysis in Crisis Situations: Can Social Media be a Reliable Information Source for Emergency Management Services? In B. Andersson, B. Johansson, S. Carlsson, C. Barry, M. Lang, H. Linger, & C. Schneider (Chairs), *Designing Digitalization*. Symposium conducted at the meeting of Lund University, Lund, Sweden.

Lea, J. (1998). *What is a Family Business? More than You Think*. Available: http://www.bizjournals.com/ triangle/stories/1998/11/02/smallb3.html

Leeuwen, T. (2011). Multimodality and Multimodal Research. In E. Margolis & L. Pauwels (Eds.), *The SAGE Handbook of Visual Research Methods* (pp. 549–569). SAGE Publications Ltd. doi:10.4135/9781446268278.n28

Lis, B., & Korchmar, S. (2013). Die digitale Mundpropaganda (Electronic Word-of-Mouth). In B. Lis & S. Korchmar (Ed.), Digitales Empfehlungsmarketing: Konzeption, Theorien und Determinanten zur Glaubwürdigkeit des Electronic Word-of-Mouth (EWOM) (pp. 11–20). Springer Fachmedien.

Livingstone, S., & Brake, D. R. (2010). On the Rapid Rise of Social Networking Sites: New Findings and Policy Implications. *Children & Society*, *24*(1), 75–83. doi:10.1111/j.1099-0860.2009.00243.x

Mayring, P. (2001). Kombination und Integration qualitativer und quantitativer Analyse. *Forum Qualitative Social Research*, *2*(1). http://www.qualitative-research.net/index.php/fqs/rt/printerFriendly/967/2110

Mayring, P. (2015). *Qualitative Inhaltsanalyse: Grundlagen und Techniken* (12th ed.). Beltz Pädagogik.

Merten, K. (1995). *Inhaltsanalyse: Einführung in Theorie, Methode und Praxis* (2nd ed.). Springer Fachmedien.

Nadegger, M., Bosio, B., Wegerer, P., Haselwanter, S. & Untersteiner, J. (2020). #mitabstandnah - Eine Hashtag-Analyse der Tiroler Tourismusverbände in Corona-Zeiten. *Tourismuswissen Quarterly, 21*, 222-227.

Nordqvist, M., & Melin, L. (2010). Entrepreneurial families and family firms. *Entrepreneurship and Regional Development*, *22*(3), 1–29.

Oliveira, A., & Huertas, A. (2019). How do destinations use twitter to recover their images after a terrorist attack? *Journal of Destination Marketing & Management*, *12*, 46–54. doi:10.1016/j.jdmm.2019.03.002

Pikkemaat, B., & Zehrer, A. (2016). Innovation and service experiences in small tourism family firms. *International Journal of Culture, Tourism and Hospitality Research*, *10*(4), 343–360. doi:10.1108/IJCTHR-06-2016-0064

Pillmayer, M., & Scherle, N. (2018). Krisen und Krisenmanagement im Tourismus - Eine konzeptionelle Einführung. In S. Hahn & Z. Neuss (Eds.), Krisenkommunikation in Tourismusorganisationen (pp. 3-18). Springer Fachmedien.

Poza, E. (2007). *Family business* (2nd ed.). Mason.

Racherla, P., & Hu, C. (2009). A Framework for Knowledge-Based Crisis Management in the Hospitality and Tourism Industry. *Cornell Hospitality Quarterly*, *50*(4), 561–577. doi:10.1177/1938965509341633

Ramadani, V., & Hoy, F. (2015). Context and Uniqueness of Family Businesses. In L. P. Dana & V. Ramadani (Eds.), *Family Businesses in Transition Economies: Management, Succession and Internationalization* (pp. 9–37). Springer.

Rau, S. B., Schneider-Siebke, V., & Günther, C. (2019). Family Firm Values Explaining Family Firm Heterogeneity. *Family Business Review, 32*(2), 195–215. doi:10.1177/0894486519846670

Ravelli, L. J., & van Leeuwen, T. (2018). Modality in the digital age. *Visual Communication, 17*(3), 277–297. doi:10.1177/1470357218764436

Ritchie, B. W. (2004). Chaos, crises and disasters: A strategic approach to crisis management in the tourism industry. *Tourism Management, 25*(6), 669–683. doi:10.1016/j.tourman.2003.09.004

Ritchie, B. W., & Jiang, Y. (2019). A review of research on tourism risk, crisis and disaster management: Launching the annals of tourism research curated collection on tourism risk, crisis and disaster management. *Annals of Tourism Research, 79*, 79. doi:10.1016/j.annals.2019.102812

Rose, G. (2016). *Visual methodologies: An introduction to researching with visual materials* (4th ed.). Sage.

Roshan, M., Warren, M., & Carr, R. (2016). Understanding the use of social media by organisations for crisis communication. *Computers in Human Behavior, 63*, 350–361. doi:10.1016/j.chb.2016.05.016

Sano, K., & Sano, H. (2019). The effect of different crisis communication channels. *Annals of Tourism Research, 79*, 79. doi:10.1016/j.annals.2019.102804

Santana, G. (2004). Crisis Management and Tourism. *Journal of Travel & Tourism Marketing, 15*(4), 299–321. doi:10.1300/J073v15n04_05

Schwaiger, K., & Zehrer, A. (2020). Familienunternehmen: Resilienz in Krisenzeiten. *TourismusWissenQuarterly, 21*, 210–215.

Schwartz, S. H. (1992). Universals in the content and structure of values: Theoretical advances and empirical tests in 20 countries. In M. P. Zanna (Ed.), (pp. 1–65). Advances in experimental social psychology. Academic Press. doi:10.1016/S0065-2601(08)60281-6

Sharma, P., Sharma, P., Chrisman, J. J., Chrisman, J. J., Chua, J. H., & Chua, J. H. (1997). Strategic management of the family business: Past research and future challenges. *Family Business Review, 10*(1), 1–35. doi:10.1111/j.1741-6248.1997.00001.x

Small, T. A. (2011). What the Hashtag? A Content Analysis of Canadian Politics on Twitter. *Information Communication and Society, 14*(6), 872–895. doi:10.1080/1369118X.2011.554572

Sorenson, R. L. (2011). Social capital and family business. In R. L. Sorenson (Ed.), *Family Business and Social Capital* (pp. 1–30). Edward Elgar Publishing.

Top.tirol. (2019). *Die größten Hotels.* https://top.tirol/juli-2019/die-groessten-hotels

Triantafillidou, A., & Yannas, P. (2020). Social media crisis communication in racially charged crises: Exploring the effects of social media and image restoration strategies. *Computers in Human Behavior, 106*, 1–10. doi:10.1016/j.chb.2020.106269

Volo, S., & Irimiás, A. (2020). Instagram: Visual methods in tourism research. *Annals of Tourism Research*, 103098. doi:10.1016/j.annals.2020.103098

Walsh, F. (2002). A family resilience framework: Innovative practice applications. *Family Relations, 51*(2), 130–137. doi:10.1111/j.1741-3729.2002.00130.x

Ward, J. L. (2004). *Perpetuating the family business: 50 lessons from long-lasting successful families in business*. Palgrave Macmillan. doi:10.1057/9780230505995

Ward, J. L. (2011). *Keeping the Family Business Healthy – How to Plan for Continuing Growth, Profitability, and Family Leadership*. Palgrave Macmillan. doi:10.1057/9780230116122

We Are Social. (2019). *Global Digital Report 2019.* https://wearesocial.com/global-digital-report-2019

Wong, I. A., Ou, J., & Wilson, A. (2021). Evolution of hoteliers' organizational crisis communication in the time of mega disruption. *Tourism Management*, 84.

World Tourism Organization (2011). *Toolbox for Crisis Communications in Tourism: Checklist and Best Practices.* World Tourism Organization (UNWTO).

World Tourism Organization. (2020). *International Tourism and COVID-19.* World Tourism Organization (UNWTO). https://www.unwto.org/international-tourism-and-covid-19

Zehrer, A., & Haslwanter, J. (2010). Management of Change in Tourism - The Problem of Family Internal Succession in Family-run Tourism SMEs. *Electronic Journal of Family Business Studies, 4*(2), 147–162.

Zehrer, A., & Leiß, G. (2019). Family Entrepreneurial Resilience - An intergenerational learning approach. *Family Business Management.*

Zenker, S., & Kock, F. (2020). The coronavirus pandemic - A critical discussion of a tourism research agenda. *Tourism Management, 81*, 81. doi:10.1016/j.tourman.2020.104164 PMID:32518437

ENDNOTES

1 Translated, Original Quote: „♥ -LICHEN DANK für diesen wundervollen Sommer! Besonders in dieser herausfordernden Zeit habt Ihr unser Haus mit so viel Liebe, Freude und Wärme gefüllt. Diese herzlichen und unvergesslichen Begegnungen von und mit Euch erfüllen uns mit so viel Dankbarkeit und Stolz. 🥰Wir haben so gut aufeinander Acht gegeben und werden das auch zukünftig tun. Versprochen! 😉DANKE für Euren Rückhalt und ganz besonders für Euer Vertrauen ins Alpenresort Schwarz und in die vielen wundervollen Menschen, die Euren Urlaub mitgestalten.Es ist so schön, für Euch da zu sein.🙏💐Bis bald im #alpenresortschwarz 😉" (@alpenresortschwarz, 15th September 2020, Retrieved from https://www.instagram.com/p/CFKQhuNpTLx/.

2 Translated, Original Quote: "ForsthofgutGLÜCK 🍀 Glücksmomente im Forsthofgut 😍Teilt eure Lieblingsbilder von eurem Urlaub mit uns und lasst uns gemeinsam an die schönen Momente zurückdenken 🤗 Jetzt mitmachen und gewinnen! Aufenthalt für 2 Personen und 2 Übernachtungen inkl. ForsthofgutKÜCHE 😋 [...]" (@naturhoteforsthofgut, 27th March 2020, Retrieved from https://www.instagram.com/p/B-Puh4UpkVm/.)

3 Translated, Original Quote: "Guten Morgen aus dem STOCK resort! Heute haben wir wieder eine #magicmemory für euch: Ein Foto aus dem Jahr 1983 - es hat sich einiges getan seitdem und auch aktuell wird fleißig umgebaut und erweitert." (@stockresort, 29th April 2020, Retrieved from https://www.instagram.com/p/B_jda-SiJD6/.)

4 Translated. Original Quote: " +++ BREAKING GOOD NEWS!!! +++ 😃 Die Grenzen zu Deutschland, Liechtenstein, der Schweiz, der Slowakei, Slowenien, Tschechien und Ungarn sind ab morgen, dem 4. Juni, wieder für den Reiseverkehr offen! 🥳 Das bedeutet für unsere vielen lieben Gäste aus diesen Nachbarländern ein weitaus früheres Wiedersehen als gedacht. 🥰 Wir freuen uns schon riesig auf euch und wollten diese freudige Nachricht sofort mit euch teilen. Auf bald ... daheim beim Stanglwirt! ♥" (@stanglwirt, 3rd June 2020, Retrieved from https://www.instagram.com/p/CA-cQLNHPvC/.)

5 Translated, Original Quote: „Sehnsuchtsort. 🧖‍♀ Für ein sicheres Miteinander: Unser Wellnessbereich bietet ganze 50 Quadratmeter für jeden Gast, so lässt es sich entspannen. 🌿." (@deralpinekraftplatz/Wellnessresidenz Alpenrose, 15th May 2020, Retrieved from https://www.instagram.com/p/CANF3iKpA4W/.)

6 Translated, Original Quote: „Liebe Fans & Freunde! Die Gesundheit unserer Mitarbeiter, Gäste und Einheimischen hat für uns oberste Priorität. Auch wenn wir uns trotz der aktuellen Umstände weiterhin über eine enorme Resonanz und erfreuliche Buchungslage dankbar schätzen dürfen, haben wir uns dazu entschlossen das 30. Jubiläum unserer legendären Weißwurstparty, das traditionelle

Sängertreffen sowie den Stanglwirt Christkindlmarkt abzusagen bzw. zu verschieben – unabhängig davon welche Lockerungen bzw. Restriktionen seitens Regierung bis zu den Veranstaltungsterminen noch getroffen werden. Angesichts der zwingenden Notwendigkeit Infektionsketten zu unterbrechen sind wir sicher, dass ihr Verständnis für diese Entscheidung im Sinne unseres hohen Verantwortungsbewusstseins habt. Wir bedanken uns von ganzem Herzen für eure Treue und das Verständnis in diesen außergewöhnlichen Zeiten! G'sund bleib'n, eure Familie Hauser & das gesamte Stanglwirt-Team." (@stanglwirt, 30th September 2020, Retrieved from https://www.instagram.com/p/CFwjKqoHPes/.)

7 Translated, Original Quote: „...seit mehr als 250 Jahren kein Ruhetag! Wir bleiben unsrem Motto treu! Nach der „unfreiwilligen Schließzeit" freuen wir uns von Herzen, heute die Pforten unseres historischen „Gasthof Stangl" wieder zu den gewohnten Zeiten, täglich von 11-22 Uhr, öffnen zu dürfen. Mia gfrein' ins auf enk! ♥ " (@stanglwirt, 15th May 2020, Retrieved from https://www.instagram.com/p/CAM5JUrHGCs/.)

8 Translated, Original Quote: „So schee war's heute am Wiedereröffnungstag bei uns im Traditionsgasthof… auch unsere lieben Freunde Tobias und Julia Moretti schauten spontan auf an frischen Spargel vorbei und natürlich wurde – so wia's beim Stanglwirt der Brauch ist – fleißig musiziert! Wir sind glücklich und dankbar endlich wieder für unsere Gäste da sein zu dürfen… denn „das Glück des Lebens ist das Glück des Gebens" … Gast-Geber-Sein! 😊" (@stanglwirt, 15th May 2020, Retrieved from https://www.instagram.com/p/CANx0xvnRAu/.)

9 Translated, Original Quote: „Kann man sich auf traditionelle Werte besinnen und gerade dadurch eine Vorreiterrolle einnehmen? Ja, man kann! Es ist uns eine große Ehre, dass die aktuelle @voguebusiness unser „Prinzip Nestwärme" als Besonderheit hervorhebt. ✨ Wir freuen uns euch diese Nestwärme im einzigen Luxusresort mit integriertem Bio-Bauernhof schon bald wieder vermitteln zu dürfen. 💛 Das Vogue Interview mit Stanglwirt Junior-Chefin Maria Hauser findet ihr hier... (zum Lesen bitte zoomen 🤗" (@stanglwirt, 6th May 2020, Retrieved from https://www.instagram.com/p/B_2E3R1nmcl/.)

10 Translated, Original Quote: „Regional. Frisch. Wertvoll. þVom Gasslbauer in Ladis, und vielen weiteren umliegenden Höfen, bekommen wir regelmäßig saisonale Produkte. Es ist uns eine Herzensangelegenheit, dass wir durch die gute Zusammenarbeit mit den Bauern unseren Gästen etwas Heimat am Teller servieren können. 🍽 " (@schlosshotel_fiss, 9th June 2020, Retrieved from https://www.instagram.com/p/CBN_YMbHRL3/.)

Chapter 10
Travellers' Resilience to Major Travel Risks of Today:
Global Terrorism and Pandemic Disease – Opportunities and Challenges for Small Tourism Businesses

Vanda Veréb
https://orcid.org/0000-0002-8894-457X
University of Aveiro, Portugal

António Azevedo
University of Minho, Portugal

ABSTRACT

It is argued that the current times can be labelled as the era of fear, as there is always something to be afraid of. Fear causes great damage to the tourism industry, as the prerequisite of global travels is (the sense of) safety. Resilience is a promising approach to address the harmful effect of fear in tourism. Travellers' resilience, while an essential component of overall tourism resilience, is scarcely studied. This chapter investigates what makes global travellers resilient in the face of current global adversities, terrorism risk, and COVID-19, the recent game-changers in tourism. By building on psychological resilience theory and synthesizing the latest risk-specific findings, general categories of travellers' resilience are outlined. The chapter concludes by profiling each travellers' category along with communication guidelines on how to encourage each segment in troubled times.

DOI: 10.4018/978-1-7998-7352-5.ch010

INTRODUCTION

Until most recently, wide-spread terrorist attacks in Europe and in the Middle East occupied most of the news media. Today, the headlines are heavy with hourly updates on the spread of corona virus. Just before that, discussions on social media were raising red flags about a potential migration crisis due the unavoidable effects of climate change. This news stream occupied only the last couple of months, with every new post heightening further the perception of risk and uncertainty (Kapuściński and Richards 2016).

The sense of uncertainty and the impression of impending danger set off our evolutionary encoded survival mechanism and trigger instinctive fear reactions (Ledoux 2015). Our survival instinct aims for effectiveness over efficiency – survival at all cost – by inflating our risk perception over the actual level of danger. Consequently, fear reactions are often exaggerated (Thompson et al. 2017) and irrational (Taha et al. 2014), even crude (Murray and Schaller 2016) and aggressive (Wu and Chang 2012). Irrationally inflated risk perception and fear as the foundation of our survival mechanism are essential for our safety. However, if the irrationality in fear left without a conscious check, instinctive fear reactions could cause substantial societal damage. For example, Human Rights Watch (2020) warned about a significant increase of hate crimes world-wide against local Asian residents who are ethnically related the population of China where the COVID-19 pandemic started. Similarly, violent backlash against local Arabs and Muslims were reported following the 9/11 terrorist attacks across the US (Human Rights Watch 2002).

The atmosphere of uncertainty creates an immense damage particularly for the travel and tourism industry, because the central requirement of tourism is safety and stability (Richter and Waugh 1986) or, most precisely, the sense of safety and stability (Williams and Baláž 2015). Risk perception inflated over the actual level of danger was demonstrated to damage destination image (Perpiña et al. 2017) and restrain the willingness to travel (Sarman et al. 2016), in some cases, even cease international travels all together (Floyd et al. 2004).

Resilience approach is argued to have significant explanatory power and suggested as a promising approach for tourism businesses to cope with the atmosphere of uncertainty global challenges create (Luthe and Wyss 2014). Ritchie (2004) also argues that effective disaster management starts with the development of resilience. Yet, research on how to build tourism resilience in the face of complex global challenges is still limited (Innerhofer et al. 2018). Tourism resilience is described as the capacity of the tourism system to absorb disturbance without collapsing and to re-organize itself to maintain structures and functions (Innerhofer et al. 2018), while allowing change, even innovation and development, after the crisis (Hall et al. 2017). Tourism resilience has three dimensions: ecological-environmental, economic-fiscal and social-cultural (Calgaro et al. 2014). The ecological-environmental dimension represents the bio-physical characteristics of a destination, like natural resources and landscape. The economic-fiscal dimension refers to factors, like fiscal imbalances or economic crisis. The social-cultural dimension includes the strengths and weaknesses of the social system, cultural norms and perceptions and attitudes towards tourism and travelling. The dynamic interrelationship between these dimensions and the uncertainties of the global environment shapes tourism resilience (Tyrrell and Johnston 2008). In order to successfully investigate resilience, each dimension should be dully examined (Pechlaner and Volgger 2015). The socio-cultural dimension focusing on the local tourism community and travellers, while an essential building block of tourism resilience, lacks comprehensive research (Tyrrell and Johnston 2008). In particular, travellers' resilience is still an unexplored field (Prayag 2018), even though the basic marketing principle stands for the tourism industry as well: the success and the very survival of a business depends on its customer (Kotler et al. 2013), in this case, the travellers.

This chapter contributes to the intermittent literature on travellers' resilience by investigating how resilient global travellers are in the face of current global adversities, and what opportunities and challenges their resilience creates particularly for the small tourism businesses. The study focuses on how global terrorism risk and the threat of infectious diseases (i.e., COVID-19) shape travellers' resilience. These are the two most recent game-changers of the travel and tourism industry (US Department of State 2020) as well as the two most potent travel risks considered in tourism literature (Karl and Schmude 2017).

Terrorism and infectious diseases represent rather different types of travel risks. Terrorism is a man-made disaster, while an epidemic is a natural catastrophe. Terrorism risk is about (a relatively low) probability of becoming a victim of an attack (Yang and Nair 2014), but the risk of COVID-19 is about (a relatively high) possibility of contracting the disease (Garfin et al. 2020). Terrorism is a distant risk for most and the destruction it creates is mainly concentrated locally (Pizam and Smith 2000). While, COVID-19 is global threat changing the daily lives for the majority of the global population through legally required home confinement and social distancing (Vinkers et al. 2020). Nevertheless, psychology domain generally assumes that fears share the same root, uncertainty, and manifest themselves in the same fear responses, irrationally exaggerated self-protective behaviour, regardless of the actual aversive stimulus (e.g., Carleton, 2016; Ledoux, 2015; Murray and Schaller, 2016; Panksepp et al., 2011; Thompson et al., 2017). Irrational behaviour was clearly demonstrated as a fear response to both terrorism and infectious diseases risks. For example, Chang et al. (2004) have found that the exaggerated fear of being infected during the 2003 epidemic of the Severe Acute Respiratory Syndrome (SARS) caused substantial decrease in health care utilization, which irrationally lead to an increase in the number of death of all other conditions. Also, following the 9/11 attacks, most American travellers grew an irrationally strong association between terrorism threat and the Muslim religion and at disproportionally large extent avoided destinations with a high percentage of Muslim population (Araña and León 2008), while others refused to travel at all and 'cocooned' up (Clancy 2012). Several scholars studied the two types of risks, terrorism threat and the risk of infectious diseases, together in order to gain a more comprehensive view on the effect that increased risk perception has in tourism and to be able to provide more comprehensive strategies to fight its overall harmful effect. For example, Avraham (2016) contrasted the influence of terrorism and diseases risk on destination image, Rittichainuwat and Chakraborty (2009) on the willingness to travel and Thompson et al. (2017) on overall decision-making. This chapter argues along with these scholars that in order to derive overarching insights and to determine potential resilience that is universal across destinations and crisis-contexts, it is essential to aim beyond crisis-specific risk analysis.

The ultimate aim is to find *the* resilient travellers "who are less concerned with potential [travel] risk, and may thus react less negatively in troubled times" (Seabra et al. 2013, p. 503). They could be the key to survival, especially, for the vulnerable small tourism business in the current challenging times. Travellers' resilience is still an uncharted area of risk perception research (Prayag 2018), and this study starts to explore this field by focusing on what makes travellers resilient to act on the irrational sway of their instinctively exaggerated risk perception in the face adversities, and how travellers' resilience can be built up to strengthen the resilience of the entire tourism system.

The chapter starts out by providing a general overview on the theoretical background on travellers' resilience and the categorisation of travellers based on their perception of and attitude towards travel risk. Then, the chapter focuses on the latest findings on travellers' resilience that explore travellers' risk perception in a new reality of pandemic disease and global terrorism, where instead of the drive for stimulation and novelty fear dominates travel decision-making (e.g., Bude 2018; Veréb et al. 2018; Yang et al. 2020). By synthesizing these latest findings, the chapter outlines general categories of travellers' resilience overarching specific travel risks. Finally, following confirmed approaches from clinical psychology domain to build up resilience (e.g., Rees et al. 2015) and highlighting the latest findings of crisis-communication of terrorism and infectious diseases (e.g., Avraham and Ketter 2017; Garfin et al. 2020), the chapter provides practical, down-to-earth guidelines on how to strengthen resilience in each traveller category. The suggested communication strategies are specifically catered for small tourism businesses, who are the most vulnerable to tourism crisis (Gössling et al. 2020; Innerhofer et al. 2018), and advises them on how to meet the changing needs and expectations of their travellers today when fear predominantly shapes tourism demand.

THEORETICAL BACKGROUND

Travellers' Resilience

Majority of the studies focusing on the socio-cultural aspect of tourism resilience assess the resilience of the local tourism community (e.g., Ruiz-Ballesteros 2011) building on the foundations of organizational resilience (e.g., Lee et al. 2013). There is almost a complete lack of studies that focus on travellers' resilience (Prayag 2018) using individual resilience approach (e.g., Garcia-Dia et al. 2013). Individual (psychological) resilience is the ability to bounce back from and successfully adopt to adversity (Henderson et al. 1999). It is the capacity to appropriately function despite of high risk, stress or trauma (Egeland et al. 1993). Psychological resilience is also described as a stable trajectory of healthy functioning after an adverse event, a relatively brief period of disequilibrium, but otherwise continued health (Bonanno et al. 2011). The focal point of psychological resilience is an active decision to keep moving forward by learning from adversity and not succumbing to its negative effects, it rests on a conscious choice that must be constantly reconfirmed (Yehuda and Flory 2007). Psychological resilience is influenced by many inter and intrapersonal as well as environmental factors, like personality traits, social, cultural and political environment (Rees et al. 2015). The building blocks of psychological resilience can be plentiful (Garcia-Dia et al. 2013; Zolli and Healy 2013), but all the definitions agree that it manifests as the capacity of a person to maintain his integrity and usual functioning in the face of changed circumstances and that it rests on a conscious choice.

Based on psychological resilience approach, travellers' resilience could be defined as the capacity of a person to consciously uphold his original interest in and desire for travelling despite of potential travel risks and to continuously maintain or quickly return to his usual travel behaviour when a new travel risk emerges. Psychological resilience is both a progressive quality that can be built up (Yehuda and Flory 2007) and a core personal attribute that manifests across various domains of life (Bonanno et al. 2011). These notions support the quest for universal categories of traveller's resilience that are not specific to any types of travel risk and can be gradually built up.

Travellers' resilience is a new field of research that was relatively unexplored until most recently. The reason could be that the socio-cultural aspect of tourism resilience is difficult to quantify and thus investigate (Tyrrell and Johnston 2008), especially without a clear definition of travellers' resilience. One of the rare studies that attempts to define resilient travellers refers to them "as those that tend to absorb risks instead of engaging in risk avoidance strategies" and measures it by how rigidly travellers stick to their original travel plans despite of an emerging adversity (Hajibaba et al. 2015, pp. 49). But, cognitive rigidity and resistance to change are not in line with the established understanding of tourism resilience. The three cornerstones of tourism resilience is: flexibility, adaptability and learning (Cochrane 2010). Flexibility, as the capability to handle unpredictable situations (Hall et al. 2017) and to recover from it (Hartman 2018). Adaptability, as the capacity to anticipate shocks (Hartman 2016) and to develop proactive coping strategies (Hartman 2018). Learning, as the ability to decipher the impacts of change and learn from a disturbance to be better prepared for future shocks (Pechlaner and Innerhofer 2018). Hajibaba et al. (2015) counts those travellers resilient who do not cancel trips already booked despite of an emerging adverse factor. This approach, while is a great attempt to advance a new research field, shows the limitations of quantitative methodology in investigating travellers' resilience, as it can only count the number of cancelled bookings due to a crisis event. This approach disregards that a crisis event, while significantly impacts immediate tourism demand, has far more significant long-term implications (Karl and Schmude 2017). Furthermore, it disregards the potential diversity of crisis-resistant travellers by overlooking those less resilient travellers who resorts to potential risk mitigation strategies, like postponing the trip or choosing a different destination, but still travel in the face of adversity (Yang and Nair 2014).

Most Recent Research on Travellers' Resilience

Qualitative methods might be the best approach to investigate travellers' resilience. As "qualitative sense may provide tourism researchers and practitioners with improved means to conceptualize trade-offs implicit in tourism resilience [...], and a means to incorporate a guiding structure to an area of debate often characterized by a lack of theoretical, conceptual and practical clarity" (Tyrrell and Johnston 2008, pp. 23). Moreover, qualitative methods, like in-depth interviews, provide a perfect framework to study psychological resilience, because "the most important and effective way to approach resilience is to start with listening to what people have to say about their everyday lives, and find what matters the most for them and how they achieve it against all of odds" (Southwick et al. 2014, pp. 10). In line with these suggestions, more recent studies rely on qualitative methods and build on interdisciplinary foundation for a more complete understanding of resilient travellers.

Relying on quantitative approach and using grounded theory on a highly diverse international sample covering 22 countries, Veréb, Nobre and Farhangmehr (2020) suggest that the dividing line between non-resilient and resilient travellers in the face of terrorism risk is the lack of or the evidence of a conscious strive for objectivity in risk evaluation and an effortful resistance against the instinctive fear impulses in behaviour when risk, in this case, potential terrorist incident, arise. Based on how effective their effort was, judged through probing questions during the in-depth interviews, scenario analysis and examination of past travel decisions, the travellers were divided into five categories: non-resilient, striving for resilience, naturally resilient, advocates and thrill-seekers. Those travellers were considered non-resilient who struggled to keep their instinctive fear reaction under rational control and reacted more according to their instinctively exaggerated perception of risk rather than the actual level of terrorism risk. Travellers in the striving for resilience category displayed very similar behavioural patterns than non-resilient travellers, but they did not let their passion for travelling vanish. They consciously fought back negative prejudice about the troubled destination keeping their interest in and their motivation to visit the destination intact. Naturally resilient travellers, backed up by multicultural social network and relying on substantial international (travel) experience, managed to keep their terrorism risk perception into rational perspective. Not taking chances, they also cancelled their trips upon a crisis-event but resumed travelling to troubled destinations immediately after the actual danger was lifted. Advocates were found very much like naturally resilient travellers in terms of travel behaviour, but their passion for travelling overflew. They openly endorsed troubled destinations, for example, by being the first to travel after a disaster and fought back openly irrational and potentially harmful prejudice breeding out of increased terrorism risk perception. Thrill-seekers were the ones who might not have cancelled a trip upon a crisis-event or even chose to travel to a troubled destination for the thrill of experiencing risk. However, common-sense did apply here as well. Their aim was to chase thrill, overcome fear and not be harmed. They were the only resilient traveller category that are commonly considered risk-resistant in the wider travel risk perception literature as well (e.g., Buckley 2012; Lepp and Gibson 2008). Table 1 displays the main characteristics of each traveller group.

Veréb, Nobre and Farhangmehr (in press) replicated the previous study in a different risk context and examined how resilient travellers were to the fear of the COVID-19 pandemic. The authors conducted 64 in-depth interviews along with direct observation of the respondents in the first 3 months after the official declaration of COVID-19 as pandemic across 29 national realities. Disregarding external controls, like political travel restrictions, the study focused on 'internal travel restrictions', the changes in attitude towards travelling in the face of COVID-19 to uncover who and how, if at all, could maintain their original passion for travelling in the shadow of this impending risk, alias be resilient to this fear. Those were labelled resilient who strived to evaluate risk objectively and fought to keep their behaviour under rational control, away of instinctive irrationality. The study concluded that the foundation of travellers' resilience is the same across risk contexts, a conscious strive for

Table 1. Categories of travellers' resilience in the face of terrorism risk

Main Characteristics/ Traveller Categories	Non-resilient	Striving for Resilience	Naturally Resilient	Advocate	Thrill-seeker
Personal Value Preference	Safety	Safety	Novelty	Universalism	Stimulation
Travel Experience	Limited	Limited	Substantial	Substantial	Substantial & exotic
Multicultural Social Network	Not present	Present	Extensive	Extensive	Extensive
Societal Value Preference	Security over civil liberties & privacy	Balancing security & civil liberties	Civil liberties over security	Civil liberties over security	Civil liberties over security
Media Consumption	Influenced by mass-media	Discerning media consumption	Discerning & selective media consumption	Discerning media consumption & active influencing role	Discerning & selective media consumption
Reaction to Fearful Stimuli (terrorism risk)	Instinctively exaggerated self-protection	Instinctively exaggerated self-protection	Objective risk assessment	Highly objective risk assessment & support of others	Thrilled by potential danger
Bias (instinctive irrationality in behaviour)	Present but not consciously recognised	Recognised & aims to correct it	Hardly present & consistently corrected	Under constant conscious check & aims to correct it in others, too	Hardly present & consistently corrected
Travel Risk Mitigation Strategy	Risk avoidance	Risk avoidance	Selective destination choice	Selective destination choice	Propelled by risk & opts for risky destinations

objectivity and an effortful resistance to act on the instinctive (potentially irrational) fear reactions. Moreover, it found very similar categories of resilient travellers in the face of COVID-19 as the previous study in the case of terrorism risk: striving for resilience, mainstream resilience, advocates and highly resilient travellers. Thus, arguing that travellers' resilience is a general construct across travel risk contexts.

Truthfully, there were two categories that were not exactly the same in the two risk contexts. Veréb et al. (in press) did not find thrill-seekers, who would go against their instinctive fear and challenge themselves to overcome the virus. It could be because overcoming a virus does not depend on a mental or physical effort, neither it is a personal achievement or a personal failure. These authors pinpointed highly resilient travellers instead who were highly resistant to the irrational fear of COVID-19 and less restrained by it than all other resilient categories. Also, there were no one who could be considered naturally resilient when it comes to a novel infectious disease. The fear of infectious diseases is evolutionary encoded and historically reinforced. Going against it is not natural. Nevertheless, Veréb et al. (in press) also found a consistently rational group of travellers in the face of COVID-19 pandemic that was labelled mainstream resilience. This group demonstrated great objectivity in risk evaluation when judging travel risk during the pandemic similarly to naturally resilient travellers in the case of terrorism risk. The slight difference between the resilience categories could be explained by the different nature of risk-contexts terrorism and COVID-19 creates. As not only the internal psychological make-up of the traveller but also the nature of the external travel risk can shape psychological resilience and (Southwick et al. 2014). Table 2 sums up the main characteristics of each traveller group based on their resilience to the fear of COVID-19.

Table 2. Categories of travellers' resilience in the face of COVID-19 risk

Main Characteristics	Non-resilient	Striving for Resilience	Mainstream Resilience	Advocate	Highly Resilient
Personal Value Preference	Safety	Safety	Novelty	Universalism	Stimulation
Travel Experience	Limited	Limited	Substantial	Substantial	Substantial & exotic
Multicultural Social Network	Not present	Present	Extensive	Extensive	Extensive
Societal Value Preference	Security over civil liberties & privacy	Security over civil liberties & privacy	Security over civil liberties	Balancing security & civil liberties	Civil liberties over security
Media Consumption	Influenced by mass-media	Discerning media consumption	Discerning & selective media consumption	Discerning media consumption & active influencer role	Discerning & selective media consumption
Reaction to Fearful Stimuli (COVID-19)	Over exaggerated self-protection & prone to panic	Over exaggerated self-protection & effortful strive to fight back panic	Effortful strive for objectivity in risk assessment & behaviour	Effortful strive for objectivity in risk assessment & behaviour & open to take risk for others	Effortful strive for objectivity in risk assessment & behaviour & open to take risk for stimulation
Bias (instinctive irrationality in attitude & behaviour)	Present but not recognised	Recognised & effortful strive to correct it	Recognised & effortful strive to instantly correct it	Recognised & consistently corrected & strive to correct it in others, too	Recognised & consistently corrected
Travel Risk Mitigation Strategy	Additional, self-imposed restrictions & apprehension towards any social contact	Additional, self-imposed restrictions	Adherence to legally required restrictions & safety measures	Adherence to legally required restrictions & safety measures	Adherence to legally required restrictions & safety measures
Attitude to Travels	"No travelling until cure or vaccine"	Open to local travels if safety measures are strictly met	Open to international travels where necessary safety measures are met	Open to travel internationally where necessary safety measures are met	Eager to travel internationally & accept reasonable safety measures

PRACTICAL IMPLICATIONS

General Categories of Travellers Resilience and Actionable Communication Guidelines

By reviewing the latest findings in travellers resilience, this chapter argues that resilient travellers are not only those who would not cancel their trips upon a crisis event. Resilient travellers are a diverse segment. Troubled destinations and all destinations in troubled times, like now, due to COVID-19 pandemic, can segment their market and target those categories of resilient travellers that are available for them and the most resource-efficient to cater for. This chapter provides a detailed profile of each traveller category to support market segmentation and outlines practical strategies to successfully target them.

The reviewed latest studies on travellers' resilience, explore travellers' resilience by building on psychological resilience theory (e.g., Garcia-Dia et al. 2013; Henderson et al. 1999) and consider those travellers resilient who are able to keep their interest in and desire for travelling despite of emerging travel risks, consciously strive to a rational stance in risk assessment and continuously maintain or quickly return to his usual travel behaviour when a new travel risk emerges. These latest studies categorize travellers based on how objective and realistic they can keep their risk assessment and travel decision-making when facing (travel) risk as well as how effectively they can resist the overpowering fear activated by survival instinct in their travel attitude and behaviour. Synthetizing the compatible findings across the risk contexts of terrorism and COVID-19, this chapter profiles the five distinct categories of travellers that emerge from literature: non-resilient travellers, travellers striving for resilience, mainstream resilience, advocates and highly resilient travellers. Below, a detailed overview along with a visual summary (Table 3) is provided on the how travel risk (fear) is approached at each of these levels of resilience and what strategies can be useful to strengthen resilience further in each category in a post-COVID era.

Non-Resilient Travellers

- **Resilience profile**: Non-resilient travellers value security and comfort over any other travel motivation. They aim to relax and recharge during holidays. The main travel concern for non-resilient travellers is safety. Thus, they are very sensitive to any potential travel risk. When facing potential danger, they instinctively rather than rationally react to fear aiming for safety and self-preservation without compromise. Their main risk mitigation strategy is risk avoidance. When facing with uncertainty, they tend to avoid troubled destinations and their region. A crisis-event can demolish their overall interest in a destination. During troubled times, like currently due to COVID-19, they do not consider travelling.
- **Suitable marketing approach:** Non-resilient travellers prefer travelling in high protection, potentially in isolation, for example, opting for all-inclusive resorts, ocean cruise trips or remote village locations. The idea of virtual travels also comes up as alternative to global travels for this group. This group also reported to accept, even welcome, rigorous, possibly uncomfortable, security measures during travelling and at a holiday destination as a reassurance of security. During the current COVID-19 health crisis, they seem not even to consider travelling. Thus, the best strategy for this segment could be to build their resilience up for the future than to try to encourage them to travel now. Communication materials that highlight the benefits that can be gained through taking the risk (of travelling), was proven to significantly change how risk is perceived (Finucane et al. 2000). A communication program that highlight the cost of their behaviour – e.g., being grounded (Floyd et al. 2004) and emotionally isolated (Lippi et al. 2020) – and emphasise the benefits of an attitude change – e.g., gaining cultural capital by travelling (Bourdieu 1987) or preserving mental well-being in home confinement (Bavel et al. 2020) could lead them toward a more rational risk assessment. As they predominantly do not travel on the face of (travel) risk, social media that

is considered a trusted information source in judging the potentials and the safety of a destination (Kim et al. 2018), could be a good place to reach and engage them.

Travellers Striving for Resilience

- **Resilience profile:** Safety is essential for this segment as well. Their first reaction to travel risk is fear and their most used risk mitigation strategy is risk avoidance. However, after the initial emotions a crisis event stirred up, they tend to calmly reassess the situation and strive to gain an objective perspective. Striving travellers are aware of their instinctively exaggerated risk perception and biased attitude towards travelling and/or destinations, and they actively strive to correct their potentially irrational preconceptions. Nevertheless, their travel decisions seem to be swayed by travel risk. They have a passion for travelling, they do not withdraw fully from travels in the face of risk. But risk profoundly changes the way they travel. Crisis-events do not deter their interest in a destination, their positive opinion about its people, culture and landscape. But it stops them from travelling there until the destinations has fully recovered.
- **Suitable marketing approach:** Under normal circumstance, this segment enjoys novelty and diversity during holidays, nevertheless strictly within the bounds of safety and comfort. During troubled times they seem to be overcautious up to the point of not travelling, or not travelling out of their local, familiar environment. This segment needs some time after a crisis event to recompose and re-evaluate. In the meantime, they could benefit from clear fact-based and science-backed crisis communication as they actively look for reliable media sources to learn truthful information behind media news, as a mean correct their own irrational risk perception. The security concern of striving travellers may be assured by transparent and honest communication about potential destination specific risks (Rittichainuwat and Chakraborty 2012; Van Niekerk and Pizam 2015) and strategies in place to handle these risks, like contingency plans, emergency and safety procedures as well as security measures (Avraham 2015, 2016; Taylor and Toohey 2005). Visible security measures, like armed security forces, alarm systems with security cameras, metal detectors, could be reassuring for this segment at tourism sights (Taylor and Toohey 2005). The right balance is essential (Rittichainuwat 2013), nevertheless, as too much security might communicate the opposite of safety. In a post-COVID era, this segment could be a great target for local tourism. Familiarity with the locals customs and language can provide the feeling of safety, and the proximity to home can provide an increased sense

of control (Wolff and Larsen 2016). Recent studies examining travel trends in the face of COVID-19 also suggest that post-pandemic travellers prefer local travels for safety concerns and to support their own economy hit by the health crisis (Gössling et al. 2020). Local tourism business could point these travellers to remote tourist attractions, well-isolated accommodations catering for small amount of people at a time highlighting the safety and hygienic measures in place. Another emerging trend in the time of COVID-19 is untact tourism (Bae and Chang 2020). The word 'untact' is a combination of the prefix 'un' and the word 'contact' and refers to a behavioural disposition to minimize direct contact between people, like using self-service kiosks, shopping online or making contactless payments. In tourism, untact travel activities satisfy the thirst for travel while minimizing the possible risks of COVID-19. For example, outdoor camping and road trips, mountain hiking trails, local scenic driving routes, forests and parks are promoted as untact tourism destinations. Untact tourism might became a norm for all travellers, but it is the perfectly fitting travel style for the striving segment.

Travellers in the Mainstream Resilience Group

- **Resilience profile:** Travellers in this group have a conscious strive for objectivity when facing risk, which became a natural tendency. They seem capable to put the arising fear into perspective when travelling, and it does not affect out of proportion neither their interest in a destination nor their overall travel behaviour. They have a strong, deep-seated desire to travel, and emerging risks do not seem to change that. These travellers seek out of ordinary experiences, novelty, stimulation and personal challenge in travelling. They prefer activity-cantered and action-packed schedule during travelling. They have a great curiosity to discover unknown places and cultures, not only to see them but to emerge themselves in them and learn from them. As their main travel motivation is novelty and experiencing diversity, their main concern is for the overall pleasure value during travels, where safety is only one of the aspects to be met in order for a holiday to be fulfilling. These travellers appear to be able to easily recognize the media's inevitable influence, so the travel risk magnified by media attention after a crisis-event did not influence them. This segment also does not rely on generalizations when forming destination image or making travel decisions, because they can confidently depend on their personal network and/or their own experience to form a view on destinations and their crisis-management progress. They cancel their trips due to the immediate effect of the crisis impact, as per the recommendations of official travel advisories. But, their travel intention to visit the destination after crisis event recovers fast.

- **Suitable marketing approach:** This segment is well-travelled and well-experienced traveller, and they crave and expect authenticity even at the expense of convenience. They desire to interact with the host country's culture, people, learn from their traditions and experience their lifestyle. Thus, this segment values customisable travel packages where they can design their holidays with the most befitting activities. As they do not rely on travel agents planning their trips, they welcome the destination's promotional effort as an information source on potential next adventures. Moreover, destination promotion after a disaster is considered as a sign of a successfully recovered tourism industry (Avraham 2015). Security measures can be reassuring for this group as well however, too lengthy, inconvenient and restricting safety procedures could reduce the entertainment value of the destination (Rittichainuwat 2013) and damaged the feeling of authenticity this group so craves. As a tourism market, they respond well to marketing communication efforts highlighting that the local tourism businesses are aware of the local risk factors and they follow closely the experts' recommendations on health measures. As travelling is more restricted in the current troubled times of COVID-19, these travellers strive to substitute the quantity of travels with a higher quality of it. They look for more quality, real authenticity and uniqueness over cheaper mass tours, even if it means paying more for holidays. Gössling et al. (2020) argues that this line of thinking points to a new trend in the post-pandemic tourism era, where the success of the tourism industry will not be measures by its quantity but its quality.

Advocate Travellers

- **Resilience profile:** Advocate travellers are very similar to the group of mainstream resilience, but they are not simply living out resilience in the face of risk like mainstream travellers, but they also promote the social change towards global resilience. They do not only bear inconvenience for the good of others, but substantial risk, too. This segment is keenly aware of the potential travel risks, they are well-informed about the necessary precautions, the legal restrictions, even the ongoing scientific discussions on hazards when planning their holidays. They are in the position to make well-informed travel decisions unaffected by a potential public panic and point others to this direction as well. The way this group chooses destinations in the face of risk resembles the practice of mainstream resilient travellers. However, this segment often chooses destinations because they want to lead by example and destroy prejudice. They aim to show that some destinations perceived risky or unwelcoming by public opinion is indeed safe to travel and has a hospitable environment.

- **Suitable marketing approach:** This segment strives to highlight the flows in common thinking and is eager to show the dangerous consequences of popular biases. They aim to make people re-evaluate their set ways and make better decisions for themselves. Providing them an opportunity to live out and display their convictions of open-mindedness, cultural sensitivity and risk-tolerance could be well-received. An invitation (promotional message) that highlights the importance of responsible tourism and the possibility to support the recovery of the local economy could catch their attention along with social causes the destination supports. Travellers shown to appreciate when engaged and not just talked to and when given the chance to contribute and advocate and not just passively take in (Lu and Stepchenkova 2015). This seemed particularly true for this segment. Tourism entities can engage them to talk about their experiences at the destination and compel them to journal their experiences on a travel blog, share them on social media or use their personal testimonies in promotional materials. Testimonials from on the-ground tourists and familiarization tours for travel writers were used successfully to encourage first time visitors' confidence and rebuild travellers' trust in Jordan (Liu et al. 2016). Over marketing communication responsibilities, this group could also be integrated into the crisis communication plan of the destination. Destination marketers and even single tourism businesses could provide advocates with fact-based crisis information, materials on safety procedures, even crisis recovery progress, to share it in their sphere of influence. This would promote awareness of the actual situation at the destination and contribute to its overall safety image. Advocates are influencers, either a large audience or a group of friends, but nevertheless. They can be a great help for any destination at large or any tourism entity in particular not only as fearless travellers, but as advocates of resilience in the face of irrational travel risks.

Highly Resilient Travellers

- **Resilience profile:** This segment wants to explore their own limits and push them further. They want to overcome fear and beat challenges during travels, and they want to build themselves up by doing so. They want to be set apart from their peers and stand out. They are looking for excitement, stimulation and thrill but not real danger. Just as every other group, they do not want to be harmed, but more than that, they do not want to miss out on great adventures. Boredom, ordinariness, conformity and routine is a bigger concern for them than safety. This segment is the only type of travellers that the traditional tourism risk domain would label crisis-resistant (e.g., Cohen 1972; Plog 1974) or adventure traveller (Buckley 2012; Ryan et al. 2005). Travel risk does not restrain them, it can even motivate them to visit a destination. Their resilience rests in both their objective and careful risk evaluation and their inherent risk tolerance. Their desire for stimulation prompts them to travel in times when others would not travel. This group seem very well-informed about the actual travel risks of their chosen destination and well-aware of the potential risk reduction strategies to eliminate or, at least, control these risks. They are a great segment for troubled destinations and for global travels in general, but they are rarely completed to go back to the same place twice.

- **Suitable marketing approach:** As they are not looking for danger but the sense of danger, clear communication on the safety and hygienic measures in place would be appreciated by this segment as well. But, creating an overly protected and inconveniently secured environment, which would assure others, might deter this segment from choosing the destination. Travelling in troubled times, like during the current pandemic, these travellers, too, seem to be more security conscious by willing to pay more and take on additional inconveniences for effective protection at the airports, planes and other means of public transportation, accommodations, tourism sites and during customs procedures. This shows that no traveller is exempt of fear (Belhassen et al. 2014; Yang and Nair 2014), and that neither they take risks lightly, but rather view it objectively. As they do not look for familiarity during holidays, it is a challenge for any destination and tourism business to engage them for a long time or to motivate them to visit again. Offering novelty, entertainment, authenticity along with the feeling of rush, but not necessarily danger is no easy task. Approaching them with always changing, highly personalisable and innovative tours might yield results. They best respond to adventure-type tourism experiences and active experience-centred holiday packages. They can be offered unique experiences and tours that never get into common holiday brochures as they are too risky or too expensive. This segment could be a suitable target for troubled destinations, as due to their high risk-tolerance and constant strive for novelty, they are the ones who lead by example and show the rest of the travellers that riskier destinations are already safe, and unknown destinations are worth visiting (Avraham 2015; Avraham and Ketter 2017). While most of them do not have the advocates' desire for addressing world issues, some of them could be open to share their experiences on social media, travel blogs even on TV, which can contribute to build both the destination as well as other travellers up.

Table 3. General categories of travellers' resilience applicable across risk contexts

Main Characteristics	Non-resilient	Striving for Resilience	Mainstream Resilient	Advocate	Highly Resilient
Resilience Profile	• Instinctive risk assessment • Reactive (potentially irrational) behaviour to fear • No acknowledgement of bias (irrationality exaggerated risk assessment)	• Conscious strive for rational risk assessment • Reactive (potentially irrational) behaviour to fear • Recognised bias & conscious strive to correct it	• Rational risk assessment • Conscious realignment of fear reactions • Conscious & consistent correction of bias	• Rational risk assessment • Conscious realignment of fear reactions • Proactive correction of bias by modelling rational risk behaviour to building up others to overcome their own biases • Open to take risk for others	• Rational risk assessment • Conscious realignment of fear reactions • Proactive correction of bias by challenging (building up) themselves to overcome their own bias • Open to take risk for stimulation
Travel Risk Mitigation Strategy in Post-COVID Era	• Avoidance of travelling • Considering travel alternatives, like virtual travels, ethnic restaurants	• Avoidance of international travels • Selective local travels • Fervent information research • Strict legal/hygienic measures in place are a must to travel	• Selective international travels • Discerning information research • Efficient legal/hygienic measures in place mean competitive advantage when choosing a destination	• Selective international travels • Discerning information research • Efficient legal/hygienic restrictions in place mean competitive advantage when choosing a destination	• Eager to return to 'normal' in travels • Reasonable legal/hygienic restrictions in place are acceptable • Unreasonable legal/hygienic restrictions are deterrent when choosing a destination
Suitable Marketing Approach	• Promote rational risk assessment	• Reassuring safety through facts • Offer (local) travel alternatives	• Reassuring safety through factual communication • Necessary safety measures in place	• Reassuring safety through factual communication • Necessary safety measures in place • Partnership in (safe) travel initiatives	• Reassuring safety through factual communication • Necessary safety measures in place • Customized & well-priced adventure travels with reasonable safety measures

CONCLUSION

Bude (2018) argues that the current times can be called the era of fear, as there is always something to be frightened of. Fear is constant and the only thing is changing what we are afraid of. Fear is an instinctive survival reaction aiming for self-preservation at any cost, thus it can be irrational, exaggerated, even aggressive impairing rational thinking and moral decision-making (Murray and Schaller 2016; Panksepp et al. 2011). Fear when prolonged tires out the nervous system and creates a constant state of anxiety, which increases overall risk perception and as an underlying state of mind tints every decision (Ledoux 2015). It creates a different type of risk perception and calls for different qualities to resist it (Thompson et al. 2017), where traditional tourism risk literature might not be entirely fitting. Therefore, Benjamin, Dillette and Alderman (2020) suggest that traditional approaches in tourism must be re-examined in this new era of fear. This chapter reviewed the most recent studies that aim to restart travellers' resilience research with a clean state of mind to build new theories on what it means to be a resilient traveller today in the face of the ever-present terrorism risk and the haunting threat of the novel coronavirus (e.g., Veréb et al. 2020; in press).

By building on the principle of psychological resilience and synthetizing the latest finding in travellers' resilience domain, the chapter outlines general categories of travellers' resilience that is applicable across risk-contexts. Furthermore, this chapter provides a detailed profile for each travellers' category along with suitable targeting guidelines following confirmed approaches from clinical psychology (e.g., Rees et al. 2015) and highlighting the latest findings of crisis-communication (e.g., Avraham and Ketter 2017; Garfin et al. 2020) on how to approach and encourage each segment in the current troubled times. The chapter also hints the latest directions and developments of emerging tourism trends in the post-pandemic world.

This chapter is one but a few studies on travellers' resilience, this research field has a vast and promising territory ahead to explore. This chapter focused on the changes in overall attitude towards travelling in the face impending travel risks and global adversities, but not dwelled deeper in how the changes in attitude play out in an environment with not only internal safety controls but external political restrictions. Future research should take the next steps and explore how the instinctively restraining fear and conscious strive for rational risk evaluation translates to actual travel decisions in a new era of tourism that we enter post-COVID. Also, COVID-19 has created a new reality and a 'new normal', especially for global travels and tourism. This new normal comes with new rules that needs to be explored and proven theories of 'old' has to be re-tested and if necessary, refined. Future research could also explore how the fear of new, upcoming infectious diseases foreseen by experts (e.g., WHO, 2015), in light of the severity of the current pandemic, impact travellers and make them rethink travelling and leisure tourism.

One of the limitations of the general categorisation of travellers' resilience this chapter outlined is that this categorisation is based on the available studies that examined resilience in terms of terrorism and infectious diseases risk only. Future research should extend this notion and test it in different risk contexts. For example, climate change is studied along with terrorism and infectious diseases as a main source of catastrophic risk that increases travel risk perception and alters travel decision-making (e.g., Gössling et al. 2020). The other major limitation of the current categorisation of travellers that it yet lack quantitative validation. The profiled groups of travellers, in the case of both terrorism risk and the threat of COVID-19, were given the same attention and emphasis, but to understand their correct proportion in the travellers' community and to rightly adjust a destination-wide marketing communication effort further research is required.

REFERENCES

Araña, J. E., & León, C. J. (2008). The impact of terrorism on tourism demand. *Annals of Tourism Research, 35*(2), 299–315. doi:10.1016/j.annals.2007.08.003

Avraham, E. (2015, April). Destination image repair during crisis: Attracting tourism during the Arab Spring uprisings. *Tourism Management, 47*, 224–232. doi:10.1016/j.tourman.2014.10.003

Avraham, E. (2016). Destination marketing and image repair during tourism crises: The case of Egypt. *Journal of Hospitality and Tourism Management, 28*(September), 41–48. doi:10.1016/j.jhtm.2016.04.004

Avraham, E., & Ketter, E. (2017). Destination marketing during and following crises: Combating negative images in Asia. *Journal of Travel & Tourism Marketing, 34*(6), 709–718. doi:10.1080/10548408.2016.1237926

Bae, S. Y., & Chang, P. J. (2020). The effect of coronavirus disease-19 (COVID-19) risk perception on behavioural intention towards 'untact' tourism in South Korea during the first wave of the pandemic (March 2020). *Current Issues in Tourism*. Advance online publication. doi:10.1080/13683500.2020.1798895

Bavel, J. J. V., Baicker, K., Boggio, P. S., Capraro, V., Cichocka, A., Cikara, M., Crockett, M. J., Crum, A. J., Douglas, K. M., Druckman, J. N., Drury, J., Dube, O., Ellemers, N., Finkel, E. J., Fowler, J. H., Gelfand, M., Han, S., Haslam, S. A., Jetten, J., ... Willer, R. (2020). Using social and behavioural science to support COVID-19 pandemic response. *Nature Human Behaviour, 4*(5), 460–471. doi:10.103841562-020-0884-z PMID:32355299

Belhassen, Y., Uriely, N., & Assor, O. (2014, November). The touristification of a conflict zone: The case of Bil'in. *Annals of Tourism Research, 49*, 174–189. doi:10.1016/j.annals.2014.09.007

Benjamin, S., Dillette, A., & Alderman, D. H. (2020). "We can't return to normal": Committing to tourism equity in the post-pandemic age. *Tourism Geographies, 22*(3), 476–483. doi:10.1080/14616688.2020.1759130

Bonanno, G. A., Westphal, M., & Mancini, A. D. (2011). Resilience to Loss and Potential Trauma. *Annual Review of Clinical Psychology, 7*(1), 511–535. doi:10.1146/annurev-clinpsy-032210-104526 PMID:21091190

Bourdieu, P. (1987). *Distinction: A Social Critique of the Judgement of Taste*. Harvard University Press.

Buckley, R. (2012). Rush as a key motivation in skilled adventure tourism: Resolving the risk recreation paradox. *Tourism Management, 33*(4), 961–970. doi:10.1016/j.tourman.2011.10.002

Bude, H. (2018). *Society of Fear*. Polity Press.

Calgaro, E., Dominey-Howes, D., & Lloyd, K. (2014). Application of the Destination Sustainability Framework to explore the drivers of vulnerability and resilience in Thailand following the 2004 Indian Ocean Tsunami. *Journal of Sustainable Tourism, 22*(3), 361–383. doi:10.1080/09669582.2013.826231

Carleton, R. N. (2016, April). Into the unknown: A review and synthesis of contemporary models involving uncertainty. *Journal of Anxiety Disorders, 39*, 30–43. doi:10.1016/j.janxdis.2016.02.007 PMID:26945765

Chang, H.-J., Huang, N., Lee, C.-H., Hsu, Y.-J., Hsieh, C.-J., & Chou, Y.-J. (2004). The Impact of the SARS Epidemic on the Utilization of Medical Services: SARS and the Fear of SARS. *American Journal of Public Health, 94*(4), 562–564. doi:10.2105/AJPH.94.4.562 PMID:15054005

Clancy, M. (2012). Security and the Future of Tourism. In J. Leigh, C. Webster, & S. Ivanov (Eds.), *Future Tourism: Political, Social and Economic Challenges* (pp. 37–52). Routledge.

Cochrane, J. (2010). The Sphere of Tourism Resilience. *Tourism Recreation Research, 35*(2), 173–185. doi:10.1080/02508281.2010.11081632

Cohen, E. (1972). Toward a Sociology of International Tourism. *Social Research, 39*(1), 164–182.

Egeland, B., Carlson, E., & Sroufe, L. A. (1993). Resilience as process. *Development and Psychopathology, 5*(04), 517–528. doi:10.1017/S0954579400006131

Finucane, M. L., Alhakami, A., Slovic, P., & Johnson, S. M. (2000). The affect heuristic in judgments of risks and benefits. *Journal of Behavioral Decision Making, 13*(1), 1–17. doi:10.1002/(SICI)1099-0771(200001/03)13:1<1::AID-BDM333>3.0.CO;2-S

Floyd, M. F., Gibson, H., Pennington-Gray, L. A., & Thapa, B. (2004). The Effect of Risk Perceptions on Intentions to Travel in the Aftermath of September 11, 2001. *Journal of Travel & Tourism Marketing, 15*(2–3), 19–38. doi:10.1300/J073v15n02_02

Garcia-Dia, M. J., DiNapoli, J. M., Garcia-Ona, L., Jakubowski, R., & O'Flaherty, D. (2013). Concept Analysis: Resilience. *Archives of Psychiatric Nursing, 27*(6), 264–270. doi:10.1016/j.apnu.2013.07.003 PMID:24238005

Garfin, D. R., Silver, R. C., & Holman, E. A. (2020). The Novel Coronavirus (COVID-2019) Outbreak: Amplification of Public Health Consequences by Media Exposure. *Health Psychology*, *39*(5), 355–357. doi:10.1037/hea0000875 PMID:32202824

Gössling, S., Scott, D., & Hall, C. M. (2020). Pandemics, tourism and global change: A rapid assessment of COVID-19. *Journal of Sustainable Tourism*, 1–20.

Hajibaba, H., Gretzel, U., Leisch, F., & Dolnicar, S. (2015, July). Crisis-resistant tourists. *Annals of Tourism Research*, *53*, 46–60. doi:10.1016/j.annals.2015.04.001

Hall, C. M., Prayag, G., & Amore, A. (2017). *Tourism and resilience: individual, organisational and destination perspectives* (1st ed.). Channel View Publications.

Hartman, S. (2016). Towards adaptive tourism areas? A complexity perspective to examine the conditions for adaptive capacity. *Journal of Sustainable Tourism*, *24*(2), 299–314. doi:10.1080/09669582.2015.1062017

Hartman, S. (2018). Resilient tourism destinations? Governance implications of bringing theories of resilience and adaptive capacity to tourism practice. In E. Innerhofer, M. Fontanari, & H. Pechlaner (Eds.), *Destination Resilience: Challenges and Opportunities for Destination Management and Governance* (1st ed., pp. 66–76). Routledge. doi:10.4324/9780203701904-5

Henderson, N., Benard, B., & Sharp-Light, N. (Eds.). (1999). *Resiliency in action: Practical ideas for overcoming risks and building strengths in youth, families and communities*. Resiliency in Action, Inc.

Human Rights Watch. (2002). *"WE ARE NOT THE ENEMY" Hate Crimes Against Arabs, Muslims, and Those Perceived to be Arab or Muslim after September 11*. Retrieved from http://www.hrw.org

Human Rights Watch. (2020). *Covid-19 Fueling Anti-Asian Racism and Xenophobia Worldwide | Human Rights Watch*. Retrieved from https://www.hrw.org/news/2020/05/12/covid-19-fueling-anti-asian-racism-and-xenophobia-worldwide

Innerhofer, E., Fontanari, M., & Pechlaner, H. (Eds.). (2018). *Destination Resilience: Challenges and Opportunities for Destination Management and Governance* (1st ed.). Routledge. doi:10.4324/9780203701904

Kapuściński, G., & Richards, B. (2016). News framing effects on destination risk perception. *Tourism Management*, *57*, 234–244. doi:10.1016/j.tourman.2016.06.017

Karl, M., & Schmude, J. (2017). Understanding the role of risk (perception) in destination choice: A literature review and synthesis. *Tourism (Zagreb)*, *65*(2), 138–155.

Kim, J., Rasouli, S., & Timmermans, H. J. P. (2018). Social networks, social influence and activity-travel behaviour: A review of models and empirical evidence. *Transport Reviews*, *38*(4), 499–523. doi:10.1080/01441647.2017.1351500

Kotler, P., Armstrong, G., Harris, L. C., & Piercy, N. (2013). Principles of Marketing (6th Europe). Essex, UK: Pearson Education Limited.

Ledoux, J. (2015). *Anxious: Using the Brain to Understand and Treat Fear and Anxiety*. Penguin Books.

Lee, A. V., Vargo, J., & Seville, E. (2013). Developing a Tool to Measure and Compare Organizations' Resilience. *Natural Hazards Review*, *14*(1), 29–41. doi:10.1061/(ASCE)NH.1527-6996.0000075

Lepp, A., & Gibson, H. (2008). Sensation seeking and tourism: Tourist role, perception of risk and destination choice. *Tourism Management*, *29*(4), 740–750. doi:10.1016/j.tourman.2007.08.002

Lippi, G., Sanchis-Gomar, F., & Henry, B. M. (2020). Coronavirus disease 2019 (COVID-19): The portrait of a perfect storm. *Annals of Translational Medicine*, *8*(7), 497–497. doi:10.21037/atm.2020.03.157 PMID:32395541

Lu, W., & Stepchenkova, S. (2015). User-Generated Content as a Research Mode in Tourism and Hospitality Applications: Topics, Methods, and Software. *Journal of Hospitality Marketing & Management*, *24*(2), 119–154. doi:10.1080/19368623.2014.907758

Luthe, T., & Wyss, R. (2014). Assessing and planning resilience in tourism. *Tourism Management*, *44*, 161–163. doi:10.1016/j.tourman.2014.03.011

Murray, D. R., & Schaller, M. (2016). The behavioral immune system: Implications for social cognition, social interaction, and social influence. In J. M. Olson & M. P. Zanna (Eds.), Advances in Experimental Social Psychology (Vol. 53, pp. 75–129). Academic Press Inc.

Panksepp, J., Fuchs, T., & Iacobucci, P. (2011). The basic neuroscience of emotional experiences in mammals: The case of subcortical FEAR circuitry and implications for clinical anxiety. *Applied Animal Behaviour Science*, *129*(1), 1–17. doi:10.1016/j.applanim.2010.09.014

Pechlaner, H., & Innerhofer, E. (2018). Linking destinations and resilience – challenges and perspectives. In E. Innerhofer, M. Fontanari, & H. Pechlaner (Eds.), *Destination Resilience: Challenges and Opportunities for Destination Management and Governance* (1st ed., pp. 3–13). Routledge. doi:10.4324/9780203701904-1

Pechlaner, H., & Volgger, M. (2015). Can Tourism Qualify for Interdisciplinary Research? A European View. In H. Pechlaner & E. Smeral (Eds.), *Tourism and Leisure:Current issues and perspectives of development* (pp. 3–21). Springer Fachmedien. doi:10.1007/978-3-658-06660-4_1

Perpiña, L., Camprubí, R., & Prats, L. (2017). Destination Image Versus Risk Perception. *Journal of Hospitality & Tourism Research.*

Pizam, A., & Smith, G. (2000). Tourism and terrorism: A quantitative analysis of major terrorist acts and their impact on tourism destinations. *Tourism Economics*, *6*(2), 123–138. doi:10.5367/000000000101297523

Plog, S. C. (1974). Why Destination Areas Rise and Fall in Popularity. *The Cornell Hotel and Restaurant Administration Quarterly*, *14*(4), 55–58. doi:10.1177/001088047401400409

Prayag, G. (2018). Symbiotic relationship or not? Understanding resilience and crisis management in tourism. *Tourism Management Perspectives*, *25*(January), 133–135. doi:10.1016/j.tmp.2017.11.012

Rees, C. S., Breen, L. J., Cusack, L., & Hegney, D. (2015). Understanding individual resilience in the workplace: The international collaboration of workforce resilience model. *Frontiers in Psychology*, *6*, 73. doi:10.3389/fpsyg.2015.00073 PMID:25698999

Richter, L. K., & Waugh, W. L. Jr. (1986). Terrorism and tourism as logical companions. *Tourism Management*, *7*(4), 230–238. doi:10.1016/0261-5177(86)90033-6

Ritchie, B. W. (2004). Chaos, crises and disasters: A strategic approach to crisis management in the tourism industry. *Tourism Management*, *25*(6), 669–683. doi:10.1016/j.tourman.2003.09.004

Rittichainuwat, B. N. (2013). Tourists' Perceived Risks Toward Overt Safety Measures. *Journal of Hospitality & Tourism Research (Washington, D.C.)*, *37*(2), 199–216. doi:10.1177/1096348011425494

Rittichainuwat, B. N., & Chakraborty, G. (2009). Perceived travel risks regarding terrorism and disease: The case of Thailand. *Tourism Management*, *30*(3), 410–418. doi:10.1016/j.tourman.2008.08.001

Rittichainuwat, B. N., & Chakraborty, G. (2012). Perceptions of importance and what safety is enough. *Journal of Business Research, 65*(1), 42–50. doi:10.1016/j.jbusres.2011.07.013 PMID:32287524

Ruiz-Ballesteros, E. (2011). Social-ecological resilience and community-based tourism: An approach from Agua Blanca, Ecuador. *Tourism Management, 32*(3), 655–666. doi:10.1016/j.tourman.2010.05.021

Ryan, C., Page, S., & Aicken, M. (Eds.). (2005). *Taking tourism to the limits: issues, concepts and managerial perspectives*. Elsevier Ltd.

Sarman, I., Scagnolari, S., & Maggi, R. (2016). Acceptance of Life-Threatening Hazards among Young Tourists: A Stated Choice Experiment. *Journal of Travel Research, 55*(8), 979–992. doi:10.1177/0047287515612595

Southwick, S. M., Bonanno, G. A., Masten, A. S., Panter-Brick, C., & Yehuda, R. (2014). Resilience definitions, theory, and challenges: Interdisciplinary perspectives. *European Journal of Psychotraumatology, 5*(1), 25338. doi:10.3402/ejpt.v5.25338 PMID:25317257

Taha, S. A., Matheson, K., & Anisman, H. (2014). H1N1 Was Not All That Scary: Uncertainty and stressor appraisals predict anxiety related to a coming viral threat. *Stress and Health, 30*(2), 149–157. doi:10.1002mi.2505 PMID:23818424

Taylor, T., & Toohey, K. (2005). Impacts of tourism-related safety and security at a major sport event. *Event Management, 9*(4), 199–209. doi:10.3727/152599506776771544

Thompson, R. R., Garfin, D. R., Holman, E. A., & Silver, R. C. (2017). Distress, Worry, and Functioning Following a Global Health Crisis: A National Study of Americans' Responses to Ebola. *Clinical Psychological Science, 5*(3), 513–521. doi:10.1177/2167702617692030

Toohey, K., Taylor, T., & Lee, C.-K. (2003). The FIFA World Cup 2002: The effects of terrorism on sport tourists. *Journal of Sport & Tourism, 8*(3), 186–196. doi:10.1080/14775080310001690495

Tyrrell, T. J., & Johnston, R. J. (2008). Tourism sustainability, resiliency and dynamics: Towards a more comprehensive perspective. *Tourism and Hospitality Research, 8*(1), 14–24. doi:10.1057/thr.2008.8

US Department of State. (2020). *Travel Advisories*. Retrieved September 9, 2020, from https://travel.state.gov/content/travel/en/traveladvisories/traveladvisories.html/

van Niekerk, M., & Pizam, A. (2015). How do Terrorism and Tourism Co-exist in Turbulent Times? In K. Glaser (Ed.), *Terrorism and the Economy. Impacts on the Capital Market and the Global Tourism Industry* (1st ed., pp. 109–126). Eleven International Publishing.

Veréb, V., Nobre, H., & Farhangmehr, M. (2020). Cosmopolitan tourists: the resilient segment in the face of terrorism. *Tourism Management Perspectives, 33*(September).

Veréb, V. N., Nobre, H., & Farhangmehr, M. (2018). The fear of terrorism and shift in cosmopolitan values. *International Journal of Tourism Cities, 4*(4), 452–483. doi:10.1108/IJTC-03-2018-0024

Veréb, V. N., Nobre, H., & Farhangmehr, M. (in press). Cosmopolitan tourists: The resilient segment in the face of COVID-19. *Service Business*.

Vinkers, C. H., van Amelsvoort, T., Bisson, J. I., Branchi, I., Cryan, J. F., Domschke, K., Howes, O. D., Manchia, M., Pinto, L., de Quervain, D., Schmidt, M. V., & van der Wee, N. J. A. (2020, June). Stress resilience during the coronavirus pandemic. *European Neuropsychopharmacology, 35*, 12–16. doi:10.1016/j.euroneuro.2020.05.003 PMID:32446705

WHO. (2015). *Anticipating emerging infectious diseases epidemics*. Retrieved from https://www.who.int/csr/disease/anticipating_epidemics/events/AE_sessionAgenda.pdf

Williams, A. M., & Baláž, V. (2015). Tourism Risk and Uncertainty: Theoretical Reflections. *Journal of Travel Research, 54*(3), 271–287. doi:10.1177/0047287514523334

Wolff, K., & Larsen, S. (2016). Flux and permanence of risk perceptions: Tourists' perception of the relative and absolute risk for various destinations. *Scandinavian Journal of Psychology, 57*(6), 584–590. doi:10.1111jop.12326 PMID:27636718

Wu, B. P., & Chang, L. (2012). The social impact of pathogen threat: How disease salience influences conformity. *Personality and Individual Differences, 53*(1), 50–54. doi:10.1016/j.paid.2012.02.023

Yang, E. C. L., & Nair, V. (2014). Tourism at Risk: A Review of Risk and Perceived Risk. *Tourism Asia-Pacific Journal of Innovation in Hospitality and Tourism, 3*(2), 239–259. doi:10.760340930-014-0013-z

Yang, Y., Zhang, H., & Chen, X. (2020). Coronavirus pandemic and tourism: Dynamic stochastic general equilibrium modelling of infectious disease outbreak. *Annals of Tourism Research, 83*, 102913. doi:10.1016/j.annals.2020.102913 PMID:32292219

Yehuda, R., & Flory, J. D. (2007). Differentiating biological correlates of risk, PTSD, and resilience following trauma exposure. *Journal of Traumatic Stress*, *20*(4), 435–447. doi:10.1002/jts.20260 PMID:17721957

Zolli, A., & Healy, A. M. (2013). *Resilience: Why things bounce back* (1st ed.). Simon & Schuster Paperbacks.

KEY TERMS AND DEFINITIONS

Cosmopolitanism: A belief that all people are entitled to equal respect and consideration, no matter what their citizenship status or other affiliations happen to be.

Destination Image: The perception of travellers about a tourist destination based on a combination of their knowledge, emotions, and intentions regarding a place.

Fear: An emotional reflex response to the threat of danger.

Pandemic: An outbreak of a disease prevalent over the whole world or a wide geographic area (such as multiple countries or continents) and affecting a significant proportion of the global population.

Psychological (Individual) Resilience: The capacity of a person to maintain his/her integrity and proper functioning in the face of adversity.

Risk Perception: The subjective judgement that people make about the characteristics and severity of a hazard.

Terrorism: The use or threat of violence to create fear and to intimidate a government or any segment of the civilian population to advance political or social objectives.

Chapter 11
Dealing With the Risks of Information Overload:
Promoting a Mindful Surprise Culture on Farm Holidays

Vincent P. Magnini
Longwood University, USA

Erika Quendler
Federal Institute of Agricultural Economics, Rural, and Mountain Research, Austria

ABSTRACT

While guests go on farm holidays in the hope of receiving authentic and memorable experiences, information overload is a factor that is increasingly detracting from such experiences. For instance, the steady stream of emails, text messages, and news received through social media channels tends to distract from the main experience. To cope with such overload, script theory suggests that guests are increasingly 'satisficing' their information intake and subconsciously tuning-out much incoming stimuli from their surroundings. This subconscious satisficing makes it more difficult for farm holiday providers to win the guest's full attention in order to create memorable and authentic experiences. Therefore, this chapter synthesizes theories and empirical research from several disciplines with a view to gaining a better understanding of how this modern information overload serves as a formidable threat to farm holiday providers, who typically operate from small family farms. This synthesis is then employed as the basis for a number of practical and research recommendations.

DOI: 10.4018/978-1-7998-7352-5.ch011

INTRODUCTION

Holidays are frequently considered '*occasions of escape and liminal leisure*' (Voase, 2018, p.384). This notwithstanding, the existing flood of information can consume too much guest attention, thus jeopardizing the enjoyment of the holiday experience (Ayeh, 2018). The steady stream of emails, text messages and news received through social media channels tends to detract from the main experience (Benselin and Ragsdell, 2016; Park, 2019; Roetzel, 2019). This is inherently connected to the development of information and communication technology. Technological devices, e.g. a mobile and apps, are an undeniable part of our daily lives. Across the family of apps, Facebook now has more than three, WhatsApp two, Messenger 1.3 and Instagram 1 billion 'monthly active people'[1] (Richter, 2020). Moreover, digital advertising, driven by the number of Internet users, connected devices, bandwith, and targeting capabilities, will continue to gain market share (Jakob, 2020). The use of (mobile) technological devices is associated with an increasing amount of information at the user's disposal (Bawden and Robinson, 2009; Floridi, 2014). Potentially, this leads to information overload (Schroder et al., 1967; Eppler and Mengis, 2008), choice overload (Park and Jang, 2013) or technostress (Lee, Chang, Lin, and Cheng, 2014) that is harmful to physical health and mental well-being. To cope with this overload, script theory suggests that guests are increasingly 'satisficing' their information intake and subconsciously tuning-out much incoming stimuli from their surroundings (Campos, Pinto, and Scott, 2019; Magnini and Dallinger, 2015).

Given this development, it is not surprising that information overload is also becoming an issue for farm holiday providers. The way the information overload felt by guests affects their escape or recreation experience has often been ignored. Obviously, farm holiday providers cannot change technology or guest behavior, but they can rethink the responsibility they hold on this matter and try to offer a new added value for guests by including elements that surprise the formers' expectations into the farm holiday experience. (cf. Stankov and Filimonau, 2019) With this in mind, the practices of integrating so-called 'mindful' surprises in order to bring a certain quality of attention to moment-by-moment experiences, are becoming an important tool for achieving well-being on a farm holiday. This in turn should lead to repeat patronage and positive word-of-mouth communications. If a guest is surprised, the provider has won her/his immediate attention. This can be valuable to the farm holiday provider. In the end, if a farm holiday provider is being judged as 'exceeding expectations' then she/he is successfully competing with everything/everyone else that is trying to occupy a portion of the guest's conscious thought. Although this area has started to receive attention from researchers (Lu, Gursoy, and Lu, 2016; Chen, Scott, and Benckendorff, 2017; Magnini and Dallinger, 2018; Stankov and Filimonau, 2019), a deeper understanding of information overload and 'mindful' surprise in the context of a farm holiday is still needed. To date, there has been no empirical evidence or proposals on how to generate and employ new and creative guest surprise tactics when providing experiences that are beneficial for the well-being of guests on holiday farms.

Thus, this chapter primarily examines the diverse segment of farm holidays, the theoretical background and practical perspectives of fostering a 'mindful' surprise culture on holiday farms in order to deliver authentic and appreciated experiences. Adopting pragmatism as a research philosophy and employing a systematic thinking approach, the authors of this chapter elaborate upon possible strategic and tactical measures and advocate for a further empirical examination of 'mindful' farm holiday surprises.

A DIVERSE SEGMENT OF FAMILY OPERATORS WITHIN THE TOURISM SECTOR: FARM HOLIDAY PROVIDERS

The farm holiday has increasingly developed in the years after the Second World War. Its importance has already been documented as synonymous to 'being in nature'. This latter is often seen as being ranked before health, friends, and family (Brämer, 2003, p.25). Logically, this section explores the evolution of a farm holiday, the complexity of its definition, the diversity of offers and the motives of guests.

Evolution of the Farm Holiday Within Tourism

Tourism, especially in rural areas, is gaining in popularity. Throughout the world, the countryside has become a common tourism destination (Ainley and Smale, 2010; Embacher, 2001; Hall et al, 2004; Müller, 2013, p.15). Among its various typifications, one of the most prominent is the farm. This kind of tourism has developed as follows:

- Changes to the agricultural sector have forced farmers to diversify and complement the economic activities of their farms e.g. having guest beds (Arnold and Staudacher, 1981, p.8f; Busby and Rendle, 2000; Cox and Fox, 2003; Dernoi, 1983; Evans and Ilbery, 1989, 1992; Frater, 1983; Ilbery et al., 1988; Nickerson et al., 2001; Pearce, 1990; Sharpley and Vass, 2006).
- In Austria, there are 113,784 guest beds on farms that account approximately for 11 percent of the tourist beds in 2019 (BMLRT, 2020a, p.60). These beds are located on around 8,467 farms or 0.05 percent of all farms (BMLRT, 2020b). In Italy about 20,174 farms are authorized for agritourism with accommodation[2], that is around 2.1 percent of the total in 2019 (Istat, 2020).

In Austria – as in Germany, France and the UK, for example, each had between 20,000 and 30,000 holiday farms in the early 1990s (Fennell and Weaver, 1997). Basically, the figures reflect the longer-term structural change in holidays on the farm towards the increased offer of holiday apartments and the structural development in agriculture.

- Moreover, there are advantages beyond the additional income for farmers and the local community, such as the generation of social benefits, e.g. health and well-being (Vujo, Petrović, and Blesic, 2015; Quendler, Magnini, and Driourech, 2021), the sustainable development of rural areas (Sonnino, 2004; Evgrafova, Ismailova, and Kalinichev, 2020), as well as the recovery of authenticity and tradition (Katila, and Ahuja, 2002; Petruzzelli, and Savino, 2015; traveloregon, 2017). So in Europe and North America, farm holiday or agritourism is used as a policy instrument to rejuvenate regional economies and preserve rural societies and landscapes (Busby and Rendle, 2000; European Parliament, 2015, 2018; Frater, 1983; Hjalager, 1996; Marsden, 1999; McEachern 1992).
- In recent years, such tourism has been so successful that it has become a 'brand' as part of a country's tourism offer, e.g. the 'farm holiday' in Austria (cf. BMNT, 2017; Embacher, 2001, 2018). It even perhaps forms one of the most important and radical product innovations that has affected agriculture, with renewed and winning business models (cf. Broccardo, Culasso, and Truant, 2017; Embacher, 2018, Roman, Roman, and Prus, 2020).
- At the same time, increasing urbanisation has created a demand, partly nostalgic in nature, for holidays in the countryside which city dwellers perceive as a relaxed, idyllic contrast to their own daily lives (Sharply, 2004; Swan, 1929).
- Finally, a holiday, spent on a farm, is a good choice in the era of the COVID-19 pandemic (Wojcieszak-Zbierska et al., 2020). The nature of travelling may change in the time of COVID-19 pandemic as tourists are generally restricted to distances that can be covered by car (Kachroo-Levine, 2020) and this may be a competitive advantage for farm holiday.

Complexity of Definitions

Apparently, the notion of a farm holiday is easy to define as tourism where guests engage in recreational activities within an agricultural setting, i.e. on the farm (Peebles, 1995; Sznajder, Przezborska, and Scrimgeour, 2009). This definition does not include the various notions, types and complexity of the activities, which have appeared in different countries. A holistic definition is missing. Labels such as 'farm holiday', 'agritourism', 'agrotourism', 'farm tourism', 'farm-based tourism', 'farm accommodation', 'vacation farms' and 'farm-stay' are often used interchangeably with each other, but have also been used explicitly to denote similar but distinct concepts (cf. Barbieri and Mshenga 2008, Flanigan et al., 2014; Kokko 2011; Phillip et al 2010) and can be exemplified as follows.

The original Austrian concept of the farm holiday ('Urlaub am Bauernhof') refers to small-scale active family farms linked to agricultural traditions with small-scale tourism accommodation (Arnold and Staudacher, 1981, p.14, 15). Well-known worldwide examples include: dude and working ranches in Canada and the USA (Weaver and Fennell, 1997); the estancias of Mexico, Argentina and Chile (Torres, 2003); the farm-stays and rural B&Bs of the U.K (Evans and Ilbery, 1989, 1992; Frater, 1983; Ilbery et al., 1998; Sharpley and Vass, 2006) and New Zealand (Opperman, 1998; Pearce, 1990); game ranches in southern Africa (Buckley, 2003); traditional small-scale family farms in Germany (Opperman, 1996) and Switzerland ('Ferien am Bauernhof' or agrotourism, Egger, Geneviève, and Passaglia, 2008) as well as agritourism in the rural Mediterranean landscapes of Italy (Santucci, 2013), Spain (Barke, 2004; Canoves et al., 2004; Perales, 2002), Greece (Dimanianos and Skuras, 1996).

These farms in some countries and regions continue to receive very few guests, while others have seen their numbers of tourists grow at a rapid pace (cf. Chatterjee and Prasad, 2019; Quendler, 2019, Van Saandt, Low, and Thilmany, 2018).

For the purposes of clarity and consistency, the term 'farm holiday' is used throughout this chapter to refer to the range of related labels, concepts and products discussed in the literature as a whole.

Variety of Offers and Motives

Whilst the archetypal holiday product is perhaps still a farm-stay on a working farm, where guests are given opportunities to experience farm life there is also room for (i) spending time with family and friends, (ii) resting and relaxing, (iii) having an adventure, (iv) improving one's health and well-being, (v) taking a break from one's everyday life, (vi) learning something new, (vii) reminiscing (nostaligia) and (viii) marking a special occasion (such as a birthday or anniversary) (Tomas and Leffew, 2016, p.3). On that note, there are now many variations of farm holiday.

In the case of Austria, nowadays, the range is vast and covers (i) catering to babies and children (ii) organic production, (iii) health and well-being, (iv) livestock, horses and riding, (v) herblore (vi) wine, (vii) barrier-free aspects, etc. (Bundesverband Urlaub am Bauernhof, 2019). This kind of holiday and its innovations aim to bring the diversity of life on farms closer to guests.

Even if motivations to go on a farm holiday vary by guest segment (Dong et al. 2013), guests are nonetheless exposed not only to a more pleasant but also more peaceful and less stressful environment (cf., Lehle, 1982; Cizek et al., 2005; Grimm, 2011; Lovell et al., 2014). The natural environment of the countryside provides valuable oases which are reflected in guests' motives such as healthy climate, relaxation, escape from the polluted environment (pers. comm, B. Grimm, Reiseanalyse 2020).

This diversity and speciality of a farm holiday are challenged by current developments. Especially information overload boosted by digital transformation comes with some risks that are outlined in the following section.

THE RISK OF INFORMATION OVERLOAD FOR FARM HOLIDAY PROVIDERS

At a basic level, information overload transpires when the volume of information received by an individual exceeds his/her information processing capacity (Schroder et al., 1967; Eppler and Mengis, 2008). Slightly more than a decade ago, it was reported that approximately two-thirds of business professionals feel that information overload has caused problems in their personal relationships, and one-third perceive such information overload as harmful to their physical health (Bawden and Robinson, 2009).

More recently, Dr. Luciano Floridi (2014) in his book titled *The Fourth Revolution: How the Infosphere is Reshaping Human Reality* details some of the profound influences that information and communication technologies are having on humankind. Using Floridi's writings as a foundational premise, Bawden and Robinson (2020, p.1) state the following:

... as society fully experiences Floridi's Fourth Revolution, and moves into hyper-history (with society dependent on, and defined by, information and communication technologies) and the infosphere (an information environment distinguished by a seamless blend of online and offline information activity), individuals and societies are dependent on, and formed by, information in an unprecedented way, information overload needs to be taken more seriously than ever.

To place the implications of the above discussions into the context of a farm holiday, the following two vignettes are provided:

- Guest A (hypothetical) at a farm holiday venue: While at the venue, she is receiving emails via her mobile device along with text messages, weather forecasts, news feeds, breaking news alerts, social media posts, as well as telephone calls.

- Guest B (hypothetical) at a farm holiday venue: While at the venue, she powers off her mobile device so that she can relax and rejuvenate. She, however, is thinking about a social media post that she saw during her journey to the holiday farm; she is reading the messaging on the shirts of the other guests visiting the farm venue; and, she is thinking about a newspaper article that she read during the current farm-stay.

In the above vignettes, research indicates that potential information overload experienced by Guest A is very much a reality. It is increasingly common for individuals to receive information about current events through social media (Gottfried and Barthel, 2018; Park, 2019). Attempting, however, to tease-out relevant/credible news nested within the large volumes of social media content mandates substantial time and mental energy (Lee, Son, and Kim, 2016; Zhang, Wu, and Mattila, 2016). Moreover, routine news updates regarding breaking news is thought to intensify perceptions of information overload (Song, Jung, and Kim, 2017; York, 2013).

On the other hand, the potential information overload experienced by Guest B is also quite plausible. Regarding absorbing printed messaging on cues present in one's environment, advances in printing technologies have made it less expensive and more commonplace to print marketing messaging on nearly all types of surfaces (Magnini, 2015). In terms of the cognitive attention described in the vignette to the newspaper, such attention is quite plausible. Research indicates that one day's newspaper in any major city contains more pieces of news than someone would have received in their entire lifespan in the 17th century (Bawden, 2001).

As depicted in Figure 1, modern information overload has led to consumers relying on mental scripts. Therefore, the next section of this chapter will discuss script theory and the need for farm holiday providers to prompt script deviations.

THE NEED FOR FARM HOLIDAY PROVIDERS TO PROMPT SCRIPT DEVIATIONS

Individuals have grown weary of such information overload described in the previous section (Spira, 2011; Aikat and Remund, 2012). Consequently, research indicates that consumers consciously and unconsciously develop defense mechanisms in attempts not to get overloaded with information (Rumbo, 2002). *Satisficing* is one of these defense mechanisms. When an individual is satisficing, s/he is taking in just enough information to solve a problem and tuning the rest out (Bawden and Robinson, 2009). Stated differently, s/he ingests just enough information to meet his/her needs during the service experience (Bawden and Robinson, 2009).

Figure 1. Farm holiday provider's resiliency model to combat guests' information overload

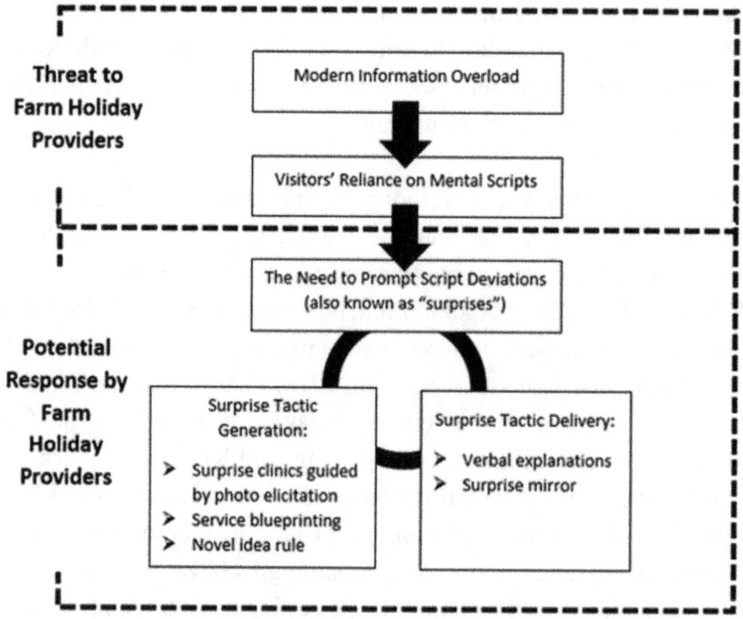

"Individuals continuously check whether their schemas match inputs coming from the surrounding environment. As soon as inputs diverge from the schema, surprise is elicited." -Joelle Vanhamme and Dirk Snelders

Within the context of the farm holiday, guests often do not pay full attention during transactions if everything transpires as they expect – in concert with their pre-formulated mental scripts (Magnini, Crotts, and Zehrer, 2011). Theory-based anchoring of this logic can be found in script theory. According to script theory, consumers interpret their experiences through particular cognitive structures in order to formulate predictions regarding how transactions will unfold; these predictions are known as scripts (Albelson, 1976). [Further detail regarding the origins of script theory is contained in Appendix A].

Consumers depend on these scripts because the scripts reduce the potential burden of continuously processing new information (Smith and Houston, 1985). Mental scripts are formed, in part, from memories of previous experiences at a particular venue or through experiences in similar settings (e.g. a different farm holiday venue). If a given person is new to staying at farm holiday venues, then a number of variables such as word-of-mouth and marketing messages aid in the composition of mental scripts (Zeithaml, Bitner, and Gremler, 2006).

A study conducted by Magnini, Crotts, and Zehrer (2011) can be used to highlight the power of script deviations within a hospitality and tourism context. The study found that:

- Nearly 100 percent of consumers who typed the phrase 'delightful surprise,' 'excellent surprise', 'pleasant surprise', or 'positive surprise' in their Internet blog posts describing a hotel stay indicated in the quantitative portion of the blog site that they would be willing to recommend the given hotel and would also be willing to stay there again.
- Blog posts that used the word 'delightful', 'excellent', 'pleasant', or 'positive', but did not use the word 'surprise' were also analyzed. Of these consumers, only three out of four would affirmatively recommend or stay again at the described hotel.
- Blogs that used the phrase 'very satisfied,' but did not mention a surprise were analyzed: in these, only 60 percent would commit to recommending or repurchasing.
- When a consumer used the word 'satisfied' without 'surprise', there was only a 1/3 likelihood that s/he would be a loyal advocate of the hotel. In other words, the person types in a blog post that s/he is 'satisfied,' but, despite this 'satisfaction,' two-thirds will not return.

The strong connection between surprise and loyalty/positive word-of-mouth and the weaker than commonly believed link between satisfaction and loyalty/positive word-of-mouth underscores the strong influence of positive script deviations: also known as 'mindful surprises'.

As previously stated, the notion that guests do not actually pay full attention until something unexpected happens can be traced to script theory (Bateson, 2002). Because the script deviation (also known as the "surprise") gains guests' full attention, it is much more probable to connect with the guest on an affective level. Surprise is a more positive and more emotional response than being great or excellent; surprise is an affect-based response that builds customer loyalty (Berman, 2005).

Furthermore, script deviations – surprises – are important because the guest's reaction is immediate (Reisenzein, 2000). When a guest is surprised, the provider has won her/his current attention and this led to a 'mindful' farm holiday experience. This immediate attention can be valuable to the provider; particularly in the era of information overload. In comparison, satisfaction judgments which do not conform to expectations are formulated after time and are not immediate. In the end, if a farm holiday provider is being judged as 'exceeding expectations' then the she/he is successfully competing with everything/everyone else that is trying to occupy a portion of the guest's conscious thought.

From a physiological perspective, a surprise spawns considerable activity in the brain (Meyer, 1997). To test the power of surprise, Berns et al. (2001) connected subjects to magnetic resonance imaging (MRI) machines and squirted a pattern of drinks into their mouths: water-juice-water-juice. When the pattern was broken with water-water or juice-juice, their brains 'lit up like Christmas trees' on the MRI machines. Due to this heightened mental activity, it is actually an accurate statement when someone says that he or she '*feels surprised*' (Reisenzein, 2000).

FOSTERING A SURPRISE (SCRIPT DEVIATION) CULTURE: TACTIC GENERATION AND DELIVERY

A guest at a holiday farm can typically only be surprised once by any one particular tactic. In addition, service delivery ideas typically cannot be patented and can, therefore, be copied by competitors. For these reasons, farm holiday providers must continuously derive new and creative guest surprise tactics (as previously depicted in Figure 1).

Photo-Elicitation at the Farm Venue to Generate Surprise Tactics

At a family farm, ideas for new surprise tactics should be derived by all family members and employees so that all feel invested in the approach and culture of surprising visitors. Therefore, those working in the holiday farm should gather periodically to participate in a 'surprise clinic' which is a session in which guest surprise ideas are brainstormed (Magnini, 2017). In such a clinic, both the volume and creativity of ideas generated can be enhanced by using photographs to guide the brainstorming. The photographs are of service points in which guests interact and the group examines the photos as a means of deriving creative surprise ideas. Before the brainstorming session begins, certain parameters can be established such as the monetary budget for the surprise ideas [e.g. all ideas must cost the farm less than two euros; or must be free]. In addition, the group should be instructed that the best surprise ideas will be the ones that reinforce the authenticity of the farm venue and its brand positioning.

The approach of utilizing photos to guide the surprise clinics is a form of photo elicitation. On a broad scope, photo elicitation can be defined as a qualitative research technique that uses a photo(s) to help guide idea generation and discussions (Harper, 2002; Haultain, 2013). There are several different forms of photo elicitation methodologies. The form utilized in the case of these surprise clinics is known as *auto driving* because the participants drive the discussion about the photographs that were provided by the session moderator (Collier 1967; Magnini, Ford, and LaTour, 2012). Auto driving is thought to be the most commonly utilized form of photo elicitation (Harper 1994) and the form most appropriate for conducting these surprise clinics at hospitality and tourism venues (Magnini, 2017).

Employing photos when attempting to generate novel and creative ideas can be useful because photo elicitation can "mine deeper shafts into a different part of the human consciousness than word-alone interviews" (Harper, 2002, p. 13). Photos also often aid the creative process because without the use of photos during a surprise clinic, the attention of the group is on the participants and such focused-attention can make some individuals timid or uncomfortable. Conversely, when photos are used to guide the brainstorming, the focus of attention is diverted from the participants themselves and on to the photos, easing the anxiety of some participants and spawning creativity (Magnini, 2006).

Service Blueprinting at the Farm Venue to Derive Surprise Tactics

Service blueprinting is a process first introduced by Shostack (1994) in the *Harvard Business Review*. Service blueprinting comes in many forms and levels of detail, but all forms entail developing a schematic of the physical servicescape to analyze nuances such as visitor and employee flow. On a service blueprint, often there are three broad areas depicted: 1) where the customer can flow; 2) where the customers cannot go but can view; and 3) the back-of-the house reserved for employees. On the service blueprint, the lines dividing these areas are frequently termed the line of interaction, line of visibility, and line of internal interaction (Bitner, Ostrom, and Morgan, 2008).

Service blueprints can vary with regard to their level of sophistication. They can be computer-generated or can simply be sketched with paper and pencil. In essence, a service blueprint is a visual depiction of a service process. Asking employees to study such visual depictions can aid in the generation of new surprise ideas at tourist attractions (Magnini, 2017). In the previous sentence the word 'study' might be misleading because the process can be fun. For instance, the family operating a holiday farm can hang the service blueprints in a back-of-the-house area and the family/employees can have a contest to see who can derive the most creative guest surprise tactics by examining the blueprints.

It is prudent to note that when a service blueprint is created, a number of copies should be produced. Multiple copies are needed because blueprinting and associated surprise tactics will vary between season of the year, day of the week, or even by time of day (for example: surprise ideas will be different at check-in versus check-out). It is posited by the authors of this chapter that the resiliency of farm holiday providers in combatting information overload of guests will be bolstered through multi-method surprise tactic generation [both photo elicitation and service blueprinting].

Implement a Novel Idea Rule

Employee buy-in has been demonstrated in a number of studies to improve branding and performance of all types of business entities including small and medium-sized enterprises (Balmer et al., 2011; Baron et al., 2009; Davies and Crane, 2010). Conceptually, buy-in can be defined as a personal and professional commitment to actively engage in a process, task, or initiative (French-Bravo and Crow, 2015, p. 2). Within the context of a family-operated holiday farm, it seems plausible that many of the service providers likely possess some level of buy-in, particularly the family members that function as service providers. It is important, however, that such service providers have specific buy-in related to the importance of generating and implementing surprise tactics to combat visitors' information overload.

One method by which to facilitate such buy-in is to have the service providers themselves derive the surprise tactic ideas. In the previously described surprise clinics driven by photo-elicitation and in the service blueprinting process, all service providers across the farm holiday entity are tasked with deriving the ideas for surprise tactics. A third mechanism for bolstering team-wide buy-in on a constant stream of surprise ideas is to create a rule that everyone on the team must bring one new surprise idea to the entity for consideration every week. More specifically, once per week, everyone on the team must write down a creative visitor surprise tactic that has never been used at the venue and anonymously place the idea in an idea collection bin. The ideas are read and one new idea each week is selected. Because the ideas are anonymous, there is no criticism of those whose ideas are not selected. Such a system in which ideas are collected once per week will aid in preventing long periods with lack of service delivery innovation (Peters, 2010).

Use of Explanations to Help Frame Expectations

One potential negative consequence associated with surprising guests with an arsenal of tactics is that the gesture/tactic might come to be expected by the given guest and, therefore, trigger disappointment during the guest's subsequent visit(s). In other words, guests' expectations might be increased to levels that are not sustainable. To circumvent this potential problem, farm holiday providers should communicate to guests as they receive surprises that they are the recipients because they are special, appreciated, and happen to be in the right place at the right time. Stated differently, it should be made apparent to the guest that this is a unique, authentic gesture that is not a typical component of routinized service delivery. This can typically be effectively communicated to the guest through a verbal or written explanation upon delivery of the given surprise (Kim and Mattila, 2013).

Foster a Surprise Mirror

In concert with the theme of this book – resiliency models for family businesses in the tourism sector – it is important to realize that resiliency can transpire on numerous levels (Quendler, 2015). Much of this chapter has addressed resilience at the meso or organizational level. Resilience, however, can also take place at the micro or individual level (cf. Alexander, 2013; Aranda et al., 2012; Quendler, 2015). Specifically, as it relates to the chapter at hand: organizational exercises such as photo elicitation and service blueprinting can only be successful if the individuals who comprise those organizations are motivated to meaningfully participate. As such, resilience – in this case, combatting information overload – is best achieved through the confluence of meso (organizational) and micro (individual) efforts. Therefore, fostering a surprise mirror within the holiday farm can be useful.

"People will forget what you said, People will forget what you did, but people will never forget how you made them feel." -Maya Angelou (American Poet)

The notion of a surprise mirror centers upon the logic that the more individuals within an organization surprise one another with random acts of kindness, the more likely each of those individuals is to surprise guests with acts of kindness (Magnini, 2015). Consequently, those within the organization should make efforts to routinely surprise one another with such gestures because the motivation to surprise visitors will be positively correlated. This concept of a surprise mirror is related to the concept of the satisfaction mirror – the notion that employee satisfaction is highly correlated with customer satisfaction (Heskett and Sasser, 2010; Vermeeren, Kuipers, and Steijn, 2011).

FURTHER RESEARCH DIRECTIONS

For farm holiday providers, surprise management is an important research direction. This should also include both technological devices owned by guests and provided by the holiday farm. These should be evaluated from the viewpoint of guests' attention. Thus, co-creation of a mindful surprise culture can be seen as the farm holiday providers' response to the growing demand for addressing the issue of well-being of guests and the viability of the farm.

Collecting empirical evidence of the effectiveness of the surprise culture on a holiday farm is an important process. In general, there are many challenges related to maintaining continuation of a mindful surprise culture while applying psychological research methods (Baer, 2011). In order to avoid the danger of becoming another fuzzy approach, further empirical examination of the feasibility of adopting a mindful surprise culture in farm holiday experiences is needed. More specifically, more research encompassing examples of its use is required, indicating how the different types of holiday farms can benefit from the broader adoption of this approach. In particular, the costs of a mindful surprise culture should be evaluated in the farm context, especially from the perspective of the implementation costs.

For academics, this piece offers a new area for research, mostly related to the issues of the guest's level of engagement in a mindful surprise culture and well-being during holidays and positive word-of-mouth communications afterwards. A mindful surprise culture may become a 'common currency'. However, it might be still seen as an unfamiliar practice, or it can even be classified as a Next Generation practice and the farm holiday consumer may be repelled by this type of product labelling as a result. This can even be the case for some sub-segments within the guest groups that consider themselves consumers of the mindful surprise culture. While the farm holiday is generally well placed to address the needs of certain guests, it might not yet be ready for the Next Generation who may seek an experience more in tune with a mindful surprise culture, but who, at the same time, sometimes refuse this epithet. These potential dissonances are some of the major concerns that need to be addressed by the farm holiday in order to create and include more mindful surprises into their day-to-day, mainstream offerings.

CONCLUSION

The specific responsibilities of a farm holiday provider to promote the well-being of their guests during vacation experiences definitely include the creation of 'mindful surprises'. Generally, farm holiday providers offer total experiences and they, therefore, should count in all the dangers that could distract the guests from achieving a desirable outcome of their farm holiday. This is not just an ethical question or question of sustainability for the holiday farm, but also a question of profitability as the effect of unsatisfactory experiences is directly reflected in the level of profit, positive online postings or lower number of guests. To this end, a mindful surprise culture, i.e. surprises, as a growing trend in the well-being movement can be a viable option for engaging guests in practices to target well-being, repeat patronage and positive word-of-mouth communications.

REFERENCES

Abelson, R. P. (1976). *Script processing in attitude formation and decision making*. Lawrence Erlbaum.

Abelson, R. P. (1981). Psychological status of the script concept. *The American Psychologist*, *36*(7), 715–729. doi:10.1037/0003-066X.36.7.715

Aikat, D., & Remund, D. (2012). Of time magazine, 24/7 media, and data deluge: The evolution of information overload theories and concepts. *Information overload: An international challenge for professional engineers and technical communicators*, 13-40.

Ainley, S., & Smale, B. (2010). A profile of Canadian agritourists and the benefits they seek. *Journal of Rural and Community Development*, *5*(1), 58–75.

Alexander, D. E. (2013). Resilience and disaster risk reduction: An etymological journey. *Natural Hazards and Earth System Sciences*, *1*, 1257–1284. doi:10.5194/nhessd-1-1257-2013

Aranda, K., Zeeman, L., Scholes, J., & Morales, A. S.-M. (2012). The resilient subject: Exploring subjectivity, identity and the body in narratives of resilience. *Health*, *16*(5), 548–563. doi:10.1177/1363459312438564 PMID:22547553

Arnold, K., & Staudacher, C. (1981). *Urlaub auf dem Bauernhof. Eine empirische Untersuchung der Struktur und Entwicklung einer spezifischen Erholungsform und ihrer Auswirkungen auf die Land- und Forstwirtschaft in Niederösterreich*. Verlag Ferdinand Hirt.

Ayeh, J. K. (2018). Distracted gaze: Problematic use of mobile technologies in vacation contexts. *Tourism Management Perspectives*, *26*, 31–38. doi:10.1016/j.tmp.2018.01.002

Baer, R. A. (2011). Measuring mindfulness. *Contemporary Buddhism*, *12*(1), 241–261. doi:10.1080/14639947.2011.564842

Balmer, J. M., Powell, S. M., Punjaisri, K., & Wilson, A. (2011). Internal branding process: Key mechanisms, outcomes and moderating factors. *European Journal of Marketing*, *45*(9/10), 1521–1537. doi:10.1108/03090561111151871

Barbieri, C., & Mshenga, P. M. (2008). The role of the firm and owner characteristics on the performance of agritourism farms. *Sociologia Ruralis*, *48*(2), 166–183. doi:10.1111/j.1467-9523.2008.00450.x

Barke, M. (2004). Rural Tourism in Spain. *International Journal of Tourism Research, 6*(3), 1367–149. doi:10.1002/jtr.480

Baron, S., Patterson, A., Oakes, S., Harris, K., Punjaisri, K., Evanschitzky, H., & Wilson, A. (2009). Internal branding: An enabler of employees' brand-supporting behaviours. *Journal of Service Management, 20*(2), 209–226. doi:10.1108/09564230910952780

Bartlett, F. C. (1932). *Remembering*. Cambridge University Press.

Bateson, J. (2002). Consumer preference and quality in services. *Managing Service Quality, 12*(4), 206–209. doi:10.1108/09604520210434811

Bawden, D. (2001). Information and digital literacies: A review of concepts. *The Journal of Documentation, 57*(2), 218–259. doi:10.1108/EUM0000000007083

Bawden, D., & Robinson, L. (2009). The dark side of information: Overload, anxiety and other paradoxes and pathologies. *Journal of Information Science, 35*(2), 180–191. doi:10.1177/0165551508095781

Bawden, D., & Robinson, L. (2020). Information Overload: An Overview. In *Oxford Encyclopedia of Political Decision Making*. Oxford University Press. doi:10.1093/acrefore/9780190228637.013.1360

Benselin, J. C., & Ragsdell, G. (2016). Information overload: The differences that age makes. *Journal of Librarianship and Information Science, 48*(3), 284–297. doi:10.1177/0961000614566341

Berman, B. (2005). How to delight your customers. *California Management Review, 48*(1), 129–151. doi:10.2307/41166331

Berns, G. S., McClure, S. M., Pagnoni, G., & Montague, P. R. (2001). Predictability modulates human brain response to reward. *The Journal of Neuroscience: The Official Journal of the Society for Neuroscience, 21*(8), 2793–2798. doi:10.1523/JNEUROSCI.21-08-02793.2001 PMID:11306631

Bitner, M. J., Ostrom, A. L., & Morgan, F. N. (2008). Service blueprinting: A practical technique for service innovation. *California Management Review, 50*(3), 66–94. doi:10.2307/41166446

BMLRT. (2020a). *Green Report*. https://gruenerbericht.at/

BMLRT. (2020b). *Nächtigungsziffern nach Unterkunftsarten 2019. Tabelle 2.6.2. Green Report*. https://gruenerbericht.at/

BMNT. (2017). *Der Cluster Urlaub am Bauernhof Österreich - Innovationsstrategie Urlaub am Bauernhof 2020*. Bundesministerium für Nachhaltigkeit und Tourismus. https://www.google.com/url?sa=t&rct=j&q=&esrc=s&source=web&cd=1&ved=2ahUKEwj8g-uf-ZffAhX5SxUIHRv-Aj8QFjAAegQIChAC&url=https%3A%2F%2Fwww.bmnt.gv.at%2Fdam%2Fjcr%3A0c0bfa96-4e61-4914-a7ff-47bdff38fb12%2FClusterstrategie%2520Urlaub%2520am%2520Bauern-hof%2520%25C3%2596sterreich%25202020.pdf&usg=AOvVaw2M6gs2O4cqZvwJbmvSJMns

Brämer, R. (2003). Zurück zur Natur? Die Wald- und Wiesen-Therapie. *Psychologie Heute*, *30*(4), 20–29.

Busby, G., & Rendle, S. (2000). The Transition from Tourism on Farms to Farm Tourism. *Tourism Management*, *21*(6), 635–642. doi:10.1016/S0261-5177(00)00011-X

Campos, A. C., Pinto, P., & Scott, N. (2019). Bottom-up factors of attention during the tourist experience: An empirical study. *Current Issues in Tourism*, 1–23. doi:10.1080/13683500.2019.1681383

Canoves, G., Villarino, M., Priestley, G. K., & Blanco, A. (2004). Rural Tourism in Spain: An analysis of recent evolution. *Geoforum*, *35*(6), 755–769. doi:10.1016/j.geoforum.2004.03.005

Chatterjee, S., & Prasad, M. V. D. (2019). The Evolution of Agri-Tourism practices in India: Some Success Stories. *Madridge J Agric Environ Sci.*, *1*(1), 19–25. doi:10.18689/mjaes-1000104

Chen, I. L., Scott, H., & Benckendorff, P. (2017). Mindful tourist experiences: A Buddhist perspective. *Annals of Tourism Research*, *64*, 1–12. doi:10.1016/j.annals.2017.01.013

Cizek, D., Schipfer, R. K., Buchebner-Ferstl, S., Geserick, M. A., Kaindl, M., Kapella, O., & Luef, C. (2005). *Urlaub am Bauernhof aus der Sicht von Gästen und Anbietern*. https://www.dafne.at/prod/dafne_plus_common/attachment_download/0748c0a56650860d343a79ba6b950cf8/Endbericht_Apr07.pdf

Collier, M. (1967). *Visual Anthropology: Photography as a Research Method*. Holt, Rinehart and Winston.

Cox, L., & Fox, M. (2003). Agricultural Based Leisure Attractions. *Journal of Tourism Studies*, *14*(1), 49–58.

Davies, I. A., & Crane, A. (2010). Corporate social responsibility in small-and medium-size enterprises: Investigating employee engagement in fair trade companies. *Business Ethics (Oxford, England)*, *19*(2), 126–139. doi:10.1111/j.1467-8608.2010.01586.x

Dernoi, L. A. (1983). Farm Tourism in Europe. *Tourism Management*, *4*(3), 155-166.

Dimianos, D., & Skuras, D. (1996). Farm business and the development of alternative farm enterprises: An empirical analysis in Greece. *Journal of Rural Studies*, *12*(3), 273–283. doi:10.1016/0743-0167(96)00017-4

Dong, E., Wang, Y., Morais, D., & Brooks, D. (2013). Segmenting the rural tourism market: The case of Potter County, Pennsylvania, USA. *Journal of Vacation Marketing*, *19*(2), 181–193. doi:10.1177/1356766712471231

Egger, T., Favre, G., & Passaglia, M. (2008). Der Agrotourismus in der Schweiz, Analyse der aktuellen Situation und Empfehlungen für die Zukunft. Schweizerische Arbeitsgemeinschaft für die Berggebiete (SAB).

Embacher, H. (2001). Österreich in Europa an der Spitze. *Der Forstliche Landwirt*, *9*, 52–53.

Embacher, H. (2018). *Agritourism and innovation*. Presentation. *1st World Congress on Agritourism*, Bolzano, Italy.

Eppler, M. J., & Mengis, J. (2008). The concept of information overload-a review of literature from organization science, accounting, marketing, mis, and related disciplines (2004). In *Kommunikationsmanagement im Wandel* (pp. 271–305). Gabler. doi:10.1007/978-3-8349-9772-2_15

European Parliament. (2015). *Report on new callenges and concepts forthe promotion of tourism in Europe* (2014/2241(INI)). https://www.europarl.europa.eu/doceo/document/A-8-2015-0258_EN.html

European Parliament. (2018). *European Parliament resolution of 3 October 2018 on addressing the specific needs of rural, mountainous and remote areas* (2018/2720(RSP). https://www.europarl.europa.eu/doceo/document/TA-8-2018-0374_EN.html

Evans, N. J., & Ilbery, B. W. (1989). A Conceptual Framework for Investigating Farm-Based Accommodation and Tourism in Britain. *Journal of Rural Studies*, *5*(3), 257–266. doi:10.1016/0743-0167(89)90004-1

Evans, N. J., & Ilbery, B. W. (1992). Farm-Based Accommodation and the Restructuring of Agriculture: Evidence from Three English Counties. *Journal of Rural Studies*, *8*(1), 85–96. doi:10.1016/0743-0167(92)90032-2

Evgrafova, L. V., Ismailova, A. Z., & Kalinichev, V. L. (2020). Agrotourism as a factor of sustainable rural development. *IOP Conference Series. Earth and Environmental Science*, *421*, 16. doi:10.1088/1755-1315/421/2/022058

Flanigan, S., Blackstock, K., & Hunter, C. (2014). Agritourism from the perspective of providers and visitors: A typology-based study. *Tourism Management*, *40*, 394–405. doi:10.1016/j.tourman.2013.07.004

Floridi, L. (2014). *The 4th Revolution: How the Infosphere is Reshaping Human Reality*. Oxford University Press.

Frater, J. M. (1983). Farm Tourism in England: Planning, Funding, Promotion and Some Lessons from Europe. *Tourism Management*, *4*(3), 167–179. doi:10.1016/0261-5177(83)90061-4

French-Bravo, M., & Crow, G. (2015). Shared governance: The role of buy-in in bringing about change. *Online Journal of Issues in Nursing*, *20*(2).

Gottfried, J., & Barthel, M. (2018). *Almost seven-in-ten Americans have news fatigue, more among Republicans*. Pew Research Center.

Grimm, B. (2011). *Urlaub auf dem Bauernhof/ Urlaub auf dem Lande 2010/11. Untersuchung der Marktsituation und der Potenziale für den Bauernhof-/Landurlaub auf Basis der Reiseanalyse 2011und einer Volumenerhebung, ergänzt durch Gruppendiskussionen mit Personen aus Haushalten ohne Kinder*. Studie für das Bundesministerium für Ernährung, Landwirtschaft und Verbraucherschutz (BMELV). http://docs.dpaq.de/882-2011_bmelv-reiseanalyse.pdf

Hall, D., Mitchell, M., & Roberts, L. (2004). Tourism and the countryside: dynamic relationships. In L. Roberts, D. Hall, & M. Morag (Eds.), *New Directions in Rural Tourism*. Routedge.

Harper, D. (1994). On the authority of the image: Visual methods at the crossroads. In N. K. Denzin & Y. S. Lincoln (Eds.), *Handbook of Qualitative Research* (pp. 403–412). Sage.

Harper, D. (1998). An argument for visual sociology. In J. Prosser (Ed.), *Image-based research: A source book for qualitative researchers* (pp. 24–41). Palmer Press.

Harper, D. (2002). Talking about pictures: A case for photo elicitation. *Visual Studies*, *17*(1), 13–26. doi:10.1080/14725860220137345

Haultain, A. (2013). A picture is worth a thousand words: The empowering potential of using photo-elicitation within research and teaching. *Conference Proceedings: CLESOL 2012*.

Heskett, J. L., & Sasser, W. E. (2010). The service profit chain. In *Handbook of Service Science* (pp. 19–29). Springer. doi:10.1007/978-1-4419-1628-0_3

Hjalager, A. M. (1996). Agricultural diversification into tourism: Evidence from a European community development programme. *Tourism Management*, *17*(2), 103–111. doi:10.1016/0261-5177(95)00113-1

Ilbery, B., Bowler, I., Clark, G., Crockett, A., & Shaw, A. (1988). Farm-Based Tourism as an Alternative Farm Enterprise: A Case Study from the Northern Pennines England. *Regional Studies Association*, *32*(4), 355–363.

Istat. (2020). *Agritourism. Agritouristc farm authorized.* http://dati.istat.it/Index.aspx?lang=en&SubSessionId=4d0df501-4308-481c-a33e-fcdc40a8ec7f

Jakob, H. (2020). *Digital Advertising: Markets, Trends, Players*. Webinar statista.

Kachroo-Levine, M. (2020). *5 Experts Weigh in on What Coronavirus Means for Your Summer Vacation Plans*. https://www.departures.com/travel/summer-vacation-coronavirus-pandemic

Kaiser, I. (2009). *Urlaub am Bauernhof. Erlebnisse aus dem Blickwinkel der Wissenschaft*. Diplomica Verlag GmbH.

Katila, R., & Ahuja, G. (2002). Something old, something new: A longitudinal study of search behavior and new product introduction. *Academy of Management Journal*, *45*, 1183–1194. https://books.google.at/books?hl=de&lr=&id=Ak_5AwAAQBAJ&oi=fnd&pg=PA1&dq=Urlaub+am+Bauernhof+Deutschland&ots=DVtj0t5Un2&sig=eB7XhY1wAcbkO-seu5PV5EcMQfw&redir_esc=y#v=onepage&q=Urlaub%20am%20Bauernhof%20Deutschland&f=false

Kim, M. G., & Mattila, A. S. (2013). Does a surprise strategy need words? The effect of explanations for a surprise strategy on customer delight and expectations. *Journal of Services Marketing*, *27*(5), 361–370. doi:10.1108/JSM-01-2012-0008

Kokko, A. (2011). *An Exploration of Agritourism: Topics, Literature and Areas for Future Research*. http://annekokko.com/resources/an%20exploration%20of%20agritourism%20-%20akokko.pdf

Lee, A. R., Son, S. M., & Kim, K. K. (2016). Information and communication technology overload and social networking service fatigue: A stress perspective. *Computers in Human Behavior*, *55*(1), 51–61. doi:10.1016/j.chb.2015.08.011

Lehle, K. (1982). *Urlaub auf dem Bauernhof. Analyse und Perspektiven d. Nebenbetriebes "Tourismus in der Landwirtschaft"*. Schriftenreihe des AID.

Libreros, M. (1998). *A conceptual framework for a tourism satellite account*. STD/NA(98)20. http://www.oecd.org/sdd/na/2681900.pdf

Lovell, R., Wheeler, B. W., Higgins, S. L., Irvine, K. N., & Depledge, M. H. (2014). A Systematic Review of the Health and Well -Being Benefits of Biodiverse Environments. *Journal of Toxicology and Environmental Health. Part B, Critical Reviews*, *17*(1), 1–20. doi:10.1080/10937404.2013.856361 PMID:24597907

Lu, A. C. C., Gursoy, D., & Lu, C. Y. R. (2016). Antecedents and outcomes of consumers' confusion in the online tourism domain. *Annals of Tourism Research*, *57*, 76–93. doi:10.1016/j.annals.2015.11.020

Magnini, V. P. (2006). Photo-elicitation as a tool to alleviate international marketing mistakes. *Journal of Global Competitiveness*, *14*(2), 75–83.

Magnini, V. P. (2015). *Surprise!: The Secret to Customer Loyalty in the Service Sector*. Business Expert Press.

Magnini, V. P. (2017). Designing Tourism Services in an era of information overload. In D. R. Fesenmaier & Z. Xiang (Eds.), Design science in tourism (pp. 161-172). Springer. doi:10.1007/978-3-319-42773-7_11

Magnini, V. P., Crotts, J. C., & Zehrer, A. (2011). Understanding customer delight: An application of travel blog analysis. *Journal of Travel Research*, *50*(5), 535–545. doi:10.1177/0047287510379162

Magnini, V. P., & Dallinger, I. (2018). Consumer information overload and the need to prompt script deviations. *Journal of Quality Assurance in Hospitality & Tourism*, *19*(3), 285–297. doi:10.1080/1528008X.2016.1230038

Magnini, V. P., Ford, J. B., & LaTour, M. S. (2012). The role of qualitative methods in tourism QOL research: A critique and future agenda. In *Handbook of Tourism and Quality-of-Life Research* (pp. 51–63). Springer. doi:10.1007/978-94-007-2288-0_4

Marsden, T. (1999). Rural Futures: The Consumption Countryside and its Regulation. *Sociologia Ruralis*, *39*(4), 501–520. doi:10.1111/1467-9523.00121

McEachern, C. (1992). Farmers and Conservation: Conflict and Accommodation in Farming Politics. *Journal of Rural Studies*, *8*(2), 159–171. doi:10.1016/0743-0167(92)90074-G

Meyer, W., Reisenzein, R., & Schützwohl, A. (1997). Towards a process analysis of emotions: The case of surprise. *Motivation and Emotion*, *21*(3), 251–274. doi:10.1023/A:1024422330338

Oppermann, M. (1996). Rural Tourism in Southern Germany. *Annals of Tourism Research, 23*(1), 86–102. doi:10.1016/0160-7383(95)00021-6

Oppermann, M. (1998). Farm Tourism in New Zealand. In R. Butler, C. M. Hall, & J. Jenkins (Eds.), *Tourism and recreation in rural areas* (pp. 225–233). Wiley.

Park, C. S. (2019). Does Too Much News on Social Media Discourage News Seeking? Mediating Role of News Efficacy Between Perceived News Overload and News Avoidance on Social Media. *Social Media + Society, 5*(3), 1–12. doi:10.1177/2056305119872956

Park, J.-Y., & Jang, S. S. (2013). Confused by too many choices? Choice overload in tourism. *Tourism Management, 35*, 1–12. doi:10.1016/j.tourman.2012.05.004

Pearce, P. L. (1990). Farm Tourism in New Zealand: A Social Situation Analysis. *Annals of Tourism Research, 17*(3), 337–352. doi:10.1016/0160-7383(90)90002-9

Peebles, M. S. (1995). Cultivating the tourist/farm tourism. *Tour. J. Tour. Soc.*, 86.

Perales, R. M. Y. (2002). Rural Tourism in Spain. *Annals of Tourism Research, 29*(4), 1104–1110.

Peters, T. (2010). *The Little Big Things: 163 Ways to Pursue Excellence*. Harper Business.

Petruzzelli, M. A., & Savino, T. (2015). Reinterpreting tradition to innovate: The case of Italian haute cuisine. *Industry and Innovation, 22*(8), 677–702. doi:10.1080/13662716.2015.1122512

Phillip, S., Hunter, C., & Blackstock, K. (2010). A typology for defining agritourism. *Tourism Management, 31*(6), 754–758. doi:10.1016/j.tourman.2009.08.001

Piaget, J. (1926). *The Language and Thought of the Child*. Harcourt, Brace.

Quendler, E. (2015). Organisational resilience: building business value in a changing world. *DIEM: Dubrovnik International Economic Meeting, 2*(1), 534-553.

Quendler, E. (2019). The position of the farm holiday in Austrian tourism. *Open Agriculture, 4*(1), 697–711. doi:10.1515/opag-2019-0069

Quendler, E., Magnini, V. P., & Driouech, N. (2021). *Strategic Positioning of the Farm Holiday's Post-Pandemic Competitive Advantages: Fresh Air and Sunlight*. Academic Press.

Reisenzein, R. (2000). The subjective experience of surprise. In H. Bless & J. Forgas (Eds.), *The Message Within* (pp. 262–279). Psychology Press.

Richter, F. (2020). *Facebook Reaches 3.2 Billion People Each Month.* Statista. https://www.statista.com/chart/2183/facebooks-mobile-users/?utm_source=Statista+Global&utm_campaign=95f88a88d2-All_InfographTicker_daily_COM_PM_KW27_2020_Mo&utm_medium=email&utm_term=0_afecd219f5-95f88a88d2-308701553

Roetzel, P. G. (2019). Information overload in the information age: A review of the literature from business administration, business psychology, and related disciplines with a bibliometric approach and framework development. *Business Research*, *12*(2), 479–522. doi:10.100740685-018-0069-z

Roman, M., Roman, M., & Prus, P. (2020). Innovations in Agritourism: Evidence from a Region in Poland. *Sustainability*, *12*(12), 4858. doi:10.3390u12124858

Rumbo, J. (2002). Consumer resistance in a world of advertising clutter: The case of adbusters. *Psychology and Marketing*, *19*(2), 127–148. doi:10.1002/mar.10006

Santucci, F. M. (2013). Agritourism for Rural Development in Italy, Evolution, Situation and Perspectives. *British Journal of Economics. Management & Trade*, *3*(3), 186–200.

Schroder, H. M., Driver, M. J., & Streufert, S. (1967). *Human information processing: Individuals and groups functioning in complex social situations*. Holt, Rinehart and Winston.

Sharpley, R. (2004). Tourism and the Countryside. In A. Lew, C. M. Hall, & A. M. Williams (Eds.), *A companion to Tourism* (pp. 374-386). Blackwell Publishing. https://books.google.at/books?hl=de&lr=&id=nePEB4e3Y0EC&oi=fnd&pg=PA374&dq=holidays+in+the+countryside&ots=6324w1G7I6&sig=o46KCCUgSGJNXg-D7sCoqLYShR0&redir_esc=y#v=onepage&q=holidays%20in%20the%20countryside&f=false

Sharpley, R., & Vass, A. (2006). Tourism, Farming and Diversification: An Attitudinal Study. *Tourism Management*, *27*(5), 1040–1052. doi:10.1016/j.tourman.2005.10.025

Shostack, G. (1984). Designing Services that Deliver. *Harvard Business Review*, *62*(1), 133–139.

Simpkins, M. D. (1976). *An experimental investigation of script theory in transactional analysis* (Doctoral dissertation, Oklahoma State University). ProQuest, UMI Dissertation Publishing.

Smith, R. A., & Houston, M. (1983). Script-based evaluations of satisfaction with services. In Emerging Perspectives on Services Marketing (pp. 59, 62). American Marketing.

Song, H., Jung, J., & Kim, Y. (2017). Perceived news overload and its cognitive and attitudinal consequences for news usage in South Korea. *Journalism & Mass Communication Quarterly*, *94*(4), 1172–1190. doi:10.1177/1077699016679975

Sonnino, R. (2004). For a "piece of bread"? Interpreting sustainable development through agritourism in Southern Tuscany. *Sociologia Ruralis*, *44*(3), 285–300. doi:10.1111/j.1467-9523.2004.00276.x

Spira, J. B. (2011). *Overload! How too much information is hazardous to your organization*. John Wiley & Sons.

Stankov, U., & Filimonau, V. (2019). Co-creating "Mindful" Holiday Resort Experience for Guests' Digital Well-Being. In J. Pesonen & J. Neidhardt (Eds.), *Information and Communication Technologies in Tourism 2019*. Springer., doi:10.1007/978-3-030-05940-8_16

Swan, R. (1929). Summer holiday and the English countriyside. *Fortnightly Review*, *126*, 259-268.

Sznajder, M., Przezborska, L., & Scrimgeour, F. (2009). *Agritourism*. CABI Publishing. doi:10.1079/9781845934828.0000

Tomas, S., & Leffew, M. B. (2016). The customer experience. Tips for agritourism enterpreneurs. In Small-Scale Tourism in Rural Areas – Trends and Research in the Nordic Countries (pp. 15-23). Nordregio Working Paper 2013:3.

Urlaub am Bauernhof, B. (2019). *Wozu eine Kategorisiserung der Höfe? Academic Press.*

Van Saandt, A., Low, S. A., & Thilmany, D. (2018, December). Exploring Regional Patterns of Agritourism in the U.S.: What's Driving Clusters of Enterprises? *Agricultural and Resource Economics Review*, *47*(3), 592–609. doi:10.1017/age.2017.36

Vanhamme, J., & Snelders, D. (2001). The role of surprise in satisfaction judgments. *Journal of Consumer Satisfaction, Dissatisfaction & Complaining Behavior*, *14*, 27–45.

Vermeeren, B., Kuipers, B., & Steijn, B. (2011). Two faces of the satisfaction mirror: A study of work environment, job satisfaction, and customer satisfaction in Dutch municipalities. *Review of Public Personnel Administration*, *31*(2), 171–189. doi:10.1177/0734371X11408569

Voase, R. (2018). Holidays under the hegemony of hyper-connectivity: Getting away, but unable to escape? *Leisure Studies*, *37*(4), 384–395. doi:10.1080/02614367.2018.1475503

Vujo, A., Petrović, I., & Blesic, M. D. (2015). Leisure time in the countryside: The health aspects of agritourism activities. *International Journal Scientific and Applicative Papers*, *8*(1), 131–136.

Weaver, D. B., & Fennell, D. A. (1997). The vacation farm sector in Saskatchewan: A profile of the operations. *Tourism Management*, *18*(6), 357–365. doi:10.1016/S0261-5177(97)00039-3

Wojcieszak-Zbierska, M. M., Jęczmyk, A., Zawadka, J., & Uglis, J. (2020). Agritourism in the Era of the Coronavirus (COVID-19): A Rapid Assessment from Poland. *Agriculture*, *10*(9), 397. doi:10.3390/agriculture10090397

York, C. (2013). Overloaded by the news: Effects of news exposure and enjoyment on reporting information overload. *Communication Research Reports*, *30*(4), 282–292. doi:10.1080/08824096.2013.836628

Zeithaml, V., Bitner, M., & Gremler, D. (2006). *Services Marketing: Integrating Customer Focus Across the Firm*. McGraw-Hill Irwin.

Zhang, L., Wu, L., & Mattila, A. S. (2016). Online reviews: The role of information load and peripheral factors. *Journal of Travel Research*, *55*(3), 299–310. doi:10.1177/0047287514559032 PMID:29708106

ENDNOTES

1 'Active people' are defined as "registered and logged-in users of Facebook, Instagram, Messenger, and/or WhatsApp who visited at least one of these Family products through a mobile device application or using a web or mobile browser in the last 30 days as of the date of measurement" (Richter, 2020).

2 For the purposes of this chapter, the focus is on farms offering private accommodation in the form of guestrooms and/or holiday apartments with additional activities and services. In Italy farms authorized for agritourism are categorized in farms offering accommodations, services or catering (Istat, 2018). In the interests of comparability with the term 'farm holiday' applied in this chapter, the farms with accommodations are listed here and not all farms authorized for agritourism. The latter are 24.576 farms in 2018 (Istat, 2020).

APPENDIX A: THEORETICAL FOUNDATIONS OF SCRIPT THEORY (ADAPTED FROM: MAGNINI AND DALLINGER, 2018)

The central components of *script theory* date back to the early sixties in the psychiatry works of Eric Berne (Simpkins, 1976) and even four decades earlier to the work of Frederic Bartlett (1932) in experimental psychology and Jean Piaget (1926) in developmental psychology (Abelson, 1981). In the mid-seventies by Robert Abelson, who, working on the design of computer simulation models of human information processing, noted the position of the two disciplines (i.e. cognitive psychology and social psychology) as too static for the dynamic nature of the cognitive process apparently involved (Abelson, 1976). This served as a call to perhaps explore past the "classical" perspective of cognitive and social psychology. Abelson's 1976 research presented the stages of development script material goes through in entering the long-term memory of an individual and defined a *"vignette"* as the basic ingredient of a script, a "raw constituent of remembered episodes in the individual's experience" (p.34). A *script* is hence "a coherently linked chain of vignettes stored as a unit" (Abelson, 1976, p.34), a "coherent sequence of events expected by the individual, involving him either as a participant or as an observer" (Abelson, 1976, p.33). Script theory, therefore, contends that "hypothesized cognitive structures [scripts], when activated, organize comprehension of event-based situations" (Abelson, 1981, p.717). Stated differently, human beings codify their experiences through certain cognitive structures that basically record recurring patterns in order to further predict how everyday events will unfold. When a cue spawns a deviation from such cognitive structures a script deviation occurs.

Compilation of References

§7 Buschenschankgesetz. (1979). *Buschenschankgesetz*. https://www.ris.bka.gv.at/GeltendeFassung.wxe?Abfrage=LrBgld&Gesetzesnummer=10000132

Abelson, R. P. (1976). *Script processing in attitude formation and decision making*. Lawrence Erlbaum.

Abelson, R. P. (1981). Psychological status of the script concept. *The American Psychologist, 36*(7), 715–729. doi:10.1037/0003-066X.36.7.715

Acquaah, M., Amoako-Gyampah, K., & Jayaram, J. (2011). Resilience in family and nonfamily firms: An examination of the relationships between manufacturing strategy, competitive strategy and firm performance. *International Journal of Production Research, 49*(18), 5527–5544. doi: 10.1080/00207543.2011.563834

Adamu, A. A., & Mohamad, B. (2019). A reliable and valid measurement scale for assessing internal crisis communication. *Journal of Communication Management (London), 23*(2), 90–108. doi:10.1108/JCOM-07-2018-0068

Agostini, L., Filippini, R., & Nosella, A. (2015). Brand-building efforts and their association with SME sales performance. *Journal of Small Business Management, 53*, 161–173. doi:10.1111/jsbm.12185

Aguirre, B. E. (2007). Dialectics of vulnerability and resilience. *Georgetown Journal on Poverty Law & Policy, 14*(1), 39.

Ahmad, S. Z. (2015). Entrepreneurship Education in Tourism and Hospitality Programs. *Journal of Hospitality & Tourism Education, 27*(1), 20–29. doi:10.1080/10963758.2014.998764

Aidis, R., Estrin, S., & Mickiewicz, T. (2008). Institutions and Entrepreneurship Development in Russia: A Comparative Perspective. *Journal of Business Venturing, 23*(6), 656–672. doi:10.1016/j.jbusvent.2008.01.005

Aikat, D., & Remund, D. (2012). Of time magazine, 24/7 media, and data deluge: The evolution of information overload theories and concepts. *Information overload: An international challenge for professional engineers and technical communicators*, 13-40.

Ainley, S., & Smale, B. (2010). A profile of Canadian agritourists and the benefits they seek. *Journal of Rural and Community Development, 5*(1), 58–75.

Alexander, D. E. (2013). Resilience and disaster risk reduction: An etymological journey. *Natural Hazards and Earth System Sciences, 1*, 1257–1284. doi:10.5194/nhessd-1-1257-2013

Alexander, D. E. (2014). Social media in disaster risk reduction and crisis management. *Science and Engineering Ethics, 20*(3), 717–733. doi:10.100711948-013-9502-z PMID:24306994

Allan, S., & Bund, R. (2017). Passing on a wine business that your kids want to own. *Wine & Viticulture Journal, 1*, 56.

Alpeza, M., Oberman Peterka, S., & Has, M. (2018). Entrepreneurship in Croatia. In R. Palaić, L. P. Dana, & V. Ramadani (Eds.), *Entrepreneurship in former Yugoslavia: Diversity, Institutional Constraints and Prospects* (pp. 37–50). Springer. doi:10.1007/978-3-319-77634-7_3

Alsos, G. A., Carter, S., & Ljunggren, E. (2014). Kinship and business: How entrepreneurial households facilitate business growth. *Entrepreneurship and Regional Development, 26*(1-2), 97–122. doi:10.1080/08985626.2013.870235

Amann, B., & Jaussaud, J. (2012). Family and non-family business resilience in an economic downturn. *Asia Pacific Business Review, 12*(1), 202–223. doi:10.1080/13602381.2010.537057

Amit, R., & Schoemaker, P. J. (1993). Strategic assets and organizational rent. *Strategic Management Journal, 14*(1), 33–46. doi:10.1002mj.4250140105

Andersson, T., Carlsen, J., & Getz, D. (2002). Family Business Goals in the Tourism and Hospitality Sector: Case Studies and Cross-Case Analysis from Australia, Canada, and Sweden. *Family Business Review, 15*(2), 89–106. doi:10.1111/j.1741-6248.2002.00089.x

Annarelli, A., & Nonino, F. (2016). Strategic and operational management of organizational resilience: Current state of research and future directions. *Omega, 62*, 1–18. doi:10.1016/j.omega.2015.08.004

Araña, J. E., & León, C. J. (2008). The impact of terrorism on tourism demand. *Annals of Tourism Research, 35*(2), 299–315. doi:10.1016/j.annals.2007.08.003

Aranda, K., Zeeman, L., Scholes, J., & Morales, A. S.-M. (2012). The resilient subject: Exploring subjectivity, identity and the body in narratives of resilience. *Health, 16*(5), 548–563. doi:10.1177/1363459312438564 PMID:22547553

Arnold, K., & Staudacher, C. (1981). *Urlaub auf dem Bauernhof. Eine empirische Untersuchung der Struktur und Entwicklung einer spezifischen Erholungsform und ihrer Auswirkungen auf die Land- und Forstwirtschaft in Niederösterreich.* Verlag Ferdinand Hirt.

Arrondo-García, R., Fernández-Méndez, C., & Menéndez-Requejo, S. (2016). The growth and performance of family businesses during the global financial crisis: The role of the generation in control. *Journal of Family Business Strategy, 7*(4), 227–237. doi:10.1016/j.jfbs.2016.11.003

Arzubiaga, U., Kotlar, J., De Massis, A., Maseda, A., & Iturralde, T. (2018). Entrepreneurial orientation and innovation in family SMEs: Unveiling the (actual) impact of the Board of Directors. *Journal of Business Venturing*, *33*(4), 455–469. doi:10.1016/j.jbusvent.2018.03.002

Assaf, A., & Scuderi, R. (2020). COVID-19 and the recovery of the tourism industry. *Tourism Economics*, *26*(5), 731–733. doi:10.1177/1354816620933712

Aßheuer, T., Thiele-Eich, I., & Braun, B. (2013). Coping with the impacts of severe flood events in Dhaka's slums—The role of social capital. *Erdkunde*, 21–35.

Astrachan, J. H. (2010). Strategy in family business: Toward a multidimensional research agenda. *Journal of Family Business Strategy*, *1*(1), 6–14. doi:10.1016/j.jfbs.2010.02.001

Astrachan, J. H., & Shanker, M. C. (2003). Family Businesses' Contribution to the U.S. Economy: A Closer Look. *Family Business Review*, *16*(3), 211–219. doi:10.1177/08944865030160030601

Ateljevic, I., & Doorne, S. (2000). 'Staying within the fence': Lifestyle entrepreneurship in tourism. *Journal of Sustainable Tourism*, *8*(5), 378–392. doi:10.1080/09669580008667374

Ates, A., & Bititci, U. (2011). Change process: A key enabler for building resilient SMEs. *International Journal of Production Research*, *49*(18), 5601–5618. doi:10.1080/00207543.2011.563825

Avey, J. B., Wernsing, T. S., & Luthans, F. (2008). Can positive employees help positive organizational change? Impact of psychological capital and emotions on relevant attitudes and behaviors. *The Journal of Applied Behavioral Science*, *44*(1), 48–70. doi:10.1177/0021886307311470

Avraham, E. (2015, April). Destination image repair during crisis: Attracting tourism during the Arab Spring uprisings. *Tourism Management*, *47*, 224–232. doi:10.1016/j.tourman.2014.10.003

Avraham, E. (2016). Destination marketing and image repair during tourism crises: The case of Egypt. *Journal of Hospitality and Tourism Management*, *28*(September), 41–48. doi:10.1016/j.jhtm.2016.04.004

Avraham, E., & Ketter, E. (2017). Destination marketing during and following crises: Combating negative images in Asia. *Journal of Travel & Tourism Marketing*, *34*(6), 709–718. doi:10.1080/10548408.2016.1237926

Ayeh, J. K. (2018). Distracted gaze: Problematic use of mobile technologies in vacation contexts. *Tourism Management Perspectives*, *26*, 31–38. doi:10.1016/j.tmp.2018.01.002

Baer, R. A. (2011). Measuring mindfulness. *Contemporary Buddhism*, *12*(1), 241–261. doi:10.1080/14639947.2011.564842

Bae, S. Y., & Chang, P. J. (2020). The effect of coronavirus disease-19 (COVID-19) risk perception on behavioural intention towards 'untact' tourism in South Korea during the first wave of the pandemic (March 2020). *Current Issues in Tourism*. Advance online publication. doi:10.1080/13683500.2020.1798895

Bajrami, D., Terzić, A., Petrović, M. D., Radovanović, M., Tretiakova, T. N., & Hadoud, A. (2020). Will we have the same employees in hospitality after all? The impact of COVID-19 on employees' work attitudes and turnover intentions. *International Journal of Hospitality Management*, (October). Advance online publication. doi:10.1016/j.ijhm.2020.102754

Balmer, J. M., Powell, S. M., Punjaisri, K., & Wilson, A. (2011). Internal branding process: Key mechanisms, outcomes and moderating factors. *European Journal of Marketing*, *45*(9/10), 1521–1537. doi:10.1108/03090561111151871

Bamberger, P. (2008). From the editors beyond contextualization: Using context theories to narrow the micro-macro gap in management research. *Academy of Management Journal*, *51*(5), 839–846. doi:10.5465/amj.2008.34789630

Banki, M. B., & Ismail, H. N. (2015). Understanding the characteristics of family owned tourism micro businesses in mountain destinations in developing countries: Evidence from Nigeria. *Tourism Management Perspectives*, *13*, 18–32. doi:10.1016/j.tmp.2014.11.001

Barach, J. A., & Ganitsky, J. B. (1995). Successful succession in family business. *Family Business Review*, *8*, 131–155.

Barbieri, C., & Mshenga, P. M. (2008). The role of the firm and owner characteristics on the performance of agritourism farms. *Sociologia Ruralis*, *48*(2), 166–183. doi:10.1111/j.1467-9523.2008.00450.x

Barbosa, S. D., Kickul, J., & Liao-Troth, M. (2007). Development and Validation of a Multidimensional Scale of Entrepreneurial Risk Perception. *Annual Meeting of the Academy of Management*, 2007(1), 1–6. 10.5465/ambpp.2007.26508257

Bardoel, E. A., Pettit, T. M., De Cieri, H., & McMillan, L. (2014). Employee resilience: An emerging challenge for HRM. *Asia Pacific Journal of Human Resources*, *52*(3), 279–297. doi:10.1111/1744-7941.12033

Barke, M. (2004). Rural Tourism in Spain. *International Journal of Tourism Research*, *6*(3), 1367–149. doi:10.1002/jtr.480

Baron, R. M., & Kenny, D. A. (1986). The Moderator-Mediator Variable Distinction in Social Psychological Research: Conceptual, Strategic, and Statistical Considerations. *Journal of Personality and Social Psychology*, *51*(6), 1173–1182. doi:10.1037/0022-3514.51.6.1173 PMID:3806354

Baron, S., Patterson, A., Oakes, S., Harris, K., Punjaisri, K., Evanschitzky, H., & Wilson, A. (2009). Internal branding: An enabler of employees' brand-supporting behaviours. *Journal of Service Management*, *20*(2), 209–226. doi:10.1108/09564230910952780

Bartlett, F. C. (1932). *Remembering*. Cambridge University Press.

Basnyat, S., & Sharma, S. (2021). Effects of COVID-19 crisis on small and medium-sized hotel operators: Insights from Nepal. *Anatolia*, *4*(1), 1–15. doi:10.1080/13032917.2021.1879184

Bateson, J. (2002). Consumer preference and quality in services. *Managing Service Quality*, *12*(4), 206–209. doi:10.1108/09604520210434811

Batista e Silva, F., Kavalov, B., & Lavalle, C. (2019). Socio-economic regional microscope series - Territorial patterns of tourism intensity and seasonality in the EU. *Publications Office of the European Union, JRC117669*. Advance online publication. doi:10.2760/061885

Baumgartner, B. (2009). *Familienunternehmen und Zukunftsgestaltung: Schlüsselfaktoren zur erfolgreichen Unternehmensnachfolge*. Gabler Verlag. doi:10.1007/978-3-8349-8541-5

Baum, T. (2018). Sustainable human resource management as a driver in tourism policy and planning: A serious sin of omission? *Journal of Sustainable Tourism*, *26*(6), 873–889. doi:10.1080/09669582.2017.1423318

Bauweraerts, J. (2016). How Do Family Firms Manage Risky Situations? An Organizational Resilience Perspective. *International Business Research*, *9*(5), 1. doi:10.5539/ibr.v9n5p1

Bavel, J. J. V., Baicker, K., Boggio, P. S., Capraro, V., Cichocka, A., Cikara, M., Crockett, M. J., Crum, A. J., Douglas, K. M., Druckman, J. N., Drury, J., Dube, O., Ellemers, N., Finkel, E. J., Fowler, J. H., Gelfand, M., Han, S., Haslam, S. A., Jetten, J., ... Willer, R. (2020). Using social and behavioural science to support COVID-19 pandemic response. *Nature Human Behaviour*, *4*(5), 460–471. doi:10.103841562-020-0884-z PMID:32355299

Bawden, D. (2001). Information and digital literacies: A review of concepts. *The Journal of Documentation*, *57*(2), 218–259. doi:10.1108/EUM0000000007083

Bawden, D., & Robinson, L. (2009). The dark side of information: Overload, anxiety and other paradoxes and pathologies. *Journal of Information Science*, *35*(2), 180–191. doi:10.1177/0165551508095781

Bawden, D., & Robinson, L. (2020). Information Overload: An Overview. In *Oxford Encyclopedia of Political Decision Making*. Oxford University Press. doi:10.1093/acrefore/9780190228637.013.1360

Becken, S., & Hughey, K. F. (2013). Linking tourism into emergency management structures to enhance disaster risk reduction. *Tourism Management*, *36*, 77–85. doi:10.1016/j.tourman.2012.11.006

Beck, L., Janssens, W., Debruyne, M., & Lommelen, T. (2011). A Study of the Relationships Between Generation, Market Orientation, and Innovation in Family Firms. *Family Business Review*, *24*(3), 252–272. doi:10.1177/0894486511409210

Beech, N., Devins, D., Gold, J., & Beech, S. (2020). In the family way: An exploration of family business resilience. *The International Journal of Organizational Analysis*, *28*(1), 160–182. doi:10.1108/IJOA-02-2019-1674

Beirman, D. (2003). *Restoring tourism destinations in crisis: A strategic marketing approach*. CABI Publishing.

Belhassen, Y., Uriely, N., & Assor, O. (2014, November). The touristification of a conflict zone: The case of Bil'in. *Annals of Tourism Research*, *49*, 174–189. doi:10.1016/j.annals.2014.09.007

Benjamin, S., Dillette, A., & Alderman, D. H. (2020). "We can't return to normal": Committing to tourism equity in the post-pandemic age. *Tourism Geographies*, *22*(3), 476–483. doi:10.1080/14616688.2020.1759130

Benselin, J. C., & Ragsdell, G. (2016). Information overload: The differences that age makes. *Journal of Librarianship and Information Science*, *48*(3), 284–297. doi:10.1177/0961000614566341

Bentfeld, M., Huppertz, S., Khazai, B., & Girard, T. (2018). Hotel resilient – plan ahead, stay ahead. Promoting the business case for disaster and climate resilience in the tourism sector. In E. Innerhofer, H. Pechlaner, & M. Fontanari, M. (Eds.), Destination Resilience (pp. 153-163). Routledge.

Beritelli, P., Bieger, T., & Laesser, C. (2007). Destination Governance: Using Corporate Governance Theories as a Foundation for Effective Destination Management. *Journal of Travel Research*, *46*(1), 96–107. doi:10.1177/0047287507302385

Berman, B. (2005). How to delight your customers. *California Management Review*, *48*(1), 129–151. doi:10.2307/41166331

Berns, G. S., McClure, S. M., Pagnoni, G., & Montague, P. R. (2001). Predictability modulates human brain response to reward. *The Journal of Neuroscience: The Official Journal of the Society for Neuroscience*, *21*(8), 2793–2798. doi:10.1523/JNEUROSCI.21-08-02793.2001 PMID:11306631

Berrone, P., Cruz, C., & Gomez-Mejia, L. R. (2012). Socioemotional wealth in family firms: Theoretical dimensions, assessment approaches, and agenda for future research. *Family Business Review*, *25*(3), 258–279. doi:10.1177/0894486511435355

Berrone, P., Cruz, C., Gomez-Mejia, L. R., & Larraza-Kintana, M. (2010). Socioemotional Wealth and Corporate Responses to Institutional Pressures: Do Family-Controlled Firms Pollute Less? *Administrative Science Quarterly*, *55*(1), 82–113. doi:10.2189/asqu.2010.55.1.82

Bertrand, M., & Schoar, A. (2006). The Role of Family in Family Firms. *The Journal of Economic Perspectives*, *20*(2), 73–96. doi:10.1257/jep.20.2.73

Bertrand, M., & Schoar, A. (2006). The role of family in family firms. *The Journal of Economic Perspectives*, *20*(2), 73–96. https://doi.org/10.1257/jep.20.2.73

Bhamra, R., Dani, S., & Burnard, K. (2011). Resilience: The concept, a literature review and future directions. *International Journal of Production Research*, *49*(18), 5375–5393. doi:10.1080/00207543.2011.563826

Bichler, B. F., & Lösch, M. (2019). Collaborative Governance in Tourism: Empirical Insights into a Community-Oriented Destination. *Sustainability*, *11*(23), 6673. doi:10.3390u11236673

Biggs, D. (2011). Understanding resilience in a vulnerable industry: The case of reef tourism in Australia. *Ecology and Society*, *16*(1), 30. doi:10.5751/ES-03948-160130

Biggs, D., Hall, C. M., & Stoeckl, N. (2012). The resilience of formal and informal tourism enterprises to disasters: Reef tourism in Phuket, Thailand. *Journal of Sustainable Tourism, 20*(5), 645–665. doi:10.1080/09669582.2011.630080

Bigliardi, B., & Dormio, A. I. (2009). Successful generational change in family business. *Measuring Business Excellence, 13*(2), 44–50.

Binder, S. B., Baker, C. K., & Barile, J. P. (2015). Rebuild or relocate? Resilience and postdisaster decision-making after Hurricane Sandy. *American Journal of Community Psychology, 56*(1-2), 180–196. doi:10.100710464-015-9727-x PMID:25903679

Birley, S. (2002). Attitudes of Owner-Managers' Children towards Family and Business Issues. *Entrepreneurship Theory and Practice, 26*(3), 5–19. doi:10.1177/104225870202600301

Bitner, M. J., Ostrom, A. L., & Morgan, F. N. (2008). Service blueprinting: A practical technique for service innovation. *California Management Review, 50*(3), 66–94. doi:10.2307/41166446

Blackstock, K. (2005). A critical look at community based tourism. *Community Development Journal: An International Forum, 40*(1), 39–49. doi:10.1093/cdj/bsi005

Blackstock, K. L., White, V., McCrum, G., Scott, A., & Hunter, C. (2008). Measuring Responsibility: An Appraisal of a Scottish National Park's Sustainable Tourism Indicators. *Journal of Sustainable Tourism, 16*(3), 276–297. doi:10.1080/09669580802154090

Blake, A., & Sinclair, M. T. (2003). Tourism crisis management. *Annals of Tourism Research, 30*(4), 813–832. doi:10.1016/S0160-7383(03)00056-2

Blumentritt, T., Mathews, T., & Marchisio, G. (2013). Game Theory and Family Business Succession: An Introduction. *Family Business Review, 26*(1), 51–67. https://doi.org/10.1177/0894486512447811

BMLRT. (2020a). *Green Report.* https://gruenerbericht.at/

BMLRT. (2020b). *Nächtigungsziffern nach Unterkunftsarten 2019. Tabelle 2.6.2. Green Report.* https://gruenerbericht.at/

BMNT. (2017). *Der Cluster Urlaub am Bauernhof Österreich - Innovationsstrategie Urlaub am Bauernhof 2020.* Bundesministerium für Nachhaltigkeit und Tourismus. https://www.google.com/url?sa=t&rct=j&q=&esrc=s&source=web&cd=1&ved=2ahUKEwj8g-uf-ZffAhX5SxUIHRv-Aj8QFjAAegQIChAC&url=https%3A%2F%2Fwww.bmnt.gv.at%2Fdam%2Fjcr%3A0c0bfa96-4e61-4914-a7ff-47bdff38fb12%2FClusterstrategie%2520Urlaub%2520am%2520Bauern-hof%2520%25C3%2596sterreich%25202020.pdf&usg=AOvVaw2M6gs2O4cqZvwJbmvSJMns

Bohdanowicz, P. (2005). European Hoteliers' Environmental Attitudes. *The Cornell Hotel and Restaurant Administration Quarterly, 46*(2), 188–204. doi:10.1177/0010880404273891

Bonanno, G. A., Westphal, M., & Mancini, A. D. (2011). Resilience to Loss and Potential Trauma. *Annual Review of Clinical Psychology, 7*(1), 511–535. doi:10.1146/annurev-clinpsy-032210-104526 PMID:21091190

Bortz, J., & Döring, N. (2015). Forschungsmethoden und Evaluation: Für Human- und Sozialwissenschaftler (5. Aufl.). Berlin: Springer.

Boschma, R. (2015). Towards an evolutionary perspective on regional resilience. *Regional Studies*, *49*(5), 733–751. doi:10.1080/00343404.2014.959481

Bosio, B., Haselwanter, S., & Ceipek, M. (2018). Social Media Nutzung in Tiroler Destinationen. In Wandel im Tourismus. Internationalität, Demografie und Digitalisierung. Berlin: ESV.

Bourdieu, P. (1987). *Distinction: A Social Critique of the Judgement of Taste*. Harvard University Press.

Brämer, R. (2003). Zurück zur Natur? Die Wald- und Wiesen-Therapie. *Psychologie Heute*, *30*(4), 20–29.

Braun, B., & Aßheuer, T. (2011). Floods in megacity environments: Vulnerability and coping strategies of slum dwellers in Dhaka/Bangladesh. *Natural Hazards*, *58*(2), 771–787. doi:10.100711069-011-9752-5

Breier, M., Kallmuenzer, A., Clauss, T., Gast, J., Kraus, S., & Tiberius, V. (2021). The role of business model innovation in the hospitality industry during the COVID-19 crisis. *International Journal of Hospitality Management*, *92*, 102723. doi:10.1016/j.ijhm.2020.102723

Brewton, K. E., Danes, S. M., Stafford, K., & Haynes, G. W. (2010). Determinants of rural and urban family firm resilience. *Journal of Family Business Strategy*, *1*(3), 155–166. doi:10.1016/j.jfbs.2010.08.003

Brislin, R. W. (1970). Back-Translation for Cross-Cultural Research. *Journal of Cross-Cultural Psychology*, *1*(3), 185–216. doi:10.1177/135910457000100301

Brosius, H.-B., Haas, A., & Koschel, F. (2012). Methoden der empirischen Kommunikationsforschung: Eine Einführung (6., erw. und aktualisierte Aufl.). In *Studienbücher zur Kommunikations- und Medienwissenschaft*. Wiesbaden: Springer VS. Retrieved from http://gbv.eblib.com/patron/FullRecord.aspx?p=970700 doi:10.1007/978-3-531-94214-8

Brouder, P. (2020). Reset redux: Possible evolutionary pathways towards the transformation of tourism in a COVID-19 world. *Tourism Geographies*, *22*(3), 484–490. doi:10.1080/14616688.2020.1760928

Brown, C., Seville, E., & Vargo, J. (2017). Measuring the organizational resilience of critical infrastructure providers: A New Zealand case study. *International Journal of Critical Infrastructure Protection*, *18*, 37–49. doi:10.1016/j.ijcip.2017.05.002

Brustbauer, J. K., & Peters, M. (2013). Risk perception of family and non-family firm managers. *International Journal of Entrepreneurship and Small Business*, *20*(1), 96. doi:10.1504/IJESB.2013.055695

Bryman, A., & Bell, E. (2011). *Business Research Methods*. Oxford University Press.

Buckley, R. (2012). Rush as a key motivation in skilled adventure tourism: Resolving the risk recreation paradox. *Tourism Management, 33*(4), 961–970. doi:10.1016/j.tourman.2011.10.002

Bude, H. (2018). *Society of Fear*. Polity Press.

Buhalis, D., & Peters, M. (2006). SMEs in tourism. In D. Buhalis & C. Costa (Eds.), *Tourism Dynamics, Challenges and Tools: Present and Future Issues* (pp. 116–129). Butterworth-Heinemann.

Bui, H. T., Jones, T. E., Weaver, D. B., & Le, A. (2020). The adaptive resilience of living cultural heritage in a tourism destination. *Journal of Sustainable Tourism, 28*(7), 1022–1040. doi:10.1080/09669582.2020.1717503

Buliga, O., Scheiner, C. W., & Voigt, K. I. (2016). Business model innovation and organizational resilience: Towards an integrated conceptual framework. *Journal of Business Economics, 86*(6), 647–670. doi:10.100711573-015-0796-y

Burnard, K., & Bhamra, R. (2011). Organisational resilience: Development of a conceptual framework for organisational responses. *International Journal of Production Research, 49*(18), 5581–5599. doi:10.1080/00207543.2011.563827

Busby, G., & Rendle, S. (2000). The Transition from Tourism on Farms to Farm Tourism. *Tourism Management, 21*(6), 635–642. doi:10.1016/S0261-5177(00)00011-X

Buse, K., & Walt, G. (2000). Global public-private partnerships: Part I - a new development in health? *Bulletin of the World Health Organization, 78*, 549–561. PMID:10885184

Butler, R. W. (2017). *Tourism and resilience*. CABI. doi:10.1079/9781780648330.0000

Cabrera-Suárez, K., de Saá-Pérez, P., & García-Almeida, D. (2001). The Succession Process from a Resource- and Knowledge-Based View of the Family Firm. *Family Business Review, 14*(1), 37–46. doi:10.1111/j.1741-6248.2001.00037.x

Calabro, A., & Valentino, A. (2019). *STEP2019 Global Family Business Survey - REPORT The impact of changing demographics on family business succession planning and governance*. Deloitte Development LLC.

Caldarice, O., Brunetta, G., & Tollin, N. (2019). The challenge of urban resilience: Operationalization. In G. Brunetta, O. Caldarice, N. Tollin, M. Rosas-Casals, & J. Morato (Eds.), *Urban Resilience for Risk and Adaptation Governance* (pp. 1–6). Springer.

Calgaro, E., Dominey-Howes, D., & Lloyd, K. (2014). Application of the Destination Sustainability Framework to explore the drivers of vulnerability and resilience in Thailand following the 2004 Indian Ocean Tsunami. *Journal of Sustainable Tourism, 22*(3), 361–383. doi:10.1080/09669582.2013.826231

Caliandro, A., & Graham, J. (2020). Studying Instagram Beyond Selfies. *Social Media + Society, 6*(2), 205630512092477. doi:10.1177/2056305120924779

Campiranon, K., & Scott, N. (2014). Critical Success Factors for Crisis Recovery Management: A Case Study of Phuket Hotels. *Journal of Travel & Tourism Marketing*, *31*(3), 313–326. doi:10.1080/10548408.2013.877414

Campos, A. C., Pinto, P., & Scott, N. (2019). Bottom-up factors of attention during the tourist experience: An empirical study. *Current Issues in Tourism*, 1–23. doi:10.1080/13683500.2019.1681383

Canoves, G., Villarino, M., Priestley, G. K., & Blanco, A. (2004). Rural Tourism in Spain: An analysis of recent evolution. *Geoforum*, *35*(6), 755–769. doi:10.1016/j.geoforum.2004.03.005

Cao, Q., Simsek, Z., & Zhang, H. (2010). Modelling the joint impact of the CEO and the TMT on organizational ambidexterity. *Journal of Management Studies*, *47*(7), 1272–1296.

Carleton, R. N. (2016, April). Into the unknown: A review and synthesis of contemporary models involving uncertainty. *Journal of Anxiety Disorders*, *39*, 30–43. doi:10.1016/j.janxdis.2016.02.007 PMID:26945765

Carlock, R. S., & Ward, J. L. (2001). *Strategic planning for the Family Business*. Palgrave. doi:10.1057/9780230508750

Carney, M. (1998). A management capacity constraint? Obstacles to the development of the overseas Chinese family business. *Asia Pacific Journal of Management*, *15*, 137–162.

Carroll, A. B. (1991). The pyramid of corporate social responsibility: Toward the moral management of organizational stakeholders. *Business Horizons*, *34*(4), 39–48. doi:10.1016/0007-6813(91)90005-G

Casillas, J. C., Moreno, A. M., & Barbero, J. L. (2011). Entrepreneurial orientation of family firms: Family and environmental dimensions. *Journal of Family Business Strategy*, *2*(2), 90–100. doi:10.1016/j.jfbs.2011.03.002

Cassia, L., de Massis, A., & Pizzurno, E. (2012). Strategic innovation and new product development in family firms. *International Journal of Entrepreneurial Behaviour & Research*, *18*(2), 198–232. doi:10.1108/13552551211204229

Cater, J. J. III, & Justis, R. T. (2010). The development and implementation of shared leadership in multi-generational family firms. *Management Research Review*, *33*(6), 563–585. doi:10.1108/01409171011050190

Cave, J., & Dredge, D. (2020). Regenerative tourism needs diverse economic practices. *Tourism Geographies*, *22*(3), 503–513. doi:10.1080/14616688.2020.1768434

Cellini, R., & Cuccia, T. (2015). The economic resilience of tourism industry in Italy: What the 'great recession' data show. *Tourism Management Perspectives*, *16*, 346–356. doi:10.1016/j.tmp.2015.09.007

Chalfen, R. M. (1979). Photography's role in tourism: Some unexplored relationships. *Annals of Tourism Research*, *6*(4), 435–447. doi:10.1016/0160-7383(79)90006-9

Chang, C. C., Chiu, C. M., & Chen, C. A. (2010). The effect of TQM practices on employee satisfaction and loyalty in government. *Total Quality Management*, *21*(12), 1299–1314. doi:10.1080/14783363.2010.530796

Chang, C. L., McAleer, M., & Ramos, V. (2020). A charter for sustainable tourism after COVID-19. *Sustainability*, *12*(9), 3671. doi:10.3390u12093671

Chang, H.-J., Huang, N., Lee, C.-H., Hsu, Y.-J., Hsieh, C.-J., & Chou, Y.-J. (2004). The Impact of the SARS Epidemic on the Utilization of Medical Services: SARS and the Fear of SARS. *American Journal of Public Health*, *94*(4), 562–564. doi:10.2105/AJPH.94.4.562 PMID:15054005

Chatterjee, S., & Prasad, M. V. D. (2019). The Evolution of Agri-Tourism practices in India: Some Success Stories. *Madridge J Agric Environ Sci.*, *1*(1), 19–25. doi:10.18689/mjaes-1000104

Chawinga, W. D. (2017). Taking social media to a university classroom: Teaching and learning using Twitter and blogs. *International Journal of Educational Technology in Higher Education*, *14*(1), 3. doi:10.118641239-017-0041-6

Cheer, J. M. (2020). Human flourishing, tourism transformation and COVID-19: A conceptual touchstone. *Tourism Geographies*, *22*(3), 514–524. doi:10.1080/14616688.2020.1765016

Cheer, J. M., & Lew, A. A. (2018). Understanding tourism resilience: Adapting to social, political, and economic change. In J. M. Cheer & A. A. Lew (Eds.), *Routledge advances in tourism. Tourism, resilience and sustainability: Adapting to social, political and economic change* (pp. 3–17). Routledge.

Cheer, J. M., & Lew, A. A. (2018). Understanding tourism resilience: Adapting to social, political, and economic change. In J. M. Cheer & A. A. Lew (Eds.), *Tourism, resilience, and sustainability: Adapting to social, political and economic change* (pp. 3–17). Routledge.

Cheer, J. M., & Lew, A. A. (Eds.). (2017). *Tourism, resilience and sustainability: Adapting to social, political and economic change*. Routledge. doi:10.4324/9781315464053

Cheer, J. M., Milano, C., & Novelli, M. (2019). Tourism and community resilience in the Anthropocene: Accentuating temporal overtourism. *Journal of Sustainable Tourism*, *27*(4), 554–572. doi:10.1080/09669582.2019.1578363

Chen, I. L., Scott, H., & Benckendorff, P. (2017). Mindful tourist experiences: A Buddhist perspective. *Annals of Tourism Research*, *64*, 1–12. doi:10.1016/j.annals.2017.01.013

Chen, L. (2017). Power and ambivalence in intergenerational communication: Deciding to institutionalize in Shanghai. *Journal of Aging Studies*, *41*, 44–51. doi:10.1016/j.jaging.2017.03.004 PMID:28610754

Cheung, K. S., & Li, L. H. (2019). Understanding visitor–resident relations in overtourism: Developing resilience for sustainable tourism. *Journal of Sustainable Tourism*, *27*(8), 1197–1216. doi:10.1080/09669582.2019.1606815

Child, J. (1972). Organizational Structure, Environment and Performance: The Role of Strategic Choice. *Sociology*, *6*(1), 1–22. doi:10.1177/003803857200600101

Chirico, F., Sirmon, D. G., Sciascia, S., & Mazzola, P. (2011). Resource orchestration in family firms: Investigating how entrepreneurial orientation, generational involvement, and participative strategy affect performance. *Strategic Entrepreneurship Journal*, *5*(4), 307–326. doi:10.1002ej.121

Chowdhury, M., Prayag, G., Orchiston, C., & Spector, S. (2019). Postdisaster social capital, adaptive resilience and business performance of tourism organizations in Christchurch, New Zealand. *Journal of Travel Research*, *58*(7), 1209–1226. doi:10.1177/0047287518794319

Chrisman, Chua, J. H., & Steier, L. P. (2011). Resilience of Family Firms: An Introduction Entrepreneurship. *Theory into Practice*, *35*(6), 1107–1119. doi:10.1111/j.1540-6520.2011.00493.x

Chrisman, J. J., Chua, J. H., & Litz, R. A. (2004). Comparing the Agency Costs of Family and Non-Family Firms: Conceptual Issues and Exploratory Evidence. *Entrepreneurship Theory and Practice*, *28*(4), 335–354. doi:10.1111/j.1540-6520.2004.00049.x

Chrisman, J. J., Chua, J. H., & Sharma, P. (2005). Trends and Directions in the Development of a Strategic Management Theory of the Family Firm. *Entrepreneurship Theory and Practice*, *29*(5), 555–575. doi:10.1111/j.1540-6520.2005.00098.x

Chrisman, J. J., Steier, L. P., & Chua, J. H. (2008). Toward a Theoretical Basis for Understanding the Dynamics of Strategic Performance in Family Firms. *Entrepreneurship Theory and Practice*, *32*(6), 935–947. doi:10.1111/j.1540-6520.2008.00264.x

Chua, J. H., Chrisman, J. J., & Bergiel, E. B. (2009). An Agency Theoretic Analysis of the Professionalized Family Firm. *Entrepreneurship Theory and Practice*, *33*(2), 355–372. doi:10.1111/j.1540-6520.2009.00294.x

Chua, J. H., Chrisman, J. J., & Sharma, P. (1999). Defining the family business by behavior. *Entrepreneurship Theory and Practice*, *23*(4), 19–39. doi:10.1177/104225879902300402

Chua, J. H., Chrisman, J. J., & Sharma, P. (2003). Succession and nonsuccession concerns of family firms and agency relationship with nonfamily managers. *Family Business Review*, *16*(2), 89–107. doi:10.1111/j.1741-6248.2003.00089.x

Chung, C. N., & Luo, X. R. (2013). Leadership succession and firm performance in an emerging economy: Successor origin, relational embeddedness, and legitimacy. *Strategic Management Journal*, *34*, 338–357.

Civelek, M. E., Cemberci, M., & Eralp, N. E. (2016). The Role of Social Media in Crisis Communication and Crisis Management. *International Journal of Research in Business and Social Science*, *5*(3), 111–120. doi:10.20525/ijrbs.v5i3.279

Cizek, D., Schipfer, R. K., Buchebner-Ferstl, S., Geserick, M. A., Kaindl, M., Kapella, O., & Luef, C. (2005). *Urlaub am Bauernhof aus der Sicht von Gästen und Anbietern*. https://www.dafne.at/prod/dafne_plus_common/attachment_download/0748c0a56650860d343a79ba6b950cf8/Endbericht_Apr07.pdf

Clancy, M. (2012). Security and the Future of Tourism. In J. Leigh, C. Webster, & S. Ivanov (Eds.), *Future Tourism: Political, Social and Economic Challenges* (pp. 37–52). Routledge.

Cleary, P., Quinn, M., & Moreno, A. (2019). Socioemotional wealth in family firms: A longitudinal content analysis of corporate disclosures. *Journal of Family Business Strategy*, *10*(2), 119–132. doi:10.1016/j.jfbs.2018.11.002

Cochrane, J. (2010). The Sphere of Tourism Resilience. *Tourism Recreation Research*, *35*(2), 173–185. doi:10.1080/02508281.2010.11081632

Cohen, E. (1972). Toward a Sociology of International Tourism. *Social Research*, *39*(1), 164–182.

Collier, M. (1967). *Visual Anthropology: Photography as a Research Method.* Holt, Rinehart and Winston.

Conz, E., Lamb, P. W., & De Massis, A. (2020). Practicing resilience in family firms: An investigation through phenomenography. *Journal of Family Business Strategy*, *11*(2), 2. doi:10.1016/j.jfbs.2020.100355

Conz, E., & Magnani, G. (2020). A dynamic perspective on the resilience of firms: A systematic literature review and a framework for future research. *European Management Journal*, *38*(3), 400–412. doi:10.1016/j.emj.2019.12.004

Coombs, W. T. (2015). The value of communication during a crisis: Insights from strategic communication research. *Business Horizons*, *58*(2), 141–148. doi:10.1016/j.bushor.2014.10.003

Coutu, D. L. (2002). How resilience works. *Harvard Business Review*, *80*, 46–56. PMID:12024758

Cox, L., & Fox, M. (2003). Agricultural Based Leisure Attractions. *Journal of Tourism Studies*, *14*(1), 49–58.

Credit Suisse Research Institute. (2018). *The CS Family 1000 in 2018. Credit Suisse AG.* Research Institute.

Credit Suisse Research Institute. (2020). *The Family 1000: Post the pandemic. Credit Suisse AG.* Research Institute.

Cruz, C., Justo, R., & de Castro, J. O. (2012). Does family employment enhance MSEs performance? *Journal of Business Venturing*, *27*(1), 62–76. doi:10.1016/j.jbusvent.2010.07.002

Curry, G. D., Leflar, J. J., Glasser, M., Loyear, R., Grey, B., Tim, J., Ong, L., Preining, W., & Sobron, J. M. (2015). How Social Media is Transforming Crisis Management and Business Continuity. *Journal of Physical Security*, *8*(2), 15–36.

Cutter, S. L. (2016). Resilience to what? Resilience for whom? *The Geographical Journal*, *182*(2), 110–113. doi:10.1111/geoj.12174

d'Angella, F., & Go, F. M. (2009). Tale of two cities' collaborative tourism marketing: Towards a theory of destination stakeholder assessment. *Tourism Management*, *30*(3), 429–440. doi:10.1016/j.tourman.2008.07.012

Dahles, H., & Susilowati, T. P. (2015). Business resilience in times of growth and crisis. *Annals of Tourism Research*, *51*, 34–50. doi:10.1016/j.annals.2015.01.002

Dalgaard-Nielsen, A. (2017). Organizational resilience in national security bureaucracies: Realistic and practicable? *Journal of Contingencies and Crisis Management*, *25*(4), 341–349. doi:10.1111/1468-5973.12164

Dana, L. E., & Smyrnios, K. X. (2010). Family business best practices: Where from and where to? *Journal of Family Business Strategy*, *1*(1), 40–53. doi:10.1016/j.jfbs.2009.12.002

Dana, L. P., Gurau, C., & Lasch, F. (2014). Entrepreneurship, tourism and regional development: A tale of two villages. *Entrepreneurship and Regional Development*, *26*(3-4), 357–374. doi:10.1080/08985626.2014.918182

Dana, L. P., Palaić, R., & Ramadani, V. (2018). Entrepreneurship in former Yugoslavia: An Introduction. In R. Palaić, L. P. Dana, & V. Ramadani (Eds.), *Entrepreneurship in former Yugoslavia: Diversity, Institutional Constraints and Prospects* (pp. 1–8). Springer. doi:10.1007/978-3-319-77634-7_1

Danes, S. M., Lee, J., Amarapurkar, S., Stafford, K., Haynes, G., & Brewton, K. E. (2009). Determinants of family business resilience after a natural disaster by gender of business owner. *Journal of Developmental Entrepreneurship*, *14*(04), 333–354. doi:10.1142/S1084946709001351

Danler & Zehrer. (2017). *Die Wahrnehmung der Arbeitgebermarke von Famlienunternehmen am Beispiel von Studierenden*. Forschungsforum der Österreichischen Fachhochschulen 2017.

Dartey-Baah, K. (2015). Resilient leadership: A transformational-transactional leadership mix. *Ecology and Systematics*, *4*, 1–23.

Das, T. K., & Teng, B.-S. (2001). Article. *Journal of Business and Psychology*, *15*(3), 449–465. doi:10.1023/A:1007874701367

DataReportal. (2020). *Digital 2020: Germany 2020.* https://datareportal.com/reports/digital-2020-germany

Dávid, L. (2011). Tourism ecology: Towards the responsible, sustainable tourism future. *Worldwide Hospitality and Tourism Themes*, *3*(3), 210–216. doi:10.1108/17554211111142176

Davidson, J. L., Jacobson, C., Lyth, A., Dedekorkut-Howes, A., Baldwin, C. L., Ellison, J. C., Holbrook, N. J., Howes, M. J., Serrao-Neumann, S., Singh-Peterson, L., & Smith, T. F. (2016). Interrogating resilience: Toward a typology to improve its operationalization. *Ecology and Society*, *21*(2), 27. doi:10.5751/ES-08450-210227

Davies, I. A., & Crane, A. (2010). Corporate social responsibility in small-and medium-size enterprises: Investigating employee engagement in fair trade companies. *Business Ethics (Oxford, England)*, *19*(2), 126–139. doi:10.1111/j.1467-8608.2010.01586.x

De Massis, A., Eddleston, K. A., & Rovelli, P. (2020). Entrepreneurial by Design: How Organizational Design Affects Family and Non-family Firms' Opportunity Exploitation. *Journal of Management Studies*. Advance online publication. doi:10.1111/joms.12568

De Massis, A., Frattini, F., Majocchi, A., & Piscitello, L. (2018). Family firms in the global economy: Toward a deeper understanding of internationalization determinants, processes, and outcomes. *Global Strategy Journal*, *8*(1), 3–21. doi:10.1002/gsj.1199

De Massis, A., Kotlar, J., Campopiano, G., & Cassia, L. (2013). Dispersion of family ownership and the performance of small-to-medium size private family firms. *Journal of Family Business Strategy*, *4*(3), 166–175. doi:10.1016/j.jfbs.2013.05.001

De Massis, A., & Rondi, E. (2020). Covid-19 and the Future of Family Business Research. *Journal of Management Studies*, *57*(8), 1727–1731. doi:10.1111/joms.12632

Decrop, A. (1999). Triangulation in qualitative tourism research. *Tourism Management*, *20*(1), 157–161. doi:10.1016/S0261-5177(98)00102-2

Della Piana, B., Vecchi, A., Marino, V., & Arco, D. (2019). How do family firms grow? The strategic goals of innovation and internationalization. *International Journal of Business and Management*, *14*(2), 1. doi:10.5539/ijbm.v14n2p1

Deloitte. (2020). *Workforce strategies for post COVID-19 recovery*. https://www2.deloitte.com/content/dam/Deloitte/us/Documents/about-deloitte/workforce-strategies-for-post-covid-19-recovery.pdf

DeNoble, A., Ehrlich, S., & Singh, G. (2007). Toward the Development of a Family Business Self-Efficacy Scale: A Resource-Based Perspective. *Family Business Review*, *20*(2), 127–140. doi:10.1111/j.1741-6248.2007.00091.x

Dernoi, L. A. (1983). Farm Tourism in Europe. *Tourism Management*, *4*(3), 155-166.

Devesa, M. J. S., & Esteban, L. P. (2011). Spanish hotel industry: Indebtedness determinants. *Applied Economics*, *43*(28), 4227–4238. doi:10.1080/00036846.2010.491446

Dickson, P. R., & Giglierano, J. J. (1986). Missing the Boat and Sinking the Boat: A Conceptual Model of Entrepreneurial Risk. *Journal of Marketing*, *50*(3), 58–70. doi:10.1177/002224298605000305

Dillman, D. A., Smyth, J. D., & Christian, L. M. (2009). *Internet, mail, and mixed-mode surveys: The tailored design method*. John Wiley.

Dimianos, D., & Skuras, D. (1996). Farm business and the development of alternative farm enterprises: An empirical analysis in Greece. *Journal of Rural Studies*, *12*(3), 273–283. doi:10.1016/0743-0167(96)00017-4

Dodds, R., & Joppe, M. (2009). The demand for, and participation in corporate social responsibility and sustainable tourism: Implications for the Caribbean. *Ara: Journal of Tourism Research*, *2*(1), 1–11.

Doern, R., Williams, N., & Vorley, T. (2019). Special issue on entrepreneurship and crises: Business as usual? An introduction and review of the literature. *Entrepreneurship and Regional Development, 31*(5–6), 400–412. doi:10.1080/08985626.2018.1541590

Dogru, T., Marchio, E. A., Bulut, U., & Suess, C. (2019). Climate change: Vulnerability and resilience of tourism and the entire economy. *Tourism Management, 72*, 292–305. doi:10.1016/j.tourman.2018.12.010

Dong, E., Wang, Y., Morais, D., & Brooks, D. (2013). Segmenting the rural tourism market: The case of Potter County, Pennsylvania, USA. *Journal of Vacation Marketing, 19*(2), 181–193. doi:10.1177/1356766712471231

Donthu, N., & Gustafsson, A. (2020). Effects of COVID-19 on business and research. *Journal of Business Research, 117*, 284–289. doi:10.1016/j.jbusres.2020.06.008 PMID:32536736

Dörflinger, C., Dörflinger, A., Gavac, K., & Vogl, B. (2013). Familienunternehmen in Österreich – Status quo 2013. Wien.

Dredge, D., Phi, G. T. L., Mahadevan, R., Meehan, E., & Popescu, E. (2018). *Digitalisation in Tourism: In-depth analysis of challenges and opportunities. Low Value procedure GRO-SME-17-C- 091-A for Executive Agency for Small and Medium-sized Enterprises (EASME) Virtual Tourism Observatory*. Aalborg University.

Duncan, R. B. (1976). The ambidextrous organization: Designing dual structures for innovation. *The Management of Organization, 1*(1), 167-188.

Dunn-Cavelty, M., & Suter, M. (2009). Public–Private Partnerships are no silver bullet: An expanded governance model for Critical Infrastructure Protection. *International Journal of Critical Infrastructure Protection, 2*(4), 179–187. doi:10.1016/j.ijcip.2009.08.006

Dupeyras, A., & MacCallum, N. (2013). *Indicators for Measuring Competitiveness in Tourism: A Guidance Document*. OECD Tourism Papers, 2013/02, OECD Publishing. https://www.oecd.org/cfe/tourism/indicators%20for%20measuring%20competitiveness%20in%20tourism.pdf

Duran, P., Kammerlander, N., van Essen, M., & Zellweger, T. (2016). Doing more with less: Innovation input and output in family firms. *Academy of Management Journal, 59*(4), 1224–1264. doi:10.5465/amj.2014.0424

Dyer, W. G., & Whetten, D. A. (2006). Family firms and social responsibility: Preliminary evidence from the S&P 500. *Entrepreneurship Theory and Practice, 30*(6), 785–802. doi:10.1111/j.1540-6520.2006.00151.x

DZS. (1991). *Statisticki ljetopis*. Drzavni Zavod za Statistiku.

Edworthy, J., Hellier, E., Newbold, L., & Titchener, K. (2015). Passing crisis and emergency risk communications: The effects of communication channel, information type, and repetition. *Applied Ergonomics, 48*, 252–262. doi:10.1016/j.apergo.2014.12.009 PMID:25683552

Efendioglu, A. M., & Muscat, E. (2009). Internal disruptions in family business succession: Death, Divorce, and Disability. *The Journal of Industrial Relations and Human Resources*, *11*(1), 17–31.

Egeland, B., Carlson, E., & Sroufe, L. A. (1993). Resilience as process. *Development and Psychopathology*, *5*(04), 517–528. doi:10.1017/S0954579400006131

Egger, T., Favre, G., & Passaglia, M. (2008). Der Agrotourismus in der Schweiz, Analyse der aktuellen Situation und Empfehlungen für die Zukunft. Schweizerische Arbeitsgemeinschaft für die Berggebiete (SAB).

Eichelberger, S., & Peters, M. (2020). Responsible Tourism Entrepreneurship nach COVID-19: Neue Wege im Tourismus. In H. Pechlaner & S. Speer (Eds.), *Responsible Entrepreneurship* (pp. 237–260). Springer Fachmedien Wiesbaden., doi:10.1007/978-3-658-31616-7_11

Eisenhardt, K. (1989). Building theories from case study research. *Academy of Management Review*, *14*(4), 532–550. doi:10.5465/amr.1989.4308385

Elgonemy, A. R. (2002). Debt-Financing Alternatives. *The Cornell Hotel and Restaurant Administration Quarterly*, *43*(3), 7–21. doi:10.1177/0010880402433001

Embacher, H. (2001). Österreich in Europa an der Spitze. *Der Forstliche Landwirt*, *9*, 52–53.

Embacher, H. (2018). *Agritourism and innovation*. Presentation. *1st World Congress on Agritourism*, Bolzano, Italy.

Engeset, A. B. (2020). For better or for worse"–the role of family ownership in the resilience of rural hospitality firms. *Scandinavian Journal of Hospitality and Tourism*, *20*(1), 68–84. doi:10.1080/15022250.2020.1717600

Eppler, M. J., & Mengis, J. (2008). The concept of information overload-a review of literature from organization science, accounting, marketing, mis, and related disciplines (2004). In *Kommunikationsmanagement im Wandel* (pp. 271–305). Gabler. doi:10.1007/978-3-8349-9772-2_15

European Commission. (2001). *Working together for the future of European tourism*. COM(2001) 665 final. https://eur-lex.europa.eu/LexUriServ/LexUriServ.do?uri=COM:2001:0665:FIN:en:PDF

European Commission. (2007). *Agenda for a sustainable and competitive European tourism*. COM(2007) 621 final. https://eur-lex.europa.eu/legal-content/EN/TXT/PDF/?uri=CELEX:52007DC0621&from=EN

European Commission. (2009). *Overview of family–business–relevant issues: research, networks, policy measures and existing studies*. Final report of the expert group.

European Commission. (2010). *Europe, the world's No 1 tourist destination – a new political framework for tourism in Europe*. COM(2010) 352 final. https://eur-lex.europa.eu/legal-content/EN/TXT/PDF/?uri=CELEX:52010DC0352&from=EN

European Commission. (2012). *Implementation and development of the common visa policy to spur growth in the EU.* COM(2012) 649 final. https://eur-lex.europa.eu/legal-content/EN/TXT/PDF/?uri=CELEX:52012DC0649&from=EN

European Commission. (2014). *A European Strategy for more Growth and Jobs in Coastal and Maritime Tourism.* COM(2014) 86 final. https://eur-lex.europa.eu/LexUriServ/LexUriServ.do?uri=COM:2014:0086:FIN:EN:PDF

European Commission. (2020a). *Europe's moment: Repair and Prepare for the Next Generation.* COM(2020) 456 final. https://eur-lex.europa.eu/legal-content/EN/TXT/PDF/?uri=CELEX:52020DC0456&from=EN

European Commission. (2020b). *Tourism and transport in 2020 and beyond.* COM(2020) 550 final. https://eur-lex.europa.eu/legal-content/EN/TXT/PDF/?uri=CELEX:52020DC0550&from=EN

European Commission. (2020c). *Commission welcomes the political agreement on REACT-EU.* European Commission - Press release. https://ec.europa.eu/commission/presscorner/detail/en/IP_20_2140

European Commission. (2020d). *The EU helps reboot Europe's tourism.* Factsheet, FS/20/851. https://ec.europa.eu/commission/presscorner/detail/en/FS_20_851

European Commission. (2020e). *Study on the Contribution of Tourism to Local and Regional Development - Evidence from the European Structural and Investment Funds 2012-2018 EASME/COSME/2018/034.* Final report. Luxembourg: Publications Office of the European Union.

European Commission. (2020f). *Tourism and employment: how the severity of future coronavirus waves could impact jobs.* https://ec.europa.eu/jrc/en/news/tourism-and-employment-how-severity-future-coronavirus-waves-could-impact-jobs

European Commission. (2020g). *Annual Sustainable Growth Strategy 2021.* COM(2020) 575 final. https://eur-lex.europa.eu/legal-content/EN/TXT/PDF/?uri=CELEX:52020DC0575&from=EN

European Parliament. (2015). *Report on new callenges and concepts fort he promotion of tourism in Europe* (2014/2241(INI)). https://www.europarl.europa.eu/doceo/document/A-8-2015-0258_EN.html

European Parliament. (2018). *European Parliament resolution of 3 October 2018 on addressing the specific needs of rural, mountainous and remote areas* (2018/2720(RSP). https://www.europarl.europa.eu/doceo/document/TA-8-2018-0374_EN.html

European Travel Commission Quarterly Report. (2020). *European tourism: trends & prospects, Quarterly report (Q2/2020).* https://etc-corporate.org/uploads/2020/07/ETC-Quarterly-Report-Q2-2020_Final-Public.pdf

European Union. (2020). *Family business.* https://ec.europa.eu/growth/smes/supporting-entrepreneurship/family-business_en

Eurostat. (2020). *Tourism statistics at regional level, Statistics Explained.* https://ec.europa.eu/eurostat/statistics-explained/pdfscache/1945.pdf

Evans, N. J., & Ilbery, B. W. (1989). A Conceptual Framework for Investigating Farm-Based Accommodation and Tourism in Britain. *Journal of Rural Studies*, *5*(3), 257–266. doi:10.1016/0743-0167(89)90004-1

Evans, N. J., & Ilbery, B. W. (1992). Farm-Based Accommodation and the Restructuring of Agriculture: Evidence from Three English Counties. *Journal of Rural Studies*, *8*(1), 85–96. doi:10.1016/0743-0167(92)90032-2

Evenhuis, E. (2020). New directions in researching regional economic resilience and adaptation. In G. Bristow & A. Healy (Eds.), *Handbook on Regional Economic Resilience*. Edward Elgar Publishing. doi:10.4337/9781785360862.00010

Evgrafova, L. V., Ismailova, A. Z., & Kalinichev, V. L. (2020). Agrotourism as a factor of sustainable rural development. *IOP Conference Series. Earth and Environmental Science*, *421*, 16. doi:10.1088/1755-1315/421/2/022058

Faghfouri, P., Kraiczy, N. D., Hack, A., & Kellermanns, F. W. (2015). Ready for a crisis? How supervisory boards affect the formalized crisis procedures of small and medium-sized family firms in Germany. *Review of Managerial Science*, *9*(2), 317–338. doi:10.100711846-014-0158-0

Falkner, E. M., & Hiebl, M. R. W. (2015). Risk management in SMEs: A systematic review of available evidence. *The Journal of Risk Finance*, *16*(2), 122–144. doi:10.1108/JRF-06-2014-0079

Fang, H., Kotlar, J., Memili, E., Chrisman, J. J., & de Massis, A. (2018). The pursuit of international opportunities in family firms: Generational differences and the role of knowledge-based resources. *Global Strategy Journal*, *8*(1), 136–157. doi:10.1002/gsj.1197

Fang, W.-T. (2020). Responsible Tourism. In W.-T. Fang (Ed.), *Tourism in Emerging Economies* (Vol. 50, pp. 131–151). Springer Singapore., doi:10.1007/978-981-15-2463-9_6

Faulkner, B. (2001). Towards a framework for tourism disaster management. *Tourism Management*, *22*(2), 135–147.

Feltham, T. S., Feltham, G., & Barnett, J. J. (2005). The dependence of family businesses on a single decision-maker. *Journal of Small Business Management*, *43*(1), 1–15. doi:10.1111/j.1540-627X.2004.00122.x

Finucane, M. L., Alhakami, A., Slovic, P., & Johnson, S. M. (2000). The affect heuristic in judgments of risks and benefits. *Journal of Behavioral Decision Making*, *13*(1), 1–17. doi:10.1002/(SICI)1099-0771(200001/03)13:1<1::AID-BDM333>3.0.CO;2-S

Finucane, M. L., & Holup, J. L. (2006). Risk as Value: Combining Affect and Analysis in Risk Judgments. *Journal of Risk Research*, *9*(2), 141–164. doi:10.1080/13669870500166930

Fishbein, M., & Ajzen, I. (1975). *Belief, attitude, intention and behaviour: An introduction to theory and research*. Addison-Wesley Reading.

Fjäder, C. (2014). The nation-state, national security and resilience in the age of globalisation. *Resilience*, *2*(2), 114–129. doi:10.1080/21693293.2014.914771

Flanigan, S., Blackstock, K., & Hunter, C. (2014). Agritourism from the perspective of providers and visitors: A typology-based study. *Tourism Management*, *40*, 394–405. doi:10.1016/j.tourman.2013.07.004

Flick, U. (2011). *Triangulation*. Beltz Verlag. doi:10.1007/978-3-531-92864-7

Floridi, L. (2014). *The 4th Revolution: How the Infosphere is Reshaping Human Reality*. Oxford University Press.

Floyd, M. F., Gibson, H., Pennington-Gray, L. A., & Thapa, B. (2004). The Effect of Risk Perceptions on Intentions to Travel in the Aftermath of September 11, 2001. *Journal of Travel & Tourism Marketing*, *15*(2–3), 19–38. doi:10.1300/J073v15n02_02

Fotiadis, A., Polyzos, S., & Huan, T.-C. T. C. (2021). The good, the bad and the ugly on COVID-19 tourism recovery. *Annals of Tourism Research*, *87*, 103117. doi:10.1016/j.annals.2020.103117 PMID:33518847

Fournier, S., & Avery, J. (2011). The uninvited brand. *Business Horizons*, *54*(3), 193–207. doi:10.1016/j.bushor.2011.01.001

Fraj-Andrés, E., Martínez-Salinas, E., & Matute-Vallejo, J. (2009). Factors affecting corporate environmental strategy in Spanish industrial firms. *Business Strategy and the Environment*, *18*(8), 500–514. doi:10.1002/bse.611

Frank, H., Lueger, M., Nosé, L., & Suchy, D. (2010). The concept of 'Familiness': Literature review and systems theory-based reflections. *Journal of Family Business Strategy*, *1*(3), 119–130.

Frater, J. M. (1983). Farm Tourism in England: Planning, Funding, Promotion and Some Lessons from Europe. *Tourism Management*, *4*(3), 167–179. doi:10.1016/0261-5177(83)90061-4

French-Bravo, M., & Crow, G. (2015). Shared governance: The role of buy-in in bringing about change. *Online Journal of Issues in Nursing*, *20*(2).

French, D. P., Sutton, S., Kinmonth, A. L., & Marteau, T. M. (2006). Assessing Perceptions of Risks due to Multiple Hazards. *Journal of Risk Research*, *9*(6), 657–682. doi:10.1080/13669870600813837

Frey, N., & George, R. (2010). Responsible tourism management: The missing link between business owners' attitudes and behaviour in the Cape Town tourism industry. *Tourism Management*, *31*(5), 621–628. doi:10.1016/j.tourman.2009.06.017

Fröhlich, K., & Hassink, R. (2018). Regional resilience: A stretched concept? *European Planning Studies*, *26*(9), 1763–1778. doi:10.1080/09654313.2018.1494137

Fu, H., Okumus, F., Wu, K., & Köseoglu, M. A. (2019). The entrepreneurship research in hospitality and tourism. *International Journal of Hospitality Management*, *78*, 1–12. doi:10.1016/j.ijhm.2018.10.005

Galdeano-Gómez, E., Céspedes-Lorente, J., & Martínez-Del-Río, J. (2008). Environmental performance and spillover effects on productivity: Evidence from horticultural firms. *Journal of Environmental Management, 88*(4), 1552–1561. doi:10.1016/j.jenvman.2007.07.028 PMID:17825476

Garbislander, S., Haigner, S., & Jenewein, S. (2020). Wertschöpfungseffekte des *Tourismusbedingten Nachfrageausfalls für Tirol im Zuge der Corona-Pandemie.* Studie Wirtschaftskammer Tirol. Available at: https://www.wirtschaft.tirol/wp-content/uploads/2020/04/Details-zur-Studie_-Wertschöpfungseffekte-des-tourismusbedingten-Nachfrageausfalls-im-Zuge-der-Corona-Pandemie-in-Tirol.pdf

Garcia-Dia, M. J., DiNapoli, J. M., Garcia-Ona, L., Jakubowski, R., & O'Flaherty, D. (2013). Concept Analysis: Resilience. *Archives of Psychiatric Nursing, 27*(6), 264–270. doi:10.1016/j.apnu.2013.07.003 PMID:24238005

Garfin, D. R., Silver, R. C., & Holman, E. A. (2020). The Novel Coronavirus (COVID-2019) Outbreak: Amplification of Public Health Consequences by Media Exposure. *Health Psychology, 39*(5), 355–357. doi:10.1037/hea0000875 PMID:32202824

Garstenauer, R., Schwarz, U., & Tod, S. (2012). Alles unter einen Hut bringen: Bäuerliche Wirtschaftsstile in zwei Regionen Niederösterreichs 1945-1985. *Historische Anthropologie: Kultur, Gesellschaft. Alltag, 20*(3), 383–426.

Gast, J., Kallmünzer, A., Kraus, S., Gundolf, K., & Arnold, J. (2019). Coopetition of small- and medium-sized family enterprises: Insights from an IT business network. *International Journal of Entrepreneurship and Small Business, 38*(1/2). doi:10.1504/IJESB.2019.102493

Gavac, K., Heckl, E., & Petzlberger, K. (2020). *Familienunternehmen in Österreich 2019: Endbericht.* Vienna: KMU Forschung Austria. https://news.wko.at/news/oesterreich/kmu-forschung-bericht-familienunternehmen-2019.pdf

Gbadegeshin, S. A. (2013). Awareness of "Ownership Succession" and Family Business Continuity. *International Journal of Business and Management Invention, 2*(9), 72–87.

Germann Molz, J., & Paris, C. M. (2015). The Social Affordances of Flashpacking: Exploring the Mobility Nexus of Travel and Communication. *Mobilities, 10*(2), 173–192. doi:10.1080/17450101.2013.848605

Germanovich, A. G., Vasilieva, O. N., Ordynskaya, M. E., Allanina, L. M., & Gorokhova, A. E. (2020). Impact of Tourism on Sustainable Development of Rural Areas: International Experience. *Journal of Environmental Management and Tourism, 11*(4), 965–972.

Gersick, K. E. (1997). *Generation to generation: Life cycles of the family business.* Harvard Business Press.

Getz, D., Carlsen, J., & Morrison, A. (2004). Family-owned tourist attractions and wineries. In The family business in tourism and hospitality (pp. 153–160). CABI Publishing.

Getz, D., & Carlsen, J. (2000). Characteristics and goals of family and owner-operated businesses in the rural tourism and hospitality sectors. *Tourism Management*, *21*(6), 547–560. doi:10.1016/S0261-5177(00)00004-2

Getz, D., & Carlsen, J. (2005). Family business in tourism: State of the art. *Annals of Tourism Research*, *32*(1), 237–258. doi:10.1016/j.annals.2004.07.006

Getz, D., Carlsen, J., & Morrison, A. (2004). *The family business in tourism and hospitality*. CABI. doi:10.1079/9780851998084.0000

Gibb Dyer, W. Jr. (1994). Potential contributions of organizational behavior to the study of family-owned businesses. *Family Business Review*, *7*(2), 109–131. doi:10.1111/j.1741-6248.1994.00109.x

Gibson, C. B., & Birkinshaw, J. (2004). The antecedents, consequences, and mediating role of organizational ambidexterity. *Academy of Management Journal*, *47*(2), 209–226. doi:10.2307/20159573

Gilsbach, C. (2018). Krisenkommunikation im Tourismus: Kommunikationskanäle, -instrumente und Medienmix. In S. Hahn & Z. Neuss (Eds.), Krisenkommunikation in Tourismusorganisationen (pp. 61–82). Springer Fachmedien.

Glaesser, D. (2006). *Crisis management in the tourism industry*. Butterworth-Heinemann.

Glowka, G., Kallmünzer, A., & Zehrer, A. (2020). Enterprise risk management in small and medium family enterprises: The role of family involvement and CEO tenure. *The International Entrepreneurship and Management Journal*. Advance online publication. doi:10.100711365-020-00682-x

Glowka, G., & Zehrer, A. (2019). Tourism Family-Business Owners' Risk Perception: Its Impact on Destination Development. *Sustainability*, *11*(6992), 1–16.

Golicic, S. L., Flint, D. J., & Signori, P. (2017). Building business sustainability through resilience in the wine industry. *International Journal of Wine Business Research*, *29*(1), 74–97. doi:10.1108/IJWBR-02-2016-0005

Gomez-Mejia, L. R., Cruz, C., Berrone, P., & de Castro, J. (2011). The Bind that Ties: Socioemotional Wealth Preservation in Family Firms. *The Academy of Management Annals*, *5*(1), 653–707. do i:10.5465/19416520.2011.593320

Gómez-Mejía, L. R., Haynes, K. T., Núñez-Nickel, M., Jacobson, K. J. L., & Moyano-Fuentes, J. (2007). Socioemotional Wealth and Business Risks in Family-controlled Firms: Evidence from Spanish Olive Oil Mills. *Administrative Science Quarterly*, *52*(1), 106–137. doi:10.2189/asqu.52.1.106

Gómez-Mejía, L. R., Haynes, K. T., Núñez-Nickel, M., Jacobson, K. J., & Moyano-Fuentes, J. (2007). Socioemotional wealth and business risks in family-controlled firms: Evidence from Spanish olive oil mills. *Administrative Science Quarterly*, *52*(1), 106–137.

Gomez-Mejia, L. R., Makri, M., & Kintana, M. L. (2010). Diversification Decisions in Family-Controlled Firms. *Journal of Management Studies*, *47*(2), 223–252. doi:10.1111/j.1467-6486.2009.00889.x

Gong, J., Detchkhajornjaroensri, P., & Knight, D. W. (2019). Responsible tourism in Bangkok, Thailand: Resident perceptions of Chinese tourist behaviour. *International Journal of Tourism Research*, *21*(2), 221–233. doi:10.1002/jtr.2256

González-Macías, C. J., & Zizaldra-Hernández, I. (2016). La unión en empresas familiares restauranteras : Una aproximación a la sucesión generacional en empresas con actividad turística de San Cristóbal de las Casas. *Revista Iberoamericana de Turismo*, *6*, 18–31.

Good, D., & Michel, E. J. (2013). Individual ambidexterity: Exploring and exploiting in dynamic contexts. *The Journal of Psychology*, *147*(5), 435–453. doi:10.1080/00223980.2012.710663 PMID:24003589

Goodrich, J. (2002). September 11, 2001 Attack on America: A Record of the Immediate Impacts and Reactions in the USA Travel and Tourism Industry. *Tourism Management*, *23*, 573–580.

Gössling, S., Scott, D., & Michael Hall, C. (2021). Pandemics, tourism and global change: A rapid assessment of COVID-19. *Journal of Sustainable Tourism*, *29*(1), 1–20. doi:10.1080/09669582.2020.1758708

Gottfried, J., & Barthel, M. (2018). *Almost seven-in-ten Americans have news fatigue, more among Republicans*. Pew Research Center.

Graves, C., & Thomas, J. (2006). Internationalization of Australian family businesses: A managerial capabilities perspective. *Family Business Review*, *19*, 207–224.

Grimm, B. (2011). *Urlaub auf dem Bauernhof/ Urlaub auf dem Lande 2010/11. Untersuchung der Marktsituation und der Potenziale für den Bauernhof-/Landurlaub auf Basis der Reiseanalyse 2011und einer Volumenerhebung, ergänzt durch Gruppendiskussionen mit Personen aus Haushalten ohne Kinder*. Studie für das Bundesministerium für Ernährung, Landwirtschaft und Verbraucherschutz (BMELV). http://docs.dpaq.de/882-2011_bmelv-reiseanalyse.pdf

Gudmundson, D., Hartman, E. A., & Tower, C. B. (1999). Strategic Orientation: Differences between Family and Nonfamily Firms. *Family Business Review*, *12*(1), 27–39. doi:10.1111/j.1741-6248.1999.00027.x

Gueye, S., & Simon, E. (2010). Is the family business a safer type of governance in time of crisis? *Problems and Perspectives in Management*, *8*(4).

Haavik, T. K., Antonsen, S., Rosness, R., & Hale, A. (2019). HRO and RE: A pragmatic perspective. *Safety Science*, *117*, 479–489. doi:10.1016/j.ssci.2016.08.010

Habbershon, T. G., & Williams, M. L. (1999). A resource-based framework for assessing the strategic advantages of family firms. *Family Business Review*, *12*(1), 1–25.

Habbershon, T. G., & Williams, M. L. (1999). A Resource-Based Framework for Assessing the Strategic Advantages of Family Firms. *Family Business Review*, *12*(1), 1–25. doi:10.1111/j.1741-6248.1999.00001.x

Habbershon, T. G., Williams, M., & MacMillan, I. C. (2003). A unified systems perspective of family firm performance. *Journal of Business Venturing*, *18*(4), 451–465. doi:10.1016/S0883-9026(03)00053-3

Hair, J. F., Anderson, R. E., Tatham, R. L., & Black, W. C. (1995). *Multivariate Data Analysis*. Prentice-Hall Inc.

Hajibaba, H., Gretzel, U., Leisch, F., & Dolnicar, S. (2015, July). Crisis-resistant tourists. *Annals of Tourism Research*, *53*, 46–60. doi:10.1016/j.annals.2015.04.001

Hall, A. (2003). *Strategizing in the context of genuine relations*. JIBS Dissertation Series No. 08, Jönköping International Business School, Jönköping, Sweden.

Hall, A., Melin, L., & Nordqvist, M. (2001). Entrepreneurship as Radical Change in the Family Business: Exploring the Role of Cultural Patterns. *Family Business Review*, *14*(3), 193–208. doi:10.1111/j.1741-6248.2001.00193.x

Hall, A., Melin, L., & Nordqvist, M. (2006). Understanding Strategizing in Family Business Contexts. In P. Poutziouris, K. Smyrnios, & S. Klein (Eds.), *Family Business Research Handbook* (pp. 253–268). Edward Elgar. doi:10.4337/9781847204394.00025

Hall, C. M. (2010). Crisis events in tourism: Subjects of crisis in tourism. *Current Issues in Tourism*, *13*(5), 401–417.

Hall, C. M., Prayag, G., & Amore, A. (2017). *Tourism and resilience: individual, organisational and destination perspectives* (1st ed.). Channel View Publications.

Hall, D., Mitchell, M., & Roberts, L. (2004). Tourism and the countryside: dynamic relationships. In L. Roberts, D. Hall, & M. Morag (Eds.), *New Directions in Rural Tourism*. Routedge.

Handler, W. C. (1990). Succession in Family Firms: A Mutual Role Adjustment between Entrepreneur and Next-generation Family Members. *Entrepreneurship Theory and Practice*, *15*(1), 37–52. doi:10.1177/104225879001500105

Handler, W. C. (1991). Key interpersonal relationships of the next-generation family members in family firms. *Journal of Small Business Management*, *29*(3), 21–32.

Handler, W. C. (1994). Succession in Family Business: A Review of the Research. *Family Business Review*, *7*(2), 133–157. doi:10.1111/j.1741-6248.1994.00133.x

Hang, H., Aroean, L., & Chen, Z. (2020). Building emotional attaching during COVID-19. *Annals of Tourism Research*, 83. PMID:32834221

Hansen. (2018). *Family Firm Performance over the Business Cycle: A Meta-Analysis*. Research Papers in Economics No. 6/18. https://ssrn.com/abstract=3288094

Hanson, S. K., Hessel, H. M., & Danes, S. M. (2019). Relational processes in family entrepreneurial culture and resilience across generations. *Journal of Family Business Strategy*, *10*(3), 100263. doi:10.1016/j.jfbs.2018.11.001

Hardy, A., Beeton, R. J. S., & Pearson, L. (2002). Sustainable Tourism: An Overview of the Concept and its Position in Relation to Conceptualisations of Tourism. *Journal of Sustainable Tourism*, *10*(6), 475–496. doi:10.1080/09669580208667183

Harper, D. (1994). On the authority of the image: Visual methods at the crossroads. In N. K. Denzin & Y. S. Lincoln (Eds.), *Handbook of Qualitative Research* (pp. 403–412). Sage.

Harper, D. (1998). An argument for visual sociology. In J. Prosser (Ed.), *Image-based research: A source book for qualitative researchers* (pp. 24–41). Palmer Press.

Harper, D. (2002). Talking about pictures: A case for photo elicitation. *Visual Studies*, *17*(1), 13–26. doi:10.1080/14725860220137345

Harrell, F. E., Lee, K. L., Califf, R. M., Pryor, D. B., & Rosati, R. A. (1984). Regression modelling strategies for improved prognostic prediction. *Statistics in Medicine*, *3*(2), 143–152. doi:10.1002im.4780030207

Hartman, S., Wielenga, B., & Heslinga, J. H. (2020). The future of tourism destination management: building productive coalitions of actor networks for complex destination development. *Journal of Tourism Futures*.

Hartman, S. (2016). Towards adaptive tourism areas? A complexity perspective to examine the conditions for adaptive capacity. *Journal of Sustainable Tourism*, *24*(2), 299–314. doi:10.1080/09669582.2015.1062017

Hartman, S. (2018). Resilient tourism destinations? Governance implications of bringing theories of resilience and adaptive capacity to tourism practice. In E. Innerhofer, M. Fontanari, & H. Pechlaner (Eds.), *Destination Resilience: Challenges and Opportunities for Destination Management and Governance* (1st ed., pp. 66–76). Routledge. doi:10.4324/9780203701904-5

Hassink, R., & Gong, H. (2020). Regional Resilience. In A. Kobayashi, M. Boyle, N. Castree, F. Collins, J. Crampton, S. de Leeuw, E. Ho, I. Luginah, B. Maharaj, & J. Tyner (Eds.), *International Encyclopedia of Human Geography* (pp. 351–355). Elsevier. doi:10.1016/B978-0-08-102295-5.10126-X

Hatch, M. J. (1997). *Organisation Theory: Modern, Symbolic and Post-modern Perspectives.* Oxford University Press.

Hauck, J., & Prügl, R. (2015). Innovation activities during intra-family leadership succession in family firms: An empirical study from a socioemotional wealth perspective. *Journal of Family Business Strategy*, *6*(2), 104–118. doi:10.1016/j.jfbs.2014.11.002

Haultain, A. (2013). A picture is worth a thousand words: The empowering potential of using photo-elicitation within research and teaching. *Conference Proceedings: CLESOL 2012.*

Haushofer, C. (2013). *Familienunternehmen in Österreich.* Academic Press.

Hauswald, H., Hack, A., Kellermanns, F. W., & Patzelt, H. (2016). Attracting new talent to family firms: Who is attracted and under what conditions? *Entrepreneurship Theory and Practice, 40*(5), 963–989.

HCFCD. (2020). *Haut Comité Français pour la Résilience Nationale.* https://www.hcfdc.org/fr/guest/nosmembres

He, H., & Harris, L. (2020). The impact of Covid-19 pandemic on corporate social responsibility and marketing philosophy. *Journal of Business Research, 116*, 176–182. doi:10.1016/j.jbusres.2020.05.030 PMID:32457556

Heimerl, P., & Peters, M. (2019). Shaping the future of Alpine tourism destinations' next generation: An action research approach. *Tourism: An International Interdisciplinary Journal, 67*(3), 281–298.

Henderson, N., Benard, B., & Sharp-Light, N. (Eds.). (1999). *Resiliency in action: Practical ideas for overcoming risks and building strengths in youth, families and communities.* Resiliency in Action, Inc.

Henschel, K., Gruner, A., & von Freyberg, B. (2013). *Hotelmanagement (4., aktualisierte und erw. Aufl.).* Oldenbourg. doi:10.1524/9783486753882

Herbane, B. (2015). Threat orientation in small and medium-sized enterprises. *Disaster Prevention and Management: An International Journal, 24*(5), 583–595. doi:10.1108/DPM-12-2014-0272

Herbane, B. (2019). Rethinking organizational resilience and strategic renewal in SMEs. *Entrepreneurship and Regional Development, 31*(5–6), 476–495. doi:10.1080/08985626.2018.1541594

Heskett, J. L., & Sasser, W. E. (2010). The service profit chain. In *Handbook of Service Science* (pp. 19–29). Springer. doi:10.1007/978-1-4419-1628-0_3

Hiebl, Duller, C., & Neubauer, H. (2019). Enterprise Risk Management in Family Firms: Evidence from Austria and Germany. *The Journal of Risk Finance, 20*(1), 39–58. doi:10.1108/JRF-01-2018-0003

Hiebl, M. R. W. (2012). Risk aversion in family firms: What do we really know? *The Journal of Risk Finance, 14*(1), 49–70. doi:10.1108/15265941311288103

Hiebl, M. R. W. (2015). Family involvement and organiza- tional ambidexterity in later-generation family businesses: A framework for further investigation. *Management Decision, 53*(5), 1061–1082. doi:10.1108/MD-04-2014-0191

Higgins-Desbiolles, F. (2010). The elusiveness of sustainability in tourism: The culture–ideology of consumerism and its implications. *Tourism and Hospitality Research, 10*(2), 116–129. doi:10.1057/thr.2009.31

Higgins-Desbiolles, F. (2020). Socialising tourism for social and ecological justice after COVID-19. *Tourism Geographies*, *22*(3), 610–623. doi:10.1080/14616688.2020.1757748

Higgins-Desbiolles, F. (2021). The "war over tourism": Challenges to sustainable tourism in the tourism academy after COVID-19. *Journal of Sustainable Tourism*, *29*(4), 551–569. doi:10.1080/09669582.2020.1803334

Hjalager, A. M. (1996). Agricultural diversification into tourism: Evidence from a European community development programme. *Tourism Management*, *17*(2), 103–111. doi:10.1016/0261-5177(95)00113-1

Holland, P. G., & Oliver, J. E. (1992). An empirical examination of stages of development of family business. *Journal of Business & Entrepreneurship*, *4*(3), 27–38.

Holling, C. S. (1973). Resilience and stability of ecological systems. *Annual Review of Ecology and Systematics*, *4*(1), 1–23. doi:10.1146/annurev.es.04.110173.000245

Holling, C. S. (1973). Resilience and Stability of Ecological Systems. *Annual Review of Ecology and Systematics*, *4*, 1–23.

Hollnagel, E. (2011). RAG – The resilience analysis grid. In E. Hollnagel, J. Pariès, D. Woods, & J. Wreathall (Eds.), *Resilience Engineering in Practice. A Guidebook* (pp. 275–296). Ashgate.

Hopkin, P. (2018). *Fundamentals of risk management: understanding, evaluating and implementing effective risk management.* Kogan Page Publishers.

Horne, J. F. (1997). The coming age of organizational resilience. In *Business Forum* (Vol. 22, No. 2/3, pp. 24 - 28). California State University, Los Angeles, School of Business and Economics.

HotelResilient. (2020). *About us. Hotel Resilient.* https://hotelresilient.org/about-us

Huang, M., Li, P., Meschke, F., & Guthrie, J. P. (2015). Family firms, employee satisfaction, and corporate performance. *Journal of Corporate Finance*, *34*, 108–127. doi:10.1016/j.jcorpfin.2015.08.002

Human Rights Watch. (2002). *"WE ARE NOT THE ENEMY" Hate Crimes Against Arabs, Muslims, and Those Perceived to be Arab or Muslim after September 11.* Retrieved from http://www.hrw.org

Human Rights Watch. (2020). *Covid-19 Fueling Anti-Asian Racism and Xenophobia Worldwide | Human Rights Watch.* Retrieved from https://www.hrw.org/news/2020/05/12/covid-19-fueling-anti-asian-racism-and-xenophobia-worldwide

Ibanescu, B. C., Eva, M., & Gheorghiu, A. (2020). Questioning the role of tourism as an engine for resilience: The role of accessibility and economic performance. *Sustainability*, *12*(14), 5527. doi:10.3390u12145527

Ibrahim, A. B., Soufani, K., & Lam, J. (2001). A study of succession in a family firm. *Family Business Review*, *14*(3), 245–258. doi:10.1111/j.1741-6248.2001.00245.x

Ilbery, B., Bowler, I., Clark, G., Crockett, A., & Shaw, A. (1988). Farm-Based Tourism as an Alternative Farm Enterprise: A Case Study from the Northern Pennines England. *Regional Studies Association*, *32*(4), 355–363.

Innerhofer, E., Fontanari, M., & Pechlaner, H. (Eds.). (2018). *Destination Resilience: Challenges and Opportunities for Destination Management and Governance* (1st ed.). Routledge. doi:10.4324/9780203701904

International Monetary Fund. (2020). *World Economic Outlook: A Long and Difficult Ascent.* https://www.imf.org/en/Publications/WEO/Issues/2020/09/30/world-economic-outlook-october-2020#Full%20Report%20and%20Executive%20Summary

International Trade Centre. (2020). *SME Competitiveness Outlook 2020: COVID-19: The Great Lockdown and its Impact on Small Business*. ITC. https://www.intracen.org/uploadedFiles/intracenorg/Content/Publications/ITCSMECO2020.pdf

Ioannides, D., & Gyimothy, S. (2020). The COVID-19 crisis as an opportunity for escaping the unsustainable global tourism path. *Tourism Geographies*, *22*(3), 624–632. doi:10.1080/14616688.2020.1763445

Ioannides, D., & Petersen, T. (2003). Tourism 'non-entrepreneurship' in peripheral destinations: A case study of small and medium tourism enterprises on Bornholm, Denmark. *Tourism Geographies*, *5*(4), 408–435. doi:10.1080/1461668032000129146

Ioannou, S., Wójcik, D., & Dymski, G. (2019). Too-Big-To-Fail: Why Megabanks Have Not Become Smaller Since the Global Financial Crisis? *Review of Political Economy*, *31*(3), 356–381. doi:10.1080/09538259.2019.1674001

Istat. (2020). *Agritourism. Agritouristc farm authorized.* http://dati.istat.it/Index.aspx?lang=en&SubSessionId=4d0df501-4308-481c-a33e-fcdc40a8ec7f

Jacobson, G. F. (1980). Crisis Theory. *New Directions for Mental Health Services*, *6*, 1–10.

Jakob, H. (2020). *Digital Advertising: Markets, Trends, Players*. Webinar statista.

James, H. (2006). *Family capitalism: Wendels, Haniels, Falcks, and the continental European model*. Harvard University Press.

Japp, K. P. (2010). Risiko und Gefahr. Zum Problem authentischer Kommunikation. In C. Büscher & K. P. Japp (Eds.), *Ökologische Aufklärung: 25 Jahre "Ökologische Kommunikation* (1st ed., pp. 281–308). VS Verl. für Sozialwiss. doi:10.1007/978-3-531-92425-0_11

Javadi, M., & Zarea, K. (2016). Understanding thematic analysis and its pitfall. *Journal of Client Care*, *1*(1), 33–39.

Jeong, M., & Lee, S. A. (2017). Do customers care about types of hotel service recovery efforts? *Journal of Hospitality and Tourism Technology*, *8*(1), 5–18. doi:10.1108/JHTT-09-2016-0049

Jessop, B. (1999). Reflections on globalisation and its (il)logic(s). In K. Olds, P. Dicken, P. F. Kelly, L. Kong, & H. W. Yeung (Eds.), *Globalisation and the Asia-Pacific: Contested territorie* (pp. 19–38). Routledge.

Jiang, Y., Ritchie, B. W., & Verreynne, M. L. (2019). Building tourism organizational resilience to crises and disasters: A dynamic capabilities view. *International Journal of Tourism Research, 21*(6), 882–900.

Jiang, Y., Ritchie, B. W., & Verreynne, M.-L. (2019). Building tourism organizational resilience to crises and disasters: A dynamic capabilities view. *International Journal of Tourism Research, 21*(6), 882–900. doi:10.1002/jtr.2312

Johnson, N., Elliott, D., & Drake, P. (2013). Exploring the role of social capital in facilitating supply chain resilience. *Supply Chain Management, 18*(3), 324–336. doi:10.1108/SCM-06-2012-0203

Jones, L. (2019). Resilience isn't the same for all: Comparing subjective and objective approaches to resilience measurement. *Wiley Interdisciplinary Reviews: Climate Change, 10*(1), 552. doi:10.1002/wcc.552

Joppe, M. (1996). Sustainable community tourism development revisited. *Tourism Management, 17*(7), 475–479. doi:10.1016/S0261-5177(96)00065-9

Joppe, M. (2012). Migrant workers: Challenges and opportunities in addressing tourism labour shortages. *Tourism Management, 33*(3), 662–671. doi:10.1016/j.tourman.2011.07.009

Joseph, J. (2013). Resilience as embedded neoliberalism: A governmentality approach. *Resilience, 1*(1), 38–52. doi:10.1080/21693293.2013.765741

Kabanoff, B., Waldersee, R., & Cohen, M. (1995). Espoused values and organizational change themes. *Academy of Management Journal, 38*, 1075–1104.

Kachroo-Levine, M. (2020). *5 Experts Weigh in on What Coronavirus Means for Your Summer Vacation Plans.* https://www.departures.com/travel/summer-vacation-coronavirus-pandemic

Kaiser, I. (2009). *Urlaub am Bauernhof. Erlebnisse aus dem Blickwinkel der Wissenschaft.* Diplomica Verlag GmbH.

Kalisch, R., Müller, M., & Tuescher, O. (2014). A conceptual framework for the neu-robiological study of resilience. *Behavioral and Brain Sciences, 27*, 1–49.

Kallmuenzer, A., Nikolakis, W., Peters, M., & Zanon, J. (2018). Trade-offs between dimensions of sustainability: Exploratory evidence from family firms in rural tourism regions. *Journal of Sustainable Tourism, 26*(7), 1204–1221. doi:10.1080/09669582.2017.1374962

Kallmuenzer, A., & Peters, M. (2017). Exploring entrepreneurial orientation in family firms: The relevance of social embeddedness in competition. *International Journal of Entrepreneurship and Small Business, 30*(2). doi:10.1504/IJESB.2017.081436

Kammerlander, N., & Ganter, M. (2015). An attention-based view of family firm adaptation to discontinuous technolog- ical change: Exploring the role of family CEOs' noneco- nomic goals. *Journal of Product Innovation Management*, *32*(3), 361–383. doi:10.1111/jpim.12205

Kammerlander, N., Patzelt, H., Behrens, J., & Röhm, C. (2020). Organizational Ambidexterity in Family-Managed Firms: The Role of Family Involvement in Top Management. *Family Business Review*, *33*(4), 393–423. doi:10.1177/0894486520961645

Kapuściński, G., & Richards, B. (2016). News framing effects on destination risk perception. *Tourism Management*, *57*, 234–244. doi:10.1016/j.tourman.2016.06.017

Karim, W., Haque, A., Anis, Z., & Ulfy, M. A. (2020). The Movement Control Order (MCO) for COVID-19 Crisis and its Impact on Tourism and Hospitality Sector in Malaysia. *International Tourism and Hopitality Yournal*, *3*(2), 1–07.

Karl, M., & Schmude, J. (2017). Understanding the role of risk (perception) in destination choice: A literature review and synthesis. *Tourism (Zagreb)*, *65*(2), 138–155.

Katila, R., & Ahuja, G. (2002). Something old, something new: A longitudinal study of search behavior and new product introduction. *Academy of Management Journal*, *45*, 1183–1194. https://books.google.at/books?hl=de&lr=&id=Ak_5AwAAQBAJ&oi=fnd&pg=PA1&dq=Urlaub+am+Bauernhof+Deutschland&ots=DVtj0t5Un2&sig=eB7XhY1wAcbkO-seu5PV5EcMQfw&redir_esc=y#v=onepage&q=Urlaub%20am%20Bauernhof%20Deutschland&f=false

Katina, P. F., Pinto, C. A., Bradley, J. M., & Hester, P. T. (2014). Interdependency-induced risk with applications to healthcare. *International Journal of Critical Infrastructure Protection*, *7*(1), 12–26. doi:10.1016/j.ijcip.2014.01.005

Kepner, E. (1991). The Family and the Firm: A Coevolutionary Perspective. *Family Business Review*, *4*(4), 445–461. doi:10.1111/j.1741-6248.1991.00445.x

Kim, J., Rasouli, S., & Timmermans, H. J. P. (2018). Social networks, social influence and activity-travel behaviour: A review of models and empirical evidence. *Transport Reviews*, *38*(4), 499–523. doi:10.1080/01441647.2017.1351500

Kim, M. G., & Mattila, A. S. (2013). Does a surprise strategy need words? The effect of explanations for a surprise strategy on customer delight and expectations. *Journal of Services Marketing*, *27*(5), 361–370. doi:10.1108/JSM-01-2012-0008

Kim, Y. (2020). Organizational resilience and employee work-role performance after a crisis situation: Exploring the effects of organizational resilience on internal crisis communication. *Journal of Public Relations Research*, *32*(1–2), 47–75. doi:10.1080/1062726X.2020.1765368

KMU Forschung Austria. (2008). *Overview of Family Business Relevant Issues*. Final Report. SME Research Austria. https://www.kmuforschung.ac.at/wp- content/uploads/2018/02/EPS_Familienunternehmen.pdf

Kobasic, A. (1987). *Turizam u Jugoslaviji: Razvoj, stanje, perspektive*. Informator.

Kokko, A. (2011). *An Exploration of Agritourism: Topics, Literature and Areas for Future Research*. http://annekokko.com/resources/an%20exploration%20of%20agritourism%20-%20akokko.pdf

Komppula, R. (2016). The role of different stakeholders in destination development. *Tourism Review*, *71*(1), 67–76. doi:10.1108/TR-06-2015-0030

Korber, S., & McNaughton, R. B. (2018). Resilience and entrepreneurship: A systematic literature review. *International Journal of Entrepreneurial Behaviour & Research*, *24*(7), 1129–1154. doi:10.1108/IJEBR-10-2016-0356

Kornai, J. (2006). The great transformation of Central and Eastern Europe: Success and disappointment. *Economics of Transition*, *14*(2), 207–244. doi:10.1111/j.1468-0351.2006.00252.x

Kossek, E. E., & Perrigino, M. B. (2016). Resilience: A Review Using a Grounded Integrated Occupational Approach. *The Academy of Management Annals*, *10*(1), 729–797. doi:10.5465/19416520.2016.1159878

Kotler, P., Armstrong, G., Harris, L. C., & Piercy, N. (2013). Principles of Marketing (6th Europe). Essex, UK: Pearson Education Limited.

Kovac, B. (1990). Politicko-Ekonomska Uloga Poduzetnistva. In D. Njavro & V. Franicevic (Eds.), *Poduzetnistvo: Teorija, Politika, Praksa* (pp. 69–93). Privredni Vjesnik.

Kovach, K. A. (1996). *Strategic Human Resource Management*. University Press of America. Inc.

KPMG. (2019). *European family business barometer. Succession top of mind as business families eye their legacy*. https://assets.kpmg/content/dam/kpmg/xx/pdf/2019/11/european-family-business-barometer.pdf

Kraiczy, N. D., Hack, A., & Kellermanns, F. W. (2015). What makes a family firm innovative? CEO risk-taking propensity and the organizational context of family firms. *Journal of Product Innovation Management*, *32*(3), 334–348. doi:10.1111/jpim.12203

Krappe, A., Goutas, L., & von Schlippe, A. (2011). The "family business brand": an enquiry into the construction of the image of family businesses. *Journal of Family Business Management*.

Kraus, S., Clauss, T., Breier, M., Gast, J., Zardini, A., & Tiberius, V. (2020). The economics of COVID-19: Initial empirical evidence on how family firms in five European countries cope with the corona crisis. *International Journal of Entrepreneurial Behaviour & Research*, *26*(5), 1067–1092. doi:10.1108/IJEBR-04-2020-0214

Krippendorf, J. (1982). Towards new tourism policies. *Tourism Management*, *3*(3), 135–148. doi:10.1016/0261-5177(82)90063-2

Kumar, P. (2020). Impact assessment of COVID-19: In tourism perspective. *Dogo Rangsang Research Journal, 10*(6).

La Porta, R., Lopez-De-Silanes, F., & Shleifer, A. (1999). Corporate Ownership Around the World. *The Journal of Finance*, *54*(2), 471–517. doi:10.1111/0022-1082.00115

Lazreg, B. M., Chkraborthy, N. R., Stieglitz, S., Potthoff, T., Ross, B., & Majchrazk, T. (2018). Social Media Analysis in Crisis Situations: Can Social Media be a Reliable Information Source for Emergency Management Services? In B. Andersson, B. Johansson, S. Carlsson, C. Barry, M. Lang, H. Linger, & C. Schneider (Chairs), *Designing Digitalization*. Symposium conducted at the meeting of Lund University, Lund, Sweden.

Le Breton-Miller, I., & Miller, D. (2006). Why Do Some Family Businesses Out-Compete? Governance, Long-Term Orientations, and Sustainable Capability. *Entrepreneurship Theory and Practice*, *30*(6), 731–746. doi:10.1111/j.1540-6520.2006.00147.x

Le Breton-Miller, I., & Miller, D. (2016). Family firms and practices of sustainability: A contingency view. *Journal of Family Business Strategy*, *7*(1), 26–33. doi:10.1016/j.jfbs.2015.09.001

Le Breton–Miller, I., Miller, D., & Steier, L. P. (2004). Toward an integrative model of effective FOB succession. *Entrepreneurship Theory and Practice*, *28*(4), 305–328. doi:10.1111/j.1540-6520.2004.00047.x

Lea, J. (1998). *What is a Family Business? More than You Think*. Available: http://www.bizjournals.com/ triangle/stories/1998/11/02/smallb3.html

Ledoux, J. (2015). *Anxious: Using the Brain to Understand and Treat Fear and Anxiety*. Penguin Books.

Lee, A. R., Son, S. M., & Kim, K. K. (2016). Information and communication technology overload and social networking service fatigue: A stress perspective. *Computers in Human Behavior*, *55*(1), 51–61. doi:10.1016/j.chb.2015.08.011

Lee, A. V., Vargo, J., & Seville, E. (2013). Developing a tool to measure and compare organizations' resilience. *Natural Hazards Review*, *14*(1), 29–41. doi:10.1061/(ASCE)NH.1527-6996.0000075

Lee, S., & Xiao, Q. (2011). An examination of the curvilinear relationship between capital intensity and firm performance for publicly traded US hotels and restaurants. *International Journal of Contemporary Hospitality Management*, *23*(6), 862–880. doi:10.1108/09596111111153510

Leeuwen, T. (2011). Multimodality and Multimodal Research. In E. Margolis & L. Pauwels (Eds.), *The SAGE Handbook of Visual Research Methods* (pp. 549–569). SAGE Publications Ltd. doi:10.4135/9781446268278.n28

Lehle, K. (1982). *Urlaub auf dem Bauernhof. Analyse und Perspektiven d. Nebenbetriebes "Tourismus in der Landwirtschaft"*. Schriftenreihe des AID.

Leiß, G., & Zehrer, A. (2018). Intergenerational communication in family firm succession. *Journal of Family Business Management*, *8*(1), 75–90.

Lengnick–Hall, C. A., & Beck, T. E. (2009). *Resilience capacity and strategic agility: prerequisites for thriving in a dynamic environment*. Working Paper, University of Texas.

Lengnick-Hall, C. A., & Beck, T. E. (2005). Adaptive fit versus robust transformation: How organizations respond to environmental change. *Journal of Management*, *31*(5), 738–757. doi:10.1177/0149206305279367

Lengnick-Hall, C. A., Beck, T. E., & Lengnick-Hall, M. L. (2011). Developing a capacity for organizational resilience through strategic human resource management. *Human Resource Management Review*, *21*(3), 243–255. doi:10.1016/j.hrmr.2010.07.001

Lepp, A., & Gibson, H. (2008). Sensation seeking and tourism: Tourist role, perception of risk and destination choice. *Tourism Management*, *29*(4), 740–750. doi:10.1016/j.tourman.2007.08.002

Letmathe, P., & Hill, M. (2006). Strukturumbrüche der Unternehmensnachfolge. *Zeitschrift Für Betriebswirtschaft*, *76*(11), 1113–1138.

Lew, A. A. (2014). Scale, change and resilience in community tourism planning. *Tourism Geographies*, *16*(1), 14–22. doi:10.1080/14616688.2013.864325

Libreros, M. (1998). *A conceptual framework for a tourism satellite account*. STD/NA(98)20. http://www.oecd.org/sdd/na/2681900.pdf

Linnenluecke, M. K. (2017). Resilience in Business and Management Research: A Review of Influential Publications and a Research Agenda. *International Journal of Management Reviews*, *19*(1), 4–30. doi:10.1111/ijmr.12076

Linnenluecke, M. K., & Griffiths, A. (2010). Corporate sustainability and organizational culture. *Journal of World Business*, *45*(4), 357–366. doi:10.1016/j.jwb.2009.08.006

Lippi, G., Sanchis-Gomar, F., & Henry, B. M. (2020). Coronavirus disease 2019 (COVID-19): The portrait of a perfect storm. *Annals of Translational Medicine*, *8*(7), 497–497. doi:10.21037/atm.2020.03.157 PMID:32395541

Lis, B., & Korchmar, S. (2013). Die digitale Mundpropaganda (Electronic Word-of-Mouth). In B. Lis & S. Korchmar (Ed.), Digitales Empfehlungsmarketing: Konzeption, Theorien und Determinanten zur Glaubwürdigkeit des Electronic Word-of-Mouth (EWOM) (pp. 11–20). Springer Fachmedien.

Liu, Z. (2003). Sustainable Tourism Development: A Critique. *Journal of Sustainable Tourism*, *11*(6), 459–475. doi:10.1080/09669580308667216

Livingstone, S., & Brake, D. R. (2010). On the Rapid Rise of Social Networking Sites: New Findings and Policy Implications. *Children & Society*, *24*(1), 75–83. doi:10.1111/j.1099-0860.2009.00243.x

Li, Y. (2004). Exploring Community Tourism in China: The Case of Nanshan Cultural Tourism Zone. *Journal of Sustainable Tourism*, *12*(3), 175–193. doi:10.1080/09669580408667232

Lovell, R., Wheeler, B. W., Higgins, S. L., Irvine, K. N., & Depledge, M. H. (2014). A Systematic Review of the Health and Well -Being Benefits of Biodiverse Environments. *Journal of Toxicology and Environmental Health. Part B, Critical Reviews*, *17*(1), 1–20. doi:10.1080/10937404.2013.856361 PMID:24597907

Lowrance, W. W. (1976). *Of Acceptable Risk.* Walliam Kaufmann.

Lu, A. C. C., Gursoy, D., & Lu, C. Y. R. (2016). Antecedents and outcomes of consumers' confusion in the online tourism domain. *Annals of Tourism Research, 57*, 76–93. doi:10.1016/j.annals.2015.11.020

Lubatkin, M. H., Simsek, Z., Ling, Y., & Veiga, J. F. (2006). Ambidexterity and performance in small-to medium-sized firms: The pivotal role of top management team behav- ioral integration. *Journal of Management, 32*(5), 646–672. doi:10.1177/0149206306290712

Lude, M., & Prügl, R. (2020). Experimental studies in family business research. *Journal of Family Business Strategy*, 100361.

Luthe, T., & Wyss, R. (2014). Assessing and planning resilience in tourism. *Tourism Management, 44*, 161–163. doi:10.1016/j.tourman.2014.03.011

Lu, W., & Stepchenkova, S. (2015). User-Generated Content as a Research Mode in Tourism and Hospitality Applications: Topics, Methods, and Software. *Journal of Hospitality Marketing & Management, 24*(2), 119–154. doi:10.1080/19368623.2014.907758

Magnini, V. P. (2017). Designing Tourism Services in an era of information overload. In D. R. Fesenmaier & Z. Xiang (Eds.), Design science in tourism (pp. 161-172). Springer. doi:10.1007/978-3-319-42773-7_11

Magnini, V. P. (2006). Photo-elicitation as a tool to alleviate international marketing mistakes. *Journal of Global Competitiveness, 14*(2), 75–83.

Magnini, V. P. (2015). *Surprise!: The Secret to Customer Loyalty in the Service Sector.* Business Expert Press.

Magnini, V. P., Crotts, J. C., & Zehrer, A. (2011). Understanding customer delight: An application of travel blog analysis. *Journal of Travel Research, 50*(5), 535–545. doi:10.1177/0047287510379162

Magnini, V. P., & Dallinger, I. (2018). Consumer information overload and the need to prompt script deviations. *Journal of Quality Assurance in Hospitality & Tourism, 19*(3), 285–297. doi:10.1080/1528008X.2016.1230038

Magnini, V. P., Ford, J. B., & LaTour, M. S. (2012). The role of qualitative methods in tourism QOL research: A critique and future agenda. In *Handbook of Tourism and Quality-of-Life Research* (pp. 51–63). Springer. doi:10.1007/978-94-007-2288-0_4

Mallak, L. (1998). Putting organizational resilience to work. *Industrial Management*, 8-13.

Mandl, I. (2008). *Overview of Family Business Relevant Issues.* Final Report. KMU Forschung Austria.

Manente, M., Minghetti, V., & Mingotto, E. (2014). *Responsible tourism and CSR: Assessment systems for sustainable development of SMEs in tourism. CSR, Sustainability, Ethics & Governance.* Springer International Publishing., doi:10.1007/978-3-319-06308-9

Manzhynski, S., & Figge, F. (2020). Coopetition for sustainability: Between organizational benefit and societal good. *Business Strategy and the Environment*, *29*(3), 827–837. doi:10.1002/bse.2400

Marana, P., Labaka, L., & Sarriegi, J. M. (2018). A framework for public-private-people partnerships in the city resilience-building process. *Safety Science*, *110*, 39–50. doi:10.1016/j.ssci.2017.12.011

Marcher, A., Kofler, I., Innerhofer, E., & Pechlaner, H. (2020). Perceptions of and interactions between locals, migrants, and tourists in South Tyrol. *Tourism Geographies*, 1–17. Advance online publication. doi:10.1080/14616688.2020.1795709

Marić, K., Samardžić, B., & Protrka, N. (2017). Analiza turističkih pokazatelja i sigurnost u turizmu Republike Hrvatske te mogućnosti jačanja konkurentnosti hrvatskog turizma nakon ulaska u Europsku uniju. *Obrazovanje za Poduzetništvo-E4E: Znanstveno Stručni Časopis o Obrazovanju za Poduzetništvo*, *7*(1), 301-318.

Märk, S., Kraus, S., & Peters, M. (2010). Der Einfluss der Familie auf den Unternehmer: Eine qualitativ-empirische Untersuchung von Familienunternehmen auf Basis der Stewardship-Theorie. *ZfKE–Zeitschrift für KMU und Entrepreneurship*, *58*(1), 31–59.

Marques Santos, A., Madrid, C., Haegeman, K., & Rainoldi, A. (2020). *Behavioural changes in tourism in times of Covid-19*. Publications Office of the European Union. https://publications.jrc.ec.europa.eu/repository/bitstream/JRC121262/report_covid_tour_emp_final.pdf doi:10.2760/00411

Marsden, T. (1999). Rural Futures: The Consumption Countryside and its Regulation. *Sociologia Ruralis*, *39*(4), 501–520. doi:10.1111/1467-9523.00121

Martínez-Ros, E., & Orfila-Sintes, F. (2012). Training plans, manager's characteristics and innovation in the accommodation industry. *International Journal of Hospitality Management*, *31*(3), 686–694. doi:10.1016/j.ijhm.2011.09.004

Martin, R. (2012). Regional economic resilience, hysteresis and recessionary shocks. *Journal of Economic Geography*, *12*(1), 1–32. doi:10.1093/jeg/lbr019

Martin, R., & Sunley, P. (2015). On the notion of regional economic resilience: Conceptualization and explanation. *Journal of Economic Geography*, *15*(1), 1–42. doi:10.1093/jeg/lbu015

Martín-Santana, J. D., Cabrera-Suárez, M. K., & La Déniz-Déniz, M. C. (2020). Familiness, social capital and market orientation in the family firm. *European Journal of Marketing*, *54*(7), 1731–1760. doi:10.1108/EJM-04-2018-0274

Mathew, P. V., & Sreejesh, S. (2017). Impact of responsible tourism on destination sustainability and quality of life of community in tourism destinations. *Journal of Hospitality and Tourism Management*, *31*, 83–89. doi:10.1016/j.jhtm.2016.10.001

Matteucci, X., & Gnoth, J. (2017). Elaborating on grounded theory in tourism research. *Annals of Tourism Research*, *65*, 49–59.

Mayring, P. (2001). Kombination und Integration qualitativer und quantitativer Analyse. *Forum Qualitative Social Research*, *2*(1). http://www.qualitative-research.net/index.php/fqs/rt/printerFriendly/967/2110

Mayring, P. (2015). *Qualitative Inhaltsanalyse- Grundlagen und Techniken* (12th ed.). Beltz.

Mayring, P. (2015). *Qualitative Inhaltsanalyse: Grundlagen und Techniken* (12th ed.). Beltz Pädagogik.

McEachern, C. (1992). Farmers and Conservation: Conflict and Accommodation in Farming Politics. *Journal of Rural Studies*, *8*(2), 159–171. doi:10.1016/0743-0167(92)90074-G

McKercher, B., & Hui, E. L. L. (2004). Terrorism, economic uncertainty and outbound travel from Hong Kong. In C. M. Hall, D. J. Timothy, & D. T. Duval (Eds.), *Safety and Security in Tourism: Relationships, Management and Marketing* (pp. 99–116). Haworth Press.

McKinsey Global Institute. (2020a). *How European businesses can position themselves for recovery*. https://www.mckinsey.com/industries/public-and-social-sector/our-insights/how-european-businesses-can-position-themselves-for-recovery

McKinsey Global Institute. (2020b). *Leadership in a crisis: Responding to the coronavirus outbreak and future challenges*. https://www.mckinsey.com/business-functions/organization/our-insights/leadership-in-a-crisis-responding-to-the-coronavirus-outbreak-and-future-challenges

McKinsey Global Institute. (2020c). *Reimagining the $9 trillion tourism economy—what will it take?* https://www.mckinsey.com/industries/travel-logistics-and-transport-infrastructure/our-insights/reimagining-the-9-trillion-tourism-economy-what-will-it-take

McKinsey Global Institute. (2020d). *Meeting the future: Dynamic risk management for uncertain times*. https://www.mckinsey.com/~/media/McKinsey/Business%20Functions/Risk/Our%20Insights/Meeting%20the%20future%20Dynamic%20risk%20management%20for%20uncertain%20times/Meeting-the-future-Dynamic-risk-management-for-uncertain-times.pdf?shouldIndex=false

McManus, S., Seville, E., Vargo, J., & Brunsdon, D. (2008). Facilitated process for improving organizational resilience. *Natural Hazards Review*, *9*(2), 81–90.

McWilliams, A., & Siegel, D. (2001). Corporate Social Responsibility: A Theory of the Firm Perspective. *Academy of Management Review*, *26*(1), 117–127. doi:10.5465/amr.2001.4011987

Mehmetoglu, M., & Altinay, L. (2006). Examination of grounded theory analysis with an application to hospitality research. *International Journal of Hospitality Management*, *25*(1), 12–33.

Mehr Grün im weissen Skiparadies - PET-Recycling Schweiz. (2019, February 24). *Petrecycling.Ch*. https://www.petrecycling.ch/de/entdecken/detail/mehr-gruen-im-weissen-skiparadies

Memili, E., Fang, H., Koç, B., Yildirim-Öktem, Ö., & Sonmez, S. (2018). Sustainability practices of family firms: The interplay between family ownership and long-term orientation. *Journal of Sustainable Tourism*, *26*(1), 9–28. doi:10.1080/09669582.2017.1308371

Memili, E., Patel, P. C., Koç, B., & Yazıcıoğlu, İ. (2020). The interplay between socioemotional wealth and family firm psychological capital in influencing firm performance in hospitality and tourism. *Tourism Management Perspectives*, *34*, 100651. doi:10.1016/j.tmp.2020.100651

Memili, E., Welsh, D. H., & Luthans, F. (2013). Going beyond research on goal setting: A proposed role for organizational psychological capital of family firms. *Entrepreneurship Theory and Practice*, *37*(6), 1289–1296. doi:10.1111/etap.12066

Merten, K. (1995). *Inhaltsanalyse: Einführung in Theorie, Methode und Praxis* (2nd ed.). Springer Fachmedien.

Meyer, A. D. (1982). Adapting to Environmental Jolts. *Administrative Science Quarterly*, *27*(4), 515–537. doi:10.2307/2392528 PMID:10257768

Meyer, W., Reisenzein, R., & Schützwohl, A. (1997). Towards a process analysis of emotions: The case of surprise. *Motivation and Emotion*, *21*(3), 251–274. doi:10.1023/A:1024422330338

Mihalic, T. (2016). Sustainable-responsible tourism discourse – Towards 'responsustable' tourism. *Journal of Cleaner Production*, *111*, 461–470. doi:10.1016/j.jclepro.2014.12.062

Miles, M. B., & Huberman, A. M. (1994). Qualitative data analysis: An expanded sourcebook. *Sage (Atlanta, Ga.)*.

Miller, D., & Le Breton–Miller, I. (2003). Challenge versus advantage in family business. *Strategic Organization*, *1*(1), 127–134. doi:10.1177/1476127003001001222

Mitchell, R., Morse, E., & Sharma, P. (2003). The transacting cognitions of nonfamily employees in the family businesses setting. *Journal of Business Venturing*, *18*(4), 533–551. doi:10.1016/S0883-9026(03)00059-4

Molly, V., Laveren, E., & Jorissen, A. (2012). Intergenerational Differences in Family Firms: Impact on Capital Structure and Growth Behavior. *Entrepreneurship Theory and Practice*, *36*(4), 703–725. doi:10.1111/j.1540-6520.2010.00429.x

Mom, T. J., Chang, Y. Y., Cholakova, M., & Jansen, J. J. (2019). A multilevel integrated framework of firm HR practices, individual ambidexterity, and organizational ambidexterity. *Journal of Management*, *45*(7), 3009–3034. doi:10.1177/0149206318776775

Mooney, S. K., Harris, C., & Ryan, I. (2016). Long hospitality careers–a contradiction in terms? *International Journal of Contemporary Hospitality Management*.

Morgan, T., Anokhin, S., Ofstein, L., & Friske, W. (2020). SME response to major exogenous shocks: The bright and dark sides of business model pivoting. *International Small Business Journal: Researching Entrepreneurship*, *38*(5), 369–379. doi:10.1177/0266242620936590

Morrison, A., & Teixeira, R. (2004). Small business performance: A tourism sector focus. *Journal of Small Business and Enterprise Development*, *11*(2), 166–173. doi:10.1108/14626000410537100

Müller, H. (2006). Qualitätsmanagement als Triebfeder von Innovationsprozessen und ihre Grenzen. In B. Pikkemaat, M. Peters, & K. Weiermair (Eds.), *Innovationen im Tourismus* (pp. 109–117). Erich Schmidt Verlag.

Mullins, J. W., & Forlani, D. (2005). Missing the boat or sinking the boat: A study of new venture decision making. *Journal of Business Venturing*, *20*(1), 47–69. doi:10.1016/j.jbusvent.2003.08.001

Muñoz-Bullon, F., Sanchez-Bueno, M. J., & Suárez-González, I. (2018). Diversification decisions among family firms: The role of family involvement and generational stage. *BRQ Business Research Quarterly*, *21*(1), 39–52. doi:10.1016/j.brq.2017.11.001

Murray, D. R., & Schaller, M. (2016). The behavioral immune system: Implications for social cognition, social interaction, and social influence. In J. M. Olson & M. P. Zanna (Eds.), Advances in Experimental Social Psychology (Vol. 53, pp. 75–129). Academic Press Inc.

Muskat, B., & Zehrer, A. (2017). A power perspective on knowledge transfer in internal succession of small family businesses. *Journal of Small Business and Entrepreneurship*, *29*(5), 333–350. https://doi.org/10.1080/08276331.2017.1345208

Mzid, I., Khachlouf, N., & Soparnot, R. (2019). How does family capital influence the resilience of family firms? *Journal of International Entrepreneurship*, *17*(2), 249–277. doi:10.100710843-018-0226-7

Nadegger, M., Bosio, B., Wegerer, P., Haselwanter, S. & Untersteiner, J. (2020). #mitabstandnah - Eine Hashtag-Analyse der Tiroler Tourismusverbände in Corona-Zeiten. *Tourismuswissen Quarterly, 21*, 222-227.

Niewiadomski, P. (2020). COVID-19: From temporary de-globalisation to a re-discovery of tourism? *Tourism Geographies*, *22*(3), 651–656. doi:10.1080/14616688.2020.1757749

NÖ Energie- und Umweltagentur. (2019). *Niederösterreichische Heurigenkultur.*

Nordqvist, M., & Jack, S. (2020). *Family businesses can tell us a lot about what makes a company resilient in times of crisis.* Stockholm School of Economics Institute for Research. https://www.hhs.se/contentassets/12721b1ec98e41f3b9ff347d02b18b3e/a40.pdf

Nordqvist, M., & Melin, L. (2010). Entrepreneurial families and family firms. *Entrepreneurship and Regional Development*, *22*(3), 1–29.

Nordqvist, M., Sharma, P., & Chirico, F. (2014). Family firm heterogeneity and governance: A configuration approach. *Journal of Small Business Management*, *52*(2), 192–209. doi:10.1111/jsbm.12096

OECD. (2020a). *Coronavirus (COVID-19): SME Policy Responses.* https://read.oecd-ilibrary.org/view/?ref=119_119680-di6h3qgi4x&title=Covid-19_SME_Policy_Responses

OECD. (2020b). *OECD Tourism Trends and Policies 2020.* OECD Publishing., doi:10.1787/6b47b985-

Ogundimu, E. O., Altman, D. G., & Collins, G. S. (2016). Adequate sample size for developing prediction models is not simply related to events per variable. *Journal of Clinical Epidemiology*, *76*, 175–182.

Okumus, F., Altinay, M., & Arasli, H. (2005). The impact of Turkey's economic crisis of February 2001 on the tourism industry in Northern Cyprus. *Tourism Management*, *26*(1), 95–104. doi:10.1016/j.tourman.2003.08.013

Okumus, F., & Karamustafa, K. (2005). Impact of an economic crisis: Evidence from Turkey. *Annals of Tourism Research*, *32*(4), 942–961. doi:10.1016/j.annals.2005.04.001 PMID:32572284

Oliveira, A., & Huertas, A. (2019). How do destinations use twitter to recover their images after a terrorist attack? *Journal of Destination Marketing & Management*, *12*, 46–54. doi:10.1016/j.jdmm.2019.03.002

Omar, A. R. C. (2020). The impact of Covid-19 Movement Control Order on SMEs' businesses and survival strategies. *Geografia OnlineTM Malaysian Journal of Society and Space*, *16*(2), 139–150. doi:10.17576/geo-2020-1602-11

Oppermann, M. (1996). Rural Tourism in Southern Germany. *Annals of Tourism Research*, *23*(1), 86–102. doi:10.1016/0160-7383(95)00021-6

Oppermann, M. (1998). Farm Tourism in New Zealand. In R. Butler, C. M. Hall, & J. Jenkins (Eds.), *Tourism and recreation in rural areas* (pp. 225–233). Wiley.

Orchiston, C. (2013). Tourism business preparedness, resilience and disaster planning in a region of high seismic risk: The case of the Southern Alps, New Zealand. *Current Issues in Tourism*, *16*(5), 477–494. doi:10.1080/13683500.2012.741115

Orchiston, C., Prayag, G., & Brown, C. (2016). Organizational resilience in the tourism sector. *Annals of Tourism Research*, *56*, 145–148. doi:10.1016/j.annals.2015.11.002

ORF. (2020, May 11). *Buschenschanken bis Jahresende geöffnet*. https://wien.orf.at/stories/3048113/

Orsini, K., & Ostojić, V. (2018). *Croatia's Tourism Industry: Beyond the Sun and Sea*. Economic Brief 036. Luxembourg: Publications Office of the European Union. https://ec.europa.eu/info/sites/info/files/economy-finance/eb036_en.pdf

Ossenbrink, J., Hoppmann, J., & Hoffmann, V. H. (2019). Hybrid ambidexterity: How the environment shapes incumbents' use of structural and contextual approaches. In Organization Science (Vol. 30). doi:10.1287/orsc.2019.1286

Österreich Wein. (2020). *Weinbaugebiete Burgenland*. https://www.oesterreichwein.at/unser-wein/weinbaugebiete/burgenland

Ozili, P. K., & Arun, T. (2020). *Spillover of COVID-19: impact on the Global Economy*. https://mpra.ub.uni-muenchen.de/99850/

Palliam, R., Cader, H. A., & Chiemeke, C. (2011). Succession Issues among Family Entrepreneurs in Countries of the Gulf. *International Journal of Business Administration*, *2*(2), 25–34.

Panksepp, J., Fuchs, T., & Iacobucci, P. (2011). The basic neuroscience of emotional experiences in mammals: The case of subcortical FEAR circuitry and implications for clinical anxiety. *Applied Animal Behaviour Science*, *129*(1), 1–17. doi:10.1016/j.applanim.2010.09.014

Pappas, N., & Papatheodorou, A. (2017). Tourism and the refugee crisis in Greece: Perceptions and decision-making of accommodation providers. *Tourism Management*, *63*, 31–41. doi:10.1016/j.tourman.2017.06.005

Park, C. S. (2019). Does Too Much News on Social Media Discourage News Seeking? Mediating Role of News Efficacy Between Perceived News Overload and News Avoidance on Social Media. *Social Media + Society*, *5*(3), 1–12. doi:10.1177/2056305119872956

Park, J.-Y., & Jang, S. S. (2013). Confused by too many choices? Choice overload in tourism. *Tourism Management*, *35*, 1–12. doi:10.1016/j.tourman.2012.05.004

Paul, J. J., Winter, M., Miller, N. J., & Fitzgerald, M. A. (2003). Cross-institutional norms for timing and sequencing and the use of adjustment strategies in families affiliated with family-owned businesses. *Marriage & Family Review*, *35*(1-2), 167–191. doi:10.1300/J002v35n01_10

Pearce, P. L. (1990). Farm Tourism in New Zealand: A Social Situation Analysis. *Annals of Tourism Research*, *17*(3), 337–352. doi:10.1016/0160-7383(90)90002-9

Pechlaner, H., Zacher, D., Eckert, C., & Petersik, L. (2019). Joint responsibility and understanding of resilience from a DMO perspective–an analysis of different situations in Bavarian tourism destinations. *International Journal of Tourism Cities, 5*(2), 146-168.

Pechlaner, H., & Innerhofer, E. (2018). Linking destinations and resilience – challenges and perspectives. In E. Innerhofer, M. Fontanari, & H. Pechlaner (Eds.), *Destination Resilience: Challenges and Opportunities for Destination Management and Governance* (1st ed., pp. 3–13). Routledge. doi:10.4324/9780203701904-1

Pechlaner, H., Kozak, M., Volgger, M., Zehrer, A., Raich, F., Siller, H., & Tschiderer, F. (2014). Leadership networks in destinations. *Tourism Review*.

Pechlaner, H., & Volgger, M. (2015). Can Tourism Qualify for Interdisciplinary Research? A European View. In H. Pechlaner & E. Smeral (Eds.), *Tourism and Leisure:Current issues and perspectives of development* (pp. 3–21). Springer Fachmedien. doi:10.1007/978-3-658-06660-4_1

Peebles, M. S. (1995). Cultivating the tourist/farm tourism. *Tour. J. Tour. Soc.*, 86.

Perales, R. M. Y. (2002). Rural Tourism in Spain. *Annals of Tourism Research*, *29*(4), 1104–1110.

Perpiña, L., Camprubí, R., & Prats, L. (2017). Destination Image Versus Risk Perception. *Journal of Hospitality & Tourism Research*.

Peters, M., & Buhalis, D. (2013). SMEs in tourism destinations. In C. Costa, E. Panyik, & D. Buhalis (Eds.), Aspects of tourism: Vol. 60. Trends in European tourism planning and organisation (pp. 93–101). Bristol: Channel View Publications. doi:10.21832/9781845414122-011

Peters, M. (2005). Succession in Tourism Family Business: The motivation of succeeding Family Members. *Tourism Review*, *60*(4), 12–18.

Peters, M., & Buhalis, D. (2004). Family hotel businesses: Strategic planning and the need for education and training. *Education + Training*, *46*(8/9), 406–415. doi:10.1108/00400910410569524

Peters, M., & Kallmuenzer, A. (2015). Entrepreneurial orientation in family firms: The case of the hospitality industry. *Current Issues in Tourism*, *21*(1), 21–40. doi:10.1080/13683500.2015.1053849

Peters, M., & Pikkemaat, B. (2006). Innovationsforschung im Tourismus: ein Ausblick. In B. Pikkemaat, M. Peters, & K. Weiermair (Eds.), *Innovationen im Tourismus. Wettbewerbsvorteil durch neue Ideen und Angebote* (pp. 303–306). Erich Schmidt Verlag.

Peters, M., Raich, M., Märk, S., & Pichler, S. (2012). The role of commitment in the succession of hospitality businesses. *Tourism Review*, *67*(2), 45–60. doi:10.1108/16605371211236169

Peters, T. (2010). *The Little Big Things: 163 Ways to Pursue Excellence*. Harper Business.

Petruzzelli, M. A., & Savino, T. (2015). Reinterpreting tradition to innovate: The case of Italian haute cuisine. *Industry and Innovation*, *22*(8), 677–702. doi:10.1080/13662716.2015.1122512

Pfefferbaum, B. J., Reissman, D. B., Pfefferbaum, R. L., Klomp, R. W., & Gurwitch, R. H. (2007). Building resilience to mass trauma events. In L. S. Doll, S. E. Bonzo, J. A. Mercy, & D. A. Sleet (Eds.), *Handbook of Injury and Violence Prevention* (pp. 347–358). Springer. doi:10.1007/978-0-387-29457-5_19

Pfefferbaum, R. L., Pfefferbaum, B., Van Horn, R. L., Klomp, R. W., Norris, F. H., & Reissman, D. B. (2013). The communities advancing resilience toolkit (CART): An intervention to build community resilience to disasters. *Journal of Public Health Management and Practice*, *19*(3), 250–258. doi:10.1097/PHH.0b013e318268aed8 PMID:23524306

Pforr, C., & Hosie, P. J. (2008). Crisis Management in Tourism: Preparing for Recovery. *Journal of Travel & Tourism Marketing*, *23*(2–4), 249–264.

Pfosser, R., & Seper, L. (2020) *Die österreichische Resilienzstrategie kommt*. http://resilienzstrategie.at/

Phillip, S., Hunter, C., & Blackstock, K. (2010). A typology for defining agritourism. *Tourism Management*, *31*(6), 754–758. doi:10.1016/j.tourman.2009.08.001

Piaget, J. (1926). *The Language and Thought of the Child*. Harcourt, Brace.

Pikkemaat, B., Peters, M., & Chan, C.-S. (2018). Needs, drivers and barriers of innovation: The case of an alpine community-model destination. *Tourism Management Perspectives*, *25*, 53–63. doi:10.1016/j.tmp.2017.11.004

Pikkemaat, B., & Zehrer, A. (2016). Innovation and service experiences in small tourism family firms. *International Journal of Culture, Tourism and Hospitality Research*, *10*(4), 343–360.

Pillmayer, M., & Scherle, N. (2018). Krisen und Krisenmanagement im Tourismus - Eine konzeptionelle Einführung. In S. Hahn & Z. Neuss (Eds.), Krisenkommunikation in Tourismusorganisationen (pp. 3-18). Springer Fachmedien.

Pinz, A., Roudyani, N., & Thaler, J. (2018). Public–private partnerships as instruments to achieve sustainability-related objectives: The state of the art and a research agenda. *Public Management Review*, *20*(1), 1–22. doi:10.1080/14719037.2017.1293143

Pittino, Chirico, F., Henssen, B., & Broekaert, W. (2019). Does Increased Generational Involvement Foster Business Growth? The Moderating Roles of Family Involvement in Ownership and Management. *European Management Review*, *17*(3), 785–801. Advance online publication. doi:10.1111/emre.12366

Pizam, A., & Smith, G. (2000). Tourism and Terrorism: A quantitative Analysis of major Terrorist Acts and their Impact on Tourism Destinations. *Tourism Economics*, *6*(2), 123–138.

Pizam, A., & Smith, G. (2000). Tourism and terrorism: A quantitative analysis of major terrorist acts and their impact on tourism destinations. *Tourism Economics*, *6*(2), 123–138. doi:10.5367/000000000101297523

Pizam, A., Tarlow, P. E., & Bloom, J. (1997). Making Tourists feel safe: Whose Responsibility is it? *Journal of Travel Research*, *36*(1), 23–28.

Plog, S. C. (1974). Why Destination Areas Rise and Fall in Popularity. *The Cornell Hotel and Restaurant Administration Quarterly*, *14*(4), 55–58. doi:10.1177/001088047401400409

Ponomarov, S. Y., & Holcomb, M. C. (2009). Understanding the concept of supply chain resilience. *International Journal of Logistics Management*, *20*(1), 124–143. doi:10.1108/09574090910954873

Poza, E. (2007). *Family business* (2nd ed.). Mason.

Pratt, M. G., Kaplan, S., & Whittington, R. (2020). Editorial essay: The tumult over transparency: Decoupling transparency from replication in establishing trustworthy qualitative research. *Administrative Science Quarterly*, *65*(1), 1–19. doi:10.1177/0001839219887663

Prayag, G. (2018). Symbiotic relationship or not? Understanding resilience and crisis management in tourism. *Tourism Management Perspectives*, *25*(January), 133–135. doi:10.1016/j.tmp.2017.11.012

Prayag, G. (2020). Time for Reset? Covid-19 and Tourism Resilience. *Tourism Review International*, *24*(2), 179–184. doi:10.3727/154427220X15926147793595

Prayag, G., Spector, S., Orchiston, C., & Chowdhury, M. (2020). Psychological resilience, organizational resilience and life satisfaction in tourism firms: Insights from the Canterbury earthquakes. *Current Issues in Tourism*, *23*(10), 1216–1233. doi:10.1080/13683500.2019.1607832

Presas, P., Guia, J., & Munos, D. (2014). Customer's Perception of Familiness in Travel Experiences. *Journal of Travel & Tourism Marketing*, *31*(2), 147–161.

Presas, P., Muñoz, D., & Guia, J. (2011). Branding familiness in tourism family firms. *Journal of Brand Management*, *18*(4-5), 274–284. doi:10.1057/bm.2010.41

Prideaux, B. (2003). The need to use disaster planning frameworks to respond to major tourism disasters: Analysis of Australia's response to tourism disasters in 2001. *Journal of Travel & Tourism Marketing*, *15*(4), 281–298. doi:10.1300/J073v15n04_04

PWC Family Business Survey. (2016). *The 'missing middle': Bridging the strategy gap in family firms.* www.pwc.com/fambizsurvey2016

Quendler, E. (2015). Organisational resilience: building business value in a changing world. *DIEM: Dubrovnik International Economic Meeting*, *2*(1), 534-553.

Quendler, E., Magnini, V. P., & Driouech, N. (2021). *Strategic Positioning of the Farm Holiday's Post-Pandemic Competitive Advantages: Fresh Air and Sunlight*. Academic Press.

Quendler, E. (2019). The position of the farm holiday in Austrian tourism. *Open Agriculture*, *4*(1), 697–711. doi:10.1515/opag-2019-0069

Racherla, P., & Hu, C. (2009). A Framework for Knowledge-Based Crisis Management in the Hospitality and Tourism Industry. *Cornell Hospitality Quarterly*, *50*(4), 561–577. doi:10.1177/1938965509341633

Ramadani, V., & Hoy, F. (2015). Context and Uniqueness of Family Businesses. In L. P. Dana & V. Ramadani (Eds.), *Family Businesses in Transition Economies: Management, Succession and Internationalization* (pp. 9–37). Springer.

Rau, S. B., Schneider-Siebke, V., & Günther, C. (2019). Family Firm Values Explaining Family Firm Heterogeneity. *Family Business Review*, *32*(2), 195–215. doi:10.1177/0894486519846670

Ravelli, L. J., & van Leeuwen, T. (2018). Modality in the digital age. *Visual Communication*, *17*(3), 277–297. doi:10.1177/1470357218764436

Rees, C. S., Breen, L. J., Cusack, L., & Hegney, D. (2015). Understanding individual resilience in the workplace: The international collaboration of workforce resilience model. *Frontiers in Psychology*, *6*, 73. doi:10.3389/fpsyg.2015.00073 PMID:25698999

Reisenzein, R. (2000). The subjective experience of surprise. In H. Bless & J. Forgas (Eds.), *The Message Within* (pp. 262–279). Psychology Press.

Resilient Europe. (2020). *Improving city resilience*. https://urbact.eu/resilient-europe

Richter, F. (2020). *Facebook Reaches 3.2 Billion People Each Month.* Statista. https://www.statista.com/chart/2183/facebooks-mobile-users/?utm_source=Statista+Global&utm_campaign=95f88a88d2-All_InfographTicker_daily_COM_PM_KW27_2020_Mo&utm_medium=email&utm_term=0_afecd219f5-95f88a88d2-308701553

Richter, L. K., & Waugh, W. L. Jr. (1986). Terrorism and tourism as logical companions. *Tourism Management, 7*(4), 230–238. doi:10.1016/0261-5177(86)90033-6

Ritchie, B. W. (2004). Chaos, crises and disasters: A strategic approach to crisis management in the tourism industry. *Tourism Management, 25*(6), 669–683.

Ritchie, B. W. (2009). *Crisis and disaster management for tourism.* Channel View Publications. doi:10.21832/9781845411077

Ritchie, B. W., & Jiang, Y. (2019). A review of research on tourism risk, crisis and disaster management: Launching the annals of tourism research curated collection on tourism risk, crisis and disaster management. *Annals of Tourism Research, 79*, 79. doi:10.1016/j.annals.2019.102812

Ritchie, J. R. B., & Crouch, G. I. (2011). *The competitive destination: A sustainable tourism perspective.* CABI Pub.

Ritchie, J., & Spencer, L. (1994). Qualitative data analysis for applied policy research. In A. Bryman & R. Burgees (Eds.), *Analyzing Qualitative Data* (pp. 173–194). Routledge. doi:10.4324/9780203413081_chapter_9

Rittichainuwat, B. N. (2013). Tourists' Perceived Risks Toward Overt Safety Measures. *Journal of Hospitality & Tourism Research (Washington, D.C.), 37*(2), 199–216. doi:10.1177/1096348011425494

Rittichainuwat, B. N., & Chakraborty, G. (2009). Perceived travel risks regarding terrorism and disease: The case of Thailand. *Tourism Management, 30*(3), 410–418. doi:10.1016/j.tourman.2008.08.001

Rittichainuwat, B. N., & Chakraborty, G. (2012). Perceptions of importance and what safety is enough. *Journal of Business Research, 65*(1), 42–50. doi:10.1016/j.jbusres.2011.07.013 PMID:32287524

Rockefeller Foundation. (2015). *Challenge Eligibility and Criteria. Who can submit an application?* http://www.100resilientcities.org/challenge-eligibility-criteria

Rodenstock, B. (2015). Wie sichere ich Familie, Unternehmen und Eigentum für die Zukunft? In A. Nagl (Ed.), *Wie regele ich meine Nachfolge? Leitfaden für Familienunternehmen* (2nd ed., pp. 37–58). Springer Gabler.

Roetzel, P. G. (2019). Information overload in the information age: A review of the literature from business administration, business psychology, and related disciplines with a bibliometric approach and framework development. *Business Research, 12*(2), 479–522. doi:10.100740685-018-0069-z

Rohrmann, B. (2008). Risk perception, risk attitude, risk communication, risk management. *International Emergency Management Society (TIEMS) Annual Conference*.

Roman, M., Roman, M., & Prus, P. (2020). Innovations in Agritourism: Evidence from a Region in Poland. *Sustainability*, *12*(12), 4858. doi:10.3390u12124858

Rose, G. (2016). *Visual methodologies: An introduction to researching with visual materials* (4th ed.). Sage.

Roshan, M., Warren, M., & Carr, R. (2016). Understanding the use of social media by organisations for crisis communication. *Computers in Human Behavior*, *63*, 350–361. doi:10.1016/j.chb.2016.05.016

Ruiz-Ballesteros, E. (2011). Social-ecological resilience and community-based tourism: An approach from Agua Blanca, Ecuador. *Tourism Management*, *32*(3), 655–666. doi:10.1016/j.tourman.2010.05.021

Rumbo, J. (2002). Consumer resistance in a world of advertising clutter: The case of adbusters. *Psychology and Marketing*, *19*(2), 127–148. doi:10.1002/mar.10006

Russell, R., & Faulkner, B. (2004). Entrepreneurship, chaos and the tourism area lifecycle. *Annals of Tourism Research*, *31*(3), 556–579. doi:10.1016/j.annals.2004.01.008

Ryan, C., Page, S., & Aicken, M. (Eds.). (2005). *Taking tourism to the limits: issues, concepts and managerial perspectives*. Elsevier Ltd.

Saarinen, J., & Gill, A. M. (Eds.). (2018). *Resilient destinations and tourism: Governance strategies in the transition towards sustainability in tourism*. Routledge. doi:10.4324/9781315162157

Sabahi, S., & Parast, M. M. (2020). Firm innovation and supply chain resilience: A dynamic capability perspective. *International Journal of Logistics Research and Applications*, *23*(3), 254–269. doi:10.1080/13675567.2019.1683522

Saiz-Álvarez, J. M., Palma-Ruiz, J. M., & Leitão, J. (2019). Introduction to Entrepreneurship and Family Business Vitality. In J. M. Saiz-Álvarez, J. M. Palma-Ruiz, & J. Leitão (Eds.), *Entrepreneurship and Family Business Vitality. Surviving and Flourishing in the long term* (pp. 1–5). Springer.

Saldaña, J. (2015). The coding manual for qualitative researchers. *Sage (Atlanta, Ga.)*.

SalvatoC.SargiacomoM.AmoreM. D.MinichilliA. (2020). Natural disasters as a source of entrepreneurial opportunity: family business resilience after an earthquake. https://ssrn.com/abstract=3693817

Sano, K., & Sano, H. (2019). The effect of different crisis communication channels. *Annals of Tourism Research*, *79*, 79. doi:10.1016/j.annals.2019.102804

Santana, G. (2004). Crisis Management and Tourism. *Journal of Travel & Tourism Marketing*, *15*(4), 299–321. doi:10.1300/J073v15n04_05

Santucci, F. M. (2013). Agritourism for Rural Development in Italy, Evolution, Situation and Perspectives. *British Journal of Economics. Management & Trade*, *3*(3), 186–200.

Saraniemi, S., & Komppula, R. (2019). The development of a destination brand identity: A story of stakeholder collaboration. *Current Issues in Tourism*, *22*(9), 1116–1132. doi:10.1080/13683500.2017.1369496

Sarman, I., Scagnolari, S., & Maggi, R. (2016). Acceptance of Life-Threatening Hazards among Young Tourists: A Stated Choice Experiment. *Journal of Travel Research*, *55*(8), 979–992. doi:10.1177/0047287515612595

Schmidt, E., Groeneveld, S., & Van de Walle, S. (2017). A change management perspective on public sector cutback management: Towards a framework for analysis. *Public Management Review*, *19*(10), 1538–1555. doi:10.1080/14719037.2017.1296488

Schönfelder, B. (2005). *The Impact of the War 1991: 1995 on the Croatian Economy: A Contribution to the Analysis of War Economies (No. 2005/14)*. Freiberger Arbeitspapiere.

Schroder, H. M., Driver, M. J., & Streufert, S. (1967). *Human information processing: Individuals and groups functioning in complex social situations*. Holt, Rinehart and Winston.

Schulze, G., & Mori, S. (1993). Increases in environmental entropy demand evolution. *Acta Biotheoretica*, *41*, 149–164. doi:10.1007/BF00712163

Schulze, W., Lubatkin, M. H., & Dino, R. N. (2003). Toward a theory of agency and altruism in family firms. *Journal of Business Venturing*, *18*(4), 473–490. doi:10.1016/S0883-9026(03)00054-5

Schumpeter, J. A. (1965). *The theory of economic development: An inquiry into profit, credit, interest, and the business cycle*. Oxford University Press.

Schwaiger, K., & Zehrer, A. (2020). *Unternehmensresilienz von Familienunternehmen in Krisenzeiten*. Eine Analyse von Resilienz-Faktoren im Gastgewerbe am Beispiel der COVID-19-Krise.

Schwaiger, K., & Zehrer, A. (2020). Familienunternehmen: Resilienz in Krisenzeiten. *TourismusWissenQuarterly*, *21*, 210–215.

Schwartz, S. H. (1992). Universals in the content and structure of values: Theoretical advances and empirical tests in 20 countries. In M. P. Zanna (Ed.), (pp. 1–65). Advances in experimental social psychology. Academic Press. doi:10.1016/S0065-2601(08)60281-6

Schweiz: größte Skigebiete - Größtes Skigebiet in der Schweiz. (n.d.). *Skiresort.De*. Retrieved December 3, 2020, from https://www.skiresort.de/skigebiete/schweiz/sortiert-nach/laenge-pisten/#:%7E:text=Gr%C3%B6%C3%9Ftes%20Skigebiet%20in%20der%20Schweiz%20(580%20km),Die%20Pistenl%C3%A4nge%20betr%C3%A4gt%20580%20km

Sciascia, S., & Mazzola, P. (2008). Family involvement in ownership and management: Exploring nonlinear effects on performance. *Family Business Review*, *21*, 331–345.

Seraphin, H., Ivanov, S., Dosquet, F., & Bourliataux-Lajoinie, S. (2020). Archetypes of locals in destinations victim of overtourism. *Journal of Hospitality and Tourism Management, 43*, 283–288. doi:10.1016/j.jhtm.2019.12.001

Serrasqueiro, Z., & Nunes, P. M. (2014). Financing Behavior of Portuguese SMEs in Hotel Industry. *International Journal of Hospitality Management, 43*, 98–107. doi:10.1016/j.ijhm.2014.09.001

Sharma, P., Chrisman, J. J., Pablo, A. L., & Chua, J. H. (2001). Determinants of Initial Satisfaction with the Succession Process in Family Firms: A Conceptual Model. *Entrepreneurship Theory and Practice*, *25*(3), 17–36. doi:10.1177/104225870102500302

Sharma, P., Sharma, P., Chrisman, J. J., Chrisman, J. J., Chua, J. H., & Chua, J. H. (1997). Strategic management of the family business: Past research and future challenges. *Family Business Review*, *10*(1), 1–35. doi:10.1111/j.1741-6248.1997.00001.x

Sharpley, R. (2004). Tourism and the Countryside. In A. Lew, C. M. Hall, & A. M. Williams (Eds.), *A companion to Tourism* (pp. 374-386). Blackwell Publishing. https://books.google.at/books?hl=de&lr=&id=nePEB4e3Y0EC&oi=fnd&pg=PA374&dq=holidays+in+the+countryside&ots=6324w1G7I6&sig=o46KCCUgSGJNXg-D7sCoqLYShR0&redir_esc=y#v=onepage&q=holidays%20in%20the%20countryside&f=false

Sharpley, R., & Vass, A. (2006). Tourism, Farming and Diversification: An Attitudinal Study. *Tourism Management*, *27*(5), 1040–1052. doi:10.1016/j.tourman.2005.10.025

Sheffi, Y., & Rice, J. B. Jr. (2005). A supply chain view of the resilient enterprise. *MIT Sloan Management Review*, *47*(1), 41–48.

Shekhar, D., & Kumar, D. N. S. (2011). *Perspectives Envisaging Employee Loyalty-A Case Analysis*. Available at SSRN 1961430.

Shin, J., Taylor, M. S., & Seo, M. G. (2012). Resources for change: The relationships of organizational inducements and psychological resilience to employees' attitudes and behaviors toward organizational change. *Academy of Management Journal*, *55*(3), 727–748. doi:10.5465/amj.2010.0325

Shostack, G. (1984). Designing Services that Deliver. *Harvard Business Review*, *62*(1), 133–139.

Siakas, K., Naaranoja, M., Vlachakis, S., & Siakas, E. (2014). Family Businesses in the New Economy: How to Survive and Develop in Times of Financial Crisis. *Procedia Economics and Finance*, *9*, 331–341. doi:10.1016/S2212-5671(14)00034-3

Siddiqui, K. (2013). Heuristics for sample size determination in multivariate statistical techniques. *World Applied Sciences Journal*, *27*(2), 285–287.

Sieger, P., Zellweger, T., & Aquino, K. (2013). Turning Agents into Psychological Principals: Aligning Interests of Non-Owners through Psychological Ownership. *Journal of Management Studies*, *50*(3), 361–388. doi:10.1111/joms.12017

Sigala, M. (2020). Tourism and COVID-19: Impacts and implications for advancing and resetting industry and research. *Journal of Business Research, 117*, 312–321. doi:10.1016/j.jbusres.2020.06.015 PMID:32546875

Simpkins, M. D. (1976). *An experimental investigation of script theory in transactional analysis* (Doctoral dissertation, Oklahoma State University). ProQuest, UMI Dissertation Publishing.

Sirmon, D. G., & Hitt, M. A. (2003). Managing Resources: Linking Unique Resources, Management, and Wealth Creation in Family Firms. *Entrepreneurship Theory and Practice*, *27*(4), 339–358. doi:10.1111/1540-8520.t01-1-00013

Sjoberg, L. (2000). Factors in Risk Perception. *Risk Analysis*, *20*(1), 1–12. doi:10.1111/0272-4332.00001

Skokic, V., Lynch, P., & Morrison, A. (2016). Hotel entrepreneurship in a turbulent environment. *International Journal of Hospitality Management*, *53*, 1–11. doi:10.1016/j.ijhm.2015.11.008

Slovic, P., & Peters, E. (2006). Risk Perception and Affect. *Current Directions in Psychological Science*, *15*(6), 322–325. doi:10.1111/j.1467-8721.2006.00461.x

Small, T. A. (2011). What the Hashtag? A Content Analysis of Canadian Politics on Twitter. *Information Communication and Society*, *14*(6), 872–895. doi:10.1080/1369118X.2011.554572

Smith, R. A., & Houston, M. (1983). Script-based evaluations of satisfaction with services. In Emerging Perspectives on Services Marketing (pp. 59, 62). American Marketing.

Smith, R. A. (1992). Beach resort evolution. *Annals of Tourism Research*, *19*(2), 304–322. doi:10.1016/0160-7383(92)90083-2

Sokro, E. (2012). Impact of employer branding on employee attraction and retention. *European Journal of Business and Management*, *4*(18), 164–173.

Solvoll, S., Alsos, G. A., & Bulanova, O. (2015). Tourism entrepreneurship–review and future directions. *Scandinavian Journal of Hospitality and Tourism*, *15*(sup1), 120–137. doi:10.1080/15022250.2015.1065592

Song, H., Jung, J., & Kim, Y. (2017). Perceived news overload and its cognitive and attitudinal consequences for news usage in South Korea. *Journalism & Mass Communication Quarterly*, *94*(4), 1172–1190. doi:10.1177/1077699016679975

Sönmez, S. F., Backman, S. J., & Allen, L. (1994). *Managing tourism crises: A guidebook.* Department of Parks, Recreation and Tourism Management, Clemson University.

Sonnino, R. (2004). For a "piece of bread"? Interpreting sustainable development through agritourism in Southern Tuscany. *Sociologia Ruralis*, *44*(3), 285–300. doi:10.1111/j.1467-9523.2004.00276.x

Sorenson, R. L. (2011). Social capital and family business. In R. L. Sorenson (Ed.), *Family Business and Social Capital* (pp. 1–30). Edward Elgar Publishing.

Southwick, S. M., Bonanno, G. A., Masten, A. S., Panter-Brick, C., & Yehuda, R. (2014). Resilience definitions, theory, and challenges: Interdisciplinary perspectives. *European Journal of Psychotraumatology*, *5*(1), 25338. doi:10.3402/ejpt.v5.25338 PMID:25317257

Spiess, T., & Zehrer, A. (2020). Employees' change-oriented and proactive behaviors in small and medium-sized family businesses. In J. M. Saiz-Àlvarez, J. Leitao, & J. M. Palma-Ruiz (Eds.), *Entrepreneurship and Family Business Vitality - Surviving and Flourishing in the Long Term* (pp. 49–64). Springer. doi:10.1007/978-3-030-15526-1_4

Spira, J. B. (2011). *Overload! How too much information is hazardous to your organization.* John Wiley & Sons.

Stainton, A., Chisholm, K., Kaiser, N., Rosen, M., Upthegrove, R., Ruhrmann, S., & Wood, S. J. (2019). Resilience as a multimodal dynamic process. *Early Intervention in Psychiatry*, *13*(4), 725–732. doi:10.1111/eip.12726 PMID:30126047

Stankov, U., & Filimonau, V. (2019). Co-creating "Mindful" Holiday Resort Experience for Guests' Digital Well-Being. In J. Pesonen & J. Neidhardt (Eds.), *Information and Communication Technologies in Tourism 2019*. Springer., doi:10.1007/978-3-030-05940-8_16

Stankov, U., Filimonau, V., & Vujičić, M. D. (2020). A mindful shift: An opportunity for mindfulness-driven tourism in a post-pandemic world. *Tourism Geographies*, *22*(3), 703–712. doi:10.1080/14616688.2020.1768432

Stanzer, M. (2012). *Rechtsfragen des Buschenschankes in der Steiermark* (Diploma thesis). Karl-Franzens-Universität Graz.

Statista. (2020). *International tourist arrivals in Europe from 2010 to 2019, by region.* https://www.statista.com/statistics/261702/international-tourist-arrivals-in-europe-by-region/

Statista. (n.d.). https://www.statista.com/

Staw, B. M., Sandelands, L. E., & Dutton, J. E. (1981). Threat Rigidity Effects in Organizational Behavior: A Multilevel Analysis. *Administrative Science Quarterly*, *26*(4), 501–524. doi:10.2307/2392337

Steiner, A., & Atterton, J. (2015). Exploring the contribution of rural enterprises to local resilience. *Journal of Rural Studies*, *40*, 30–45. doi:10.1016/j.jrurstud.2015.05.004

Stewart, A. (2003). Help one another, use one another: Toward an anthropology of family business. *Entrepreneurship Theory and Practice*, *27*(4), 383–396.

Stojanov, D. (2004). *Hungary and Bosnia and Herzegovina: A success and a failure of transition.* Institute for World Economics, Hungarian Academy of Sciences.

Stubner, S., Blarr, W. H., Brands, C., & Wulf, T. (2012). Organizational Ambidexterity and Family Firm Performance. *Journal of Small Business and Entrepreneurship*, *25*(2), 217–229. doi:10.1080/08276331.2012.10593570

Su, L., Swanson, S. R., & He, X. (2020). A scale to measure residents perceptions of destination social responsibility. *Journal of Sustainable Tourism, 28*(6), 873–897. doi:10.1080/09669582.2019.1708372

Sutcliffe, K. M., & Vogus, T. J. (2003). Organizing for resilience. In K. S. Cameron, J. E. Dutton, & R. E. Quinn (Eds.), *Positive organizational scholarship: Foundations of a new discipline* (pp. 94–110). Berret-Koehler.

Swan, R. (1929). Summer holiday and the English countriyside. *Fortnightly Review, 126*, 259-268.

Sydnor-Bousso, S. B. (2009). *Assessing the impact of industry resilience as a function of community resilience: the case of natural disasters* (Doctoral dissertation). The Ohio State University.

Sznajder, M., Przezborska, L., & Scrimgeour, F. (2009). *Agritourism.* CABI Publishing. doi:10.1079/9781845934828.0000

Szydlik, M., & Künemund, H. (2009). *Generationen: Multidisziplinäre Perspektiven; Martin Kohli zum 65. Geburtstag* (1st ed.). VS Verlag für Sozialwissenschaften / GWV Fachverlage GmbH Wiesbaden. doi:10.1007/978-3-531-91499-2

Tabor, W., Chrisman, J. J., Madison, K., & Vardaman, J. M. (2018). Nonfamily Members in Family Firms: A Review and Future Research Agenda. *Family Business Review, 31*(1), 54–79. doi:10.1177/0894486517734683

Tagiuri, R., & Davis, J. (1982). Bivalent attributes of the family firms. Working paper, Harvard Business School, Cambridge, MA. *Family Business Review, 9*(2), 199-208.

Taha, S. A., Matheson, K., & Anisman, H. (2014). H1N1 Was Not All That Scary: Uncertainty and stressor appraisals predict anxiety related to a coming viral threat. *Stress and Health, 30*(2), 149–157. doi:10.1002mi.2505 PMID:23818424

Tarlow, P. E. (2003). The Tenth Annual Las Vegas Tourism Security and Safety Conference, Las Vegas, Nevada. *International Journal of Tourism Research, 4*, 251–252.

Taylor, P. J., & Derudder, B. (2016). *World city network: a global urban analysis.* Routledge.

Taylor, T., & Toohey, K. (2005). Impacts of tourism-related safety and security at a major sport event. *Event Management, 9*(4), 199–209. doi:10.3727/152599506776771544

The Austrian Federal Economic Chamber (Ed.). (2019). *Das Serviceangebot der Wirtschaftskammer.* https://www.wko.at/service/Startseite.html

Thomas, R., Shaw, G., & Page, S. J. (2011). Understanding small firms in tourism: A perspective on research trends and challenges. *Tourism Management, 32*(5), 963–976.

Thompson, R. R., Garfin, D. R., Holman, E. A., & Silver, R. C. (2017). Distress, Worry, and Functioning Following a Global Health Crisis: A National Study of Americans' Responses to Ebola. *Clinical Psychological Science, 5*(3), 513–521. doi:10.1177/2167702617692030

Tibay, V., Miller, J., Chang-Richards, A. Y., Egbelakin, T., Seville, E., & Wilkinson, S. (2018). Business resilience: A study of Auckland hospitality sector. *Procedia Engineering, 212*, 1217–1224.

Tingsanchali, T. (2012). Urban flood disaster management. *Procedia Engineering, 32*, 25–37. doi:10.1016/j.proeng.2012.01.1233

Tobisch, K. K. (2018). *Generationenwechsel im Familienunternehmen unter dem Gesichtspunkt der wandelnden Organisation (Master thesis)*. Karl-Franzens-Universität Graz.

Tomas, S., & Leffew, M. B. (2016). The customer experience. Tips for agritourism enterpreneurs. In Small-Scale Tourism in Rural Areas – Trends and Research in the Nordic Countries (pp. 15-23). Nordregio Working Paper 2013:3.

Toohey, K., Taylor, T., & Lee, C.-K. (2003). The FIFA World Cup 2002: The effects of terrorism on sport tourists. *Journal of Sport & Tourism, 8*(3), 186–196. doi:10.1080/1477508031000169O495

Top.tirol. (2019). *Die größten Hotels*. https://top.tirol/juli-2019/die-groessten-hotels

Tran, H., Hwang, Y., Yu, C., & Yoo, S. (2018). The Effect of Destination Social Responsibility on Tourists' Satisfaction: The Mediating Role of Emotions. *Sustainability, 10*(9), 3044. doi:10.3390u10093044

TrenkerD. (2010). *Heurigenkultur*. https://www.bmlrt.gv.at/land/lebensmittel/trad-lebensmittel/spezialkulturen/heurigenkultur.html

Triantafillidou, A., & Yannas, P. (2020). Social media crisis communication in racially charged crises: Exploring the effects of social media and image restoration strategies. *Computers in Human Behavior, 106*, 1–10. doi:10.1016/j.chb.2020.106269

Tripsas, M., & Gavetti, G. (2000). Capabilities, cognition, and inertia: Evidence from digital imaging. *Strategic Management Journal, 21*(10-11), 1147–1161. doi:10.1002/1097-0266(200010/11)21:10/11<1147::AID-SMJ128>3.0.CO;2-R

Tsai, C. H., Wu, T. C., Wall, G., & Linliu, S. C. (2016). Perceptions of tourism impacts and community resilience to natural disasters. *Tourism Geographies, 18*(2), 152–173.

Tsao, C. Y., & Ni, C. C. (2016). Vulnerability, resilience, and the adaptive cycle in a crisis-prone tourism community. *Tourism Geographies, 18*(1), 80–105. doi:10.1080/14616688.2015.1116600

Tushman, M. L., & O'Reilly, C. A. III. (1996). Ambidextrous organizations: Managing evolutionary and revolutionary change. *California Management Review, 38*(4), 8–29. doi:10.2307/41165852

Twigg, J. (2015). *Disaster risk reduction*. Overseas Development Institute, Humanitarian Policy Group.

Tyrrell, T. J., & Johnston, R. J. (2008). Tourism sustainability, resiliency and dynamics: Towards a more comprehensive perspective. *Tourism and Hospitality Research, 8*(1), 14–24. doi:10.1057/thr.2008.8

UNWTO. (2020). *International tourist numbers could fall 60-80% in 2020*. UNWTO Reports. https://www.unwto.org/news/covid-19-international-tourist-numbers-could-fall-60-80-in-2020

Urlaub am Bauernhof, B. (2019). *Wozu eine Kategorisiserung der Höfe? Academic Press.*

US Department of State. (2020). *Travel Advisories*. Retrieved September 9, 2020, from https://travel.state.gov/content/travel/en/traveladvisories/traveladvisories.html/

van Niekerk, M., & Pizam, A. (2015). How do Terrorism and Tourism Co-exist in Turbulent Times? In K. Glaser (Ed.), *Terrorism and the Economy. Impacts on the Capital Market and the Global Tourism Industry* (1st ed., pp. 109–126). Eleven International Publishing.

Van Saandt, A., Low, S. A., & Thilmany, D. (2018, December). Exploring Regional Patterns of Agritourism in the U.S.: What's Driving Clusters of Enterprises? *Agricultural and Resource Economics Review*, *47*(3), 592–609. doi:10.1017/age.2017.36

Vanhamme, J., & Snelders, D. (2001). The role of surprise in satisfaction judgments. *Journal of Consumer Satisfaction, Dissatisfaction & Complaining Behavior*, *14*, 27–45.

Vasvári, T. (2015). Risk, Risk Perception, Risk Management – a Review of the Literature. *Public Finance Quarterly*, *60*(1), 29–48.

Vedeld, T., Coly, A., Ndour, N. M., & Hellevik, S. (2016). Climate adaptation at what scale? Multi-level governance, resilience, and coproduction in Saint Louis, Senegal. *Natural Hazards*, *82*(2), 173–199. doi:10.100711069-015-1875-7

Veider, V., & Kallmuenzer, A. (2016). Assessing long-term orientation among founder- and descendant-led firms. *Journal of Family Business Management*, *6*(1). Advance online publication. doi:10.1108/JFBM-07-2015-0027

Veider, V., & Matzler, K. (2016). The ability and willingness of family-controlled firms to arrive at organizational ambidexterity. *Journal of Family Business Strategy*, *7*(2), 105–116. doi:10.1016/j.jfbs.2015.10.001

Venkatarman, S. (2002). Ten principles of entrepreneurial creation. *Batten Briefings, 1*(1).

Venugopal, A., Krishnan, T. N., Kumar, M., & Upadhyayula, R. S. (2019). Strengthening organizational ambidexterity with top management team mechanisms and processes. *International Journal of Human Resource Management*, *30*(4), 586–617. doi:10.1080/09585192.2016.1277369

Veréb, V., Nobre, H., & Farhangmehr, M. (2020). Cosmopolitan tourists: the resilient segment in the face of terrorism. *Tourism Management Perspectives, 33*(September).

Veréb, V. N., Nobre, H., & Farhangmehr, M. (2018). The fear of terrorism and shift in cosmopolitan values. *International Journal of Tourism Cities*, *4*(4), 452–483. doi:10.1108/IJTC-03-2018-0024

Veréb, V. N., Nobre, H., & Farhangmehr, M. (in press). Cosmopolitan tourists: The resilient segment in the face of COVID-19. *Service Business*.

Vermeeren, B., Kuipers, B., & Steijn, B. (2011). Two faces of the satisfaction mirror: A study of work environment, job satisfaction, and customer satisfaction in Dutch municipalities. *Review of Public Personnel Administration, 31*(2), 171–189. doi:10.1177/0734371X11408569

Villalonga, B., & Amit, R. (2004). *How do family ownership, control and management affect firm value?* EFA 2004 Maastricht Meetings Paper no. 3620. Available at: https://papers.ssrn.com/sol3/papers.cfm?abstract_id=556032

Villalonga, B., & Amit, R. (2006). How do family ownership, control, and management affect firm value? *Journal of Financial Economics, 80*(2), 385–417. doi:10.1016/j.jfineco.2004.12.005

Vinkers, C. H., van Amelsvoort, T., Bisson, J. I., Branchi, I., Cryan, J. F., Domschke, K., Howes, O. D., Manchia, M., Pinto, L., de Quervain, D., Schmidt, M. V., & van der Wee, N. J. A. (2020, June). Stress resilience during the coronavirus pandemic. *European Neuropsychopharmacology, 35*, 12–16. doi:10.1016/j.euroneuro.2020.05.003 PMID:32446705

Voase, R. (2018). Holidays under the hegemony of hyper-connectivity: Getting away, but unable to escape? *Leisure Studies, 37*(4), 384–395. doi:10.1080/02614367.2018.1475503

Volo, S., & Irimiás, A. (2020). Instagram: Visual methods in tourism research. *Annals of Tourism Research*, 103098. doi:10.1016/j.annals.2020.103098

von Westphalen, F. G. (2020). Einige Vorüberlegungen zum bevorstehenden Lieferkettengesetz. *ZIP: Zeitschrift für Wirtschaftsrecht, 41*(49), 2421–2422.

Vujo, A., Petrović, I., & Blesic, M. D. (2015). Leisure time in the countryside: The health aspects of agritourism activities. *International Journal Scientific and Applicative Papers, 8*(1), 131–136.

Walker, B., & Salt, D. (2012). *Resilience thinking: sustaining ecosystems and people in a changing world.* Island Press.

Walsh, F. (2002). A family resilience framework: Innovative practice applications. *Family Relations, 51*(2), 130–137. doi:10.1111/j.1741-3729.2002.00130.x

Wang, J., & Ritchie, B. W. (2012). Understanding accommodation managers' crisis planning intention: An application of the theory of planned behaviour. *Tourism Management, 33*(5), 1057–1067. doi:10.1016/j.tourman.2011.12.006

Wang, Y.-S. (2009). The impact of crisis events and macroeconomic activity on Taiwan's international inbound tourism demand. *Tourism Management, 30*(1), 75–82. doi:10.1016/j.tourman.2008.04.010 PMID:32287727

Ward, J. L. (2004). *Perpetuating the family business: 50 lessons from long-lasting successful families in business*. Palgrave Macmillan. doi:10.1057/9780230505995

Ward, J. L. (2011). *Keeping the Family Business Healthy – How to Plan for Continuing Growth, Profitability, and Family Leadership*. Palgrave Macmillan. doi:10.1057/9780230116122

Watson, P., & Deller, S. (2021). Tourism and economic resilience. *Tourism Economics*. Advance online publication. doi:10.1177/1354816621990943

We Are Social. (2019). *Global Digital Report 2019.* https://wearesocial.com/global-digital-report-2019

Weaver, D. B., & Fennell, D. A. (1997). The vacation farm sector in Saskatchewan: A profile of the operations. *Tourism Management, 18*(6), 357–365. doi:10.1016/S0261-5177(97)00039-3

Weber, E. U., & Milliman, R. A. (1997). Perceived Risk Attitudes: Relating Risk Perception to Risky Choice. *Management Science, 43*(2), 123–144. doi:10.1287/mnsc.43.2.123

Weick, K. E., & Quinn, R. E. (1999). Organizational change and development. *Annual Review of Psychology, 50*(1), 361–386. doi:10.1146/annurev.psych.50.1.361 PMID:15012461

Weissearena. (2020, October 5). *Portrait.* WAG.

Welter, F. (2011). Contextualizing entrepreneurship—Conceptual challenges and ways forward. *Entrepreneurship Theory and Practice, 35*(1), 165–184. doi:10.1111/j.1540-6520.2010.00427.x

Welter, F., Wolter, H., & Kranzusch, P. (2020). *Preliminary assessments of the IfM Bonn on the economic impact of the coronavirus pandemic on the "Mittelstand" businesses in Germany. IFM-Background Paper.* IFM.

Wenzel, M., Stanske, S., & Lieberman, M. B. (2020). Strategic responses to crisis. *Strategic Management Journal, 42*(2), O16–O27. doi:10.1002mj.3161

Weston. (2019). *Research for TRAN Committee – European tourism: recent developments and future challenges, European Parliament.* Policy Department for Structural and Cohesion Policies.

White, I., & O'Hare, P. (2014). From rhetoric to reality: Which resilience, why resilience, and whose resilience in spatial planning? *Environment and Planning. C, Government & Policy, 32*(5), 934–950. doi:10.1068/c12117

WHO. (2015). *Anticipating emerging infectious diseases epidemics.* Retrieved from https://www.who.int/csr/disease/anticipating_epidemics/events/AE_sessionAgenda.pdf

WHO. (2020). *Coronavirus disease 2019 (COVID-19). Situation Report – 51.* Available at: https://www.who.int/docs/default-source/coronaviruse/situation-reports/20200311-sitrep-51-covid-19.pdf?sfvrsn=1ba62

Williams, A. M., & Baláž, V. (2015). Tourism Risk and Uncertainty: Theoretical Reflections. *Journal of Travel Research, 54*(3), 271–287. doi:10.1177/0047287514523334

Williams, A. M., Shaw, G., & Greenwood, J. (1989). From tourist to tourism entrepreneur, from consumption to production: Evidence from Cornwall, England. *Environment & Planning A, 21*(12), 1639–1653. doi:10.1068/a211639

Williams, C., You, J. J., & Joshua, K. (2020). Small-business resilience in a remote tourist destination: Exploring close relationship capabilities on the island of St Helena. *Journal of Sustainable Tourism, 28*(7), 937–955. doi:10.1080/09669582.2020.1712409

Williams, N., & Vorley, T. (2014). Economic resilience and entrepreneurship: Lessons from the Sheffield City Region. *Entrepreneurship and Regional Development, 26*(3-4), 257–281. doi:10.1080/08985626.2014.894129

Williams, T. A., Gruber, D. A., Sutcliffe, K. M., Shepherd, D. A., & Zhao, E. Y. (2017). Organizational response to adversity: Fusing crisis management and resilience research streams. *The Academy of Management Annals, 11*(2), 733–769. doi:10.5465/annals.2015.0134

Wimmer, R., Domayer, E., Oswald, M., & Vater, G. (2005). Familienunternehmen—Auslaufmodell oder Erfolgstyp? Familienunternehmen — Auslaufmodell oder Erfolgstyp? doi:10.1007/978-3-322-82895-8

Wirtschaftskammer Burgenland. (2019). *Das freie Gastgewerbe*. Academic Press.

Wittkowski, J. (1994). *Das Interview in der Psychologie: Interviewtechnik und Codierung von Interviewmaterial*. VS Verlag für Sozialwissenschaften. doi:10.1007/978-3-322-92472-8

Wojcieszak-Zbierska, M. M., Jęczmyk, A., Zawadka, J., & Uglis, J. (2020). Agritourism in the Era of the Coronavirus (COVID-19): A Rapid Assessment from Poland. *Agriculture, 10*(9), 397. doi:10.3390/agriculture10090397

Wolff, K., & Larsen, S. (2016). Flux and permanence of risk perceptions: Tourists' perception of the relative and absolute risk for various destinations. *Scandinavian Journal of Psychology, 57*(6), 584–590. doi:10.1111jop.12326 PMID:27636718

Wong, I. A., Ou, J., & Wilson, A. (2021). Evolution of hoteliers' organizational crisis communication in the time of mega disruption. *Tourism Management*, 84.

World Bank. (2020a). *Chapter 3: Lasting Scars of the COVID-19 Pandemic*. Global Economic Prospects. https://pubdocs.worldbank.org/en/112641588788257004/Global-Economic-Prospects-June-2020-Topical-Issue-1.pdf

World Bank. (2020b). *Global Outlook – Pandemic, Recession: The Global Economy in Crisis*. Global Economic Prospects. https://openknowledge.worldbank.org/bitstream/handle/10986/33748/211553-Ch01.pdf

World Economic Forum. (2019). *The Global Risks Report*. Retrieved from http://wef.ch/risks2019

World Economic Forum. (2019). *The Travel & Tourism Competitiveness Report 2019: Travel and Tourism at a Tipping Point*. Geneva: World Economic Forum. http://www3.weforum.org/docs/WEF_TTCR_2019.pdf

World Economic Forum. (2020). *COVID-19: These countries are most at risk from falling tourism*. https://www.weforum.org/agenda/2020/07/coronavirus-covid19-travel-tourism-gdp-economics/

World Tourism Organization (2011). *Toolbox for Crisis Communications in Tourism: Checklist and Best Practices.* World Tourism Organization (UNWTO).

World Tourism Organization. (2020). *International Tourism and COVID-19.* World Tourism Organization (UNWTO). https://www.unwto.org/international-tourism-and-covid-19

WTTC Data Gateway. (n.d.). https://tool.wttc.org/

WTTC. (2019). *Travel & Tourism continues strong growth above global GDP.* https://www.wttc.org/about/media-centre/press-releases/press-releases/2019/travel-tourism-continues-strong-growth-above-global-gdp/

WTTC. (2020). *Travel and Tourism Economic Impact: Croatia.* WTTC.

WTTC. (2020a). *Travel & Tourism: Global Economic Impact & Trends 2020.* https://wttc.org/Portals/0/Documents/Reports/2020/Global%20Economic%20Impact%20Trends%202020.pdf?ver=2021-02-25-183118-360

WTTC. (2020b). *Travel & Tourism Economic Impact from COVID-19: Global Data.* World Travel & Tourism Council. https://wttcweb.on.uat.co/Research/Economic-Impact

Wu, B. P., & Chang, L. (2012). The social impact of pathogen threat: How disease salience influences conformity. *Personality and Individual Differences*, *53*(1), 50–54. doi:10.1016/j.paid.2012.02.023

Yang, E. C. L., & Nair, V. (2014). Tourism at Risk: A Review of Risk and Perceived Risk. *Tourism Asia-Pacific Journal of Innovation in Hospitality and Tourism*, *3*(2), 239–259. doi:10.760340930-014-0013-z

Yang, Y., Zhang, H., & Chen, X. (2020). Coronavirus pandemic and tourism: Dynamic stochastic general equilibrium modelling of infectious disease outbreak. *Annals of Tourism Research*, *83*, 102913. doi:10.1016/j.annals.2020.102913 PMID:32292219

Ye, J. (2021). *4 ways family businesses can lead the pandemic recovery.* World Economic Forum. https://www.weforum.org/agenda/2021/01/4-ways-family-businesses-can-lead-the-pandemic-recovery/

Yehuda, R., & Flory, J. D. (2007). Differentiating biological correlates of risk, PTSD, and resilience following trauma exposure. *Journal of Traumatic Stress*, *20*(4), 435–447. doi:10.1002/jts.20260 PMID:17721957

York, C. (2013). Overloaded by the news: Effects of news exposure and enjoyment on reporting information overload. *Communication Research Reports*, *30*(4), 282–292. doi:10.1080/08824096.2013.836628

Youssef, C. M., & Luthans, F. (2007). Positive Organizational Behavior in the Workplace. *Journal of Management*, *33*(5), 774–800. doi:10.1177/0149206307305562

Zach, F. (2012). Partners and Innovation in American Destination Marketing Organizations. *Journal of Travel Research*, *51*(4), 412–425. doi:10.1177/0047287511426340

Zahra, S. A. (2005). Entrepreneurial Risk Taking in Family Firms. *Family Business Review*, *18*(1), 23–40. doi:10.1111/j.1741-6248.2005.00028.x

Zahra, S. A., Hayton, J. C., Neubaum, D. O., Dibrell, C., & Craig, J. (2008). Culture of Family Commitment and Strategic Flexibility: The Moderating Effect of Stewardship. *Entrepreneurship Theory and Practice*, *32*(6), 1035–1054. doi:10.1111/j.1540-6520.2008.00271.x

Zapalska, A. M., & Brozik, D. (2007). Managing family businesses in the tourism and hospitality industry: The transitional economy of Poland. *Proceedings of Rijeka Faculty of Economics: Journal of Economics and Business*, *25*(1), 141–165.

Zehrer, A. (2009). The culture of facing crisis in tourism – an empirical study on crisis competencies of future destination managers. Culture Meets Economy Conference "Culture and Crisis – opportunity or threat?"

Zehrer, A., & Halswanter, J. (2010). *Management of change in tourism - The problem of family internal succession in family-run tourism SMEs*. Academic Press.

Zehrer, A., & Leiß, G. (2019). Family Entrepreneurial Resilience - An intergenerational learning approach. *Family Business Management*.

Zehrer, A., & Leiß, G. (2019). Family entrepreneurial resilience – an intergenerational learning approach. *Journal of Family Business Management*. https://www.emerald.com/insight/content/doi/10.1108/JFBM-09-2018-0037/full/html doi:10.1108/JFBM-09-2018-0037

Zehrer, A. (2019). Structure, significance and challenges of family firms in community-type Alpine tourism destinations. In H. Pechlaner (Ed.), *Destination und Lebensraum* (pp. 25–38). Springer Gabler. doi:10.1007/978-3-658-28110-6_2

Zehrer, A., & Hallmann, K. (2015). A stakeholder perspective on policy indicators of destination competitiveness. *Journal of Destination Marketing & Management*, *4*(2), 120–126. doi:10.1016/j.jdmm.2015.03.003

Zehrer, A., & Haslwanter, J. (2010). Management of Change in Tourism - The Problem of Family Internal Succession in Family-run Tourism SMEs. *Electronic Journal of Family Business Studies*, *4*(2), 147–162.

Zehrer, A., & Leiß, G. (2020). Intergenerational communication barriers and pitfalls of business families in transition—A qualitative action research approach. *Corporate Communications*, *25*(3), 515–532. https://doi.org/10.1108/CCIJ-03-2020-0056

Zehrer, A., & Muskat, B. (2016). Der Treiber Innovation in touristischen Unternehmen. In H. Siller & A. Zehrer (Eds.), *Entrepreneurship und Tourismus: Unternehmerisches Denken und Erfolgskonzepte aus der Praxis* (2nd ed., pp. 29–46). Linde.

Zehrer, A., Pechlander, H., & Reuter, C. (2009). Innovativeness in tourism: The perception of innovation awards participants. *Journal of Tourism, 14*(1), 11–29.

Zehrer, A., & Raich, F. (2020). Corporate Social Responsibility im Tourismus - der Einsatz von CSR-Aktivitäten in familiengeführten touristischen Unternehmen. In H. Pechlaner & S. Speer (Eds.), *Responsible Entrepreneurship* (pp. 215–236). Springer Fachmedien Wiesbaden. doi:10.1007/978-3-658-31616-7_10

Zeithaml, V., Bitner, M., & Gremler, D. (2006). *Services Marketing: Integrating Customer Focus Across the Firm*. McGraw-Hill Irwin.

Zelger, J., & Oberprantacher, A. (2002, May). Processing of verbal data and knowledge representation by GABEK®-WinRelan®. *Forum Qualitative Social Research, 3*(2).

Zenker, S., & Kock, F. (2020). The coronavirus pandemic - A critical discussion of a tourism research agenda. *Tourism Management, 81*, 81. doi:10.1016/j.tourman.2020.104164 PMID:32518437

Zhang, L., Wu, L., & Mattila, A. S. (2016). Online reviews: The role of information load and peripheral factors. *Journal of Travel Research, 55*(3), 299–310. doi:10.1177/0047287514559032 PMID:29708106

Zimmermann, A., Raisch, S., & Cardinal, L. B. (2018). Managing persistent tensions on the frontline: A configurational perspective on ambidexterity. *Journal of Management Studies, 55*(5), 739–769. doi:10.1111/joms.12311

Zolli, A., & Healy, A. M. (2013). *Resilience: Why things bounce back* (1st ed.). Simon & Schuster Paperbacks.

Zurub, Ionescu, A., & Constantin, V. D. (2015). Measuring the Economic Impact of Tourism in European Emerging Markets. *Procedia Economics and Finance, 32*, 95–102. doi:10.1016/S2212-5671(15)01369-6

About the Contributors

Anita Zehrer is Head of the Family Business Center at the Management Center Innsbruck (MCI). From 2007 to 2015 she was Deputy Head of the MCI Tourism Department, from 2012 to 2017 she was Deputy Head of the Academic Council, from 2016-2019 she was Adjunct Professor at the University of Notre Dame in Sydney, Australia, from 2012-2016 she was Adjunct Professor at the University of Canberra, Australia. From 2009 to 2018 she served as Vice-President of the German Association for Tourism Research DGT, 2014 to 2017 she was Member of the Tourism Advisory Board of the Federal Ministry of Foreign Affairs and Energy, Germany, from 2016 to 2017 she was tourism expert at the Committee of Regions at the European Union. She is member of the Academy of Management (AoM), the International Association of Scientific Experts in Tourism (AIEST), the International Federation for Information Technology and Travel & Tourism (IFITT), the International Center for Research and Education in Tourism (ICRET) the Senate of the Christian Doppler Gesellschaft, the Working Group Family Firm of the FGF e.V., the International Family Enterprise Research Association (IFERA), EQUA Stiftung and FGF e.V.

Gundula Glowka is currently working as a doctoral student at the Family Business Center at Management Center Innsbruck, Austria. Before, Gundula Glowka completed her Master's degree in International Business and Economy at the University of Innsbruck, majoring in strategic management and marketing. In her recent research, she focuses on the risk behaviour of small and medium-structured family businesses. Her research interests are on how owner-managers perceive, identify and deal with risks in such informally structured businesses. Previous studies therefore explore the link of Enterprise Risk Management on performance in small and medium structured family firms and the effect of family involvement and CEO tenure as moderating variables. In tourism, Gundula Glowka examines the risk perception of family business-owners and its impact on destination development.

* * *

António Azevedo is an Assistant Professor in the School of Economics and Management of University of Minho, Braga, Portugal. His primary research topics are tourism marketing, place marketing, destination branding, celebrity branding, advertising processing, and marketing strategy amongst others. He has published several articles in international academic journals and international conferences. He is also researcher of the Lab2PT Investigation Unit - Laboratory of Landscape, Heritage and Territory (University of Minho). For more information: ORCID Nr: 0000-0002-3918-5108.

Martina Basarac Sertić is a senior research associate at the Economic Research Division of the Croatian Academy of Sciences and Arts. In 2020, she became an associate member of the Croatian Academy of Sciences and Arts. As an external associate and an assistant professor, she participated in teaching the courses Croatian Economy, Economic Policy and Economics of Education at the Department of Macroeconomics and Economic Development and the course Statistics at the Department of Statistics (at the Faculty of Economic & Business, University of Zagreb). She currently teaches the course on Sustainable Development and Natural Resources at the Faculty of Economics at the University of Zagreb. She has published about 50 scientific and professional articles and is the editor of numerous Proceedings. Her research focuses on the macroeconomics, sectoral competitiveness and sustainable development.

Laura Bechthold is post-doctoral researcher at the Friedrichshafen Institute of Family Entrepreneurship at Zeppelin University. Additionally, she is Director for Science Services at philoneos GmbH, an innovation office for family businesses located in Munich. Laura holds a BA in Management and Economics from Zeppelin University, a MSc in Sustainability Science and Policy from Maastricht University, as well as a Master of Business Research (MBR) from Ludwig-Maximilians-University in Munich. Her dissertation at the Max-Planck-Institute for Innovation and Competition focused on female entrepreneurship and her field experimental study about the impacts of female entrepreneurial role models has been awarded twice at international conferences, such as the Best Empirical Paper Award by the Entrepreneurship Division at the 2018 Academy of Management Conference.

Sarah Eichelberger is a Pre Doc University Assistant and PhD student at the Department of Strategic Management, Marketing and Tourism at the Faculty of Business and Management, University of Innsbruck. Sarah graduated with a Master's degree in Strategic Management and a Bachelor's degree in Management and Economics from the University of Innsbruck. She is a leading member of the doctoral program "Tourism and Leisure in Mountain Regions". Her research interests include quality of life in tourism, entrepreneurial ecosystems, as well as responsible and sustainable tourism. In this regard, she focuses on the impact of sustainable tourism concepts on responsible tourism behaviour.

Denise Fecker is a PhD student at the University of Innsbruck and at the Tourism Department of Management Center Innsbruck. Her research interests include Winter Sports Tourism, Leadership, Organisational Commitment and New Work.

Martina Harc is a research associate at The Institute for Scientific Research and Artistic Work in Osijek's division of the Croatian Academy of Sciences and Arts. Currently, she is a visiting scholar at the Technical University of Munich, Chair of Corporate Sustainability at the TUM School of Management. So far, her research was oriented toward SMEs, emphasizing the outstanding importance of them. Until 2020, she has published one scientific book and numerous scientific and professional articles as author or co-author. Her main fields of interest are sustainable entrepreneurship, financial management and environmental economics.

Stefanie Haselwanter is a research and teaching assistant at the Tourism Department of Management Center Innsbruck (MCI) and a doctoral students at the Catholic University of Eichstätt-Ingolstadt. Her research interests include Entrepreneurship, Leadership as well as the Governance and Management of Tourism Destinations.

Maximilian Lude is interim Professor at the TUM School of Management, TUM Campus Heilbronn. Concurrently he is Managing Partner at philoneos GmbH, a marketing and innovation office for family businesses located in Munich. He studied Management and Economics at Zeppelin University and Copenhagen Business School. His dissertation was on the perception of family firms, bridging marketing, consumer behavior and family business research. This intersection also reflects the current research focus of Maximilian. His research has been published in international peer-reviewed journals such as Entrepreneurship Theory and Practice (ETP) or Journal of Business Research. Maximilian also serves as Reviewer for several journals such as Journal of Brand Management and Entrepreneurship Theory and Practice and is Member of the Editorial Board of Journal of Family Business Strategy.

Vincent Magnini serves as faculty at Longwood University in the USA. He conducts research in the area of services marketing with specific application to hospitality and tourism settings. Dr. Magnini holds a number of editorial board appointments on well-regarded journals and has been ranked among the most prolific hospitality and tourism researchers in the world. Further, he is a U.S. Fulbright Scholar. He has published seven books and more than 250 articles and reports.

Monica Nadegger is a PhD student at the University of Innsbruck and the Management Center Innsbruck. Her research focuses on the intersection of communication, organization, and the role of materiality on digital platforms in the context of tourism. Her recent work was published in the Journal of Sustainable Tourism.

Harald Pechlaner is Professor of Tourism and Head of the Center for Entrepreneurship at the Catholic University of Eichstätt-Ingolstadt, Germany and Head of the Center for Advanced Studies at EURAC Research in Bozen/Bolzano (EURAC), Italy. Since 2014 Harald Pechlaner is Adjunct Research Professor at the School of Management and Marketing/Curtin Business School in Perth/Western Australia. His research is focused on destination governance and leadership, entrepreneurship and global challenges for regions and destinations. He is author and (co-) editor of more than 60 books and about 150 research articles. Prof. Pechlaner is President of the AIEST (Association Internationale d'Experts Scientifiques du Tourisme) and was President (2002-2012) of the German Association of Tourism Research (Deutsche Gesellschaft für Tourismuswissenschaft). Since 2016 Harald Pechlaner is a full member of the European Academy of Sciences and Arts.

Mike Peter's research interest lies in the field of entrepreneurship and small and medium-sized enterprises. In doing so, he essentially investigates the mechanisms that influence company developments. Using the example of the tourism industry, he has analysed various questions relating to entrepreneurship in small and medium-sized enterprises. The aim was to investigate why entrepreneurs are all too often sceptical about further growth, contrary to Schumpeter's growth-oriented entrepreneur. A potential explanation could be provided by entrepreneurship and quality of life research. Today, the research focus is on researching the heterogeneity of family businesses and small businesses, understanding which factors influence entrepreneurial behaviour and business success. Secondly, Peters and his team deal with relevant questions in tourism: These, mostly demand-side studies are currently analysing, for example, the specificities of food labelling and the importance of regionality for the consumer. It also records how the preferences, travel decisions and travel behaviour of the Alpine tourist change. Thirdly, together with Tyrolean business partners and international research centres, the conditions are currently being created to develop sustainable databases and networks for family business research.

Reinhard Prügl is full professor and academic director of the Friedrichshafen Institute for Family Entrepreneurship (FIF), Zeppelin University, Lake Constance, Germany. His overall research interest is focused on the intersection of marketing, innovation, and family business, particularly from the perspective of the next generation of enterprising families. His work has been published in renowned journals such as for example Journal of Management Studies, Family Business Review, Journal of Family Business Strategy, Research Policy, Journal of Product Innovation Management, or Marketing Letters.

Erika Quendler is a researcher at the Federal Institute of Agricultural Economics, Rural and Mountain Research. She holds a PhD in Agricultural Economics from the University of Natural Resources and Life Sciences in Vienna with the dissertation thesis "Small scale farming from a historical and current integrated point of view: Philosophies and Genesis; Structural Analysis (Classification) as well as Characteristics and Potentials; Discussion of possible Development-scenarios and Outlook". Her main research interests are sustainable development, quality of life, resilience, digital transformation, young farmer and family farm issues, value chain and economic accounts for agriculture. Erika Quendler has published several scientific papers and participated in many scientific international conferences.

Vlatka Skokic is an associate professor of Entrepreneurship at University of Split, Faculty of Economics, Business and Tourism, Croatia. Prior joining the University of Split, she spent six years as an assistant professor (Lecturer) at the University of Surrey, School of Hospitality and Tourism Management, UK. Vlatka holds a PhD from the University of Strathclyde, Scotland. Her research is focused on entrepreneurship, innovation and networks. Vlatka's research is published in the leading management and social science journals including Annals of Tourism Research and European Management Journal. Vlatka is experienced at teaching a range of subjects including entrepreneurship, family business and research methods (qualitative) at postgraduate and doctoral level. She is a Fellow of the Higher Education Academy (HEA).

Lara Soleder is a research and teaching assistant at Management Center Innsbruck at the Department of Tourism and Leisure Business. Her research interests are intercultural communication and management, diversity and leadership. She is responsible for the internationalisation of the tourism business studies at MCI and did her Masters in Executive Management as well as Leadership in Tourism.

Martin Tusch developed a deep interest in the tourism industry. He decided to complete an education as a tourism manager at Villa Blanka Innsbruck. Afterwards he gained work experience in the hospitality industry. Then he completed a bachelor's degree in "Tourism Business Studies" and deepened his knowledge in the master's program "Entrepreneurship & Tourism" at MCI (graduation in 2020). In his master thesis, he focuses on generational differences of risk perception in tourism family firms.

Vanda Vereb is a research fellow at GOVCOPP (Research Unit on Governance, Competitiveness and Public Policies), University of Aveiro (Portugal) and an invited professor at the National Economics University, Hanoi (Vietnam) of social, non-profit and public marketing. She also serves as a reviewer of several international journals on tourism, entrepreneurship and consumer behaviour. She holds a PhD in Strategic Marketing from the University of Minho (Portugal). Before entering academia, she worked for 10 years as a management consultant across the Middle East advising local governments on strategic management and service quality. Her research interest is in tourism, terrorism, public fear, cosmopolitanism, social marketing and strategic entrepreneurship. For more information: orcid.org/0000-0002-8894-457X.

Daniel Zacher is a research fellow at the Catholic University of Eichstätt-Ingolstadt and PhD candidate at the Chair of Tourism / Centre for Entrepreneurship based there. He also works at the transfer office of the CU where he is part of the structural development of regional cooperation with companies and civil society institutions. His research focuses on the role of tourism in the context of sustainable regional development, organisational development and resilience in tourism destinations. Daniel Zacher is (co-) author of about 20 research articles.

Index

O

P

R

S

T

www.ingramcontent.com/pod-product-compliance
Lightning Source LLC
LaVergne TN
LVHW080603300926
849162LV00005B/1037